The goal of industrial ecology is the evolution of the world's industrial activity into a sustainable and environmentally benign system. As a field of study, it requires a long-range view and a deep analysis of the environmental implications of today's industrial systems, and a creative approach to the design of services, products, and governmental policy.

This book is a wide-ranging exploration of this new approach to environmental problems. With contributions from a broad range of disciplines – environmental science, technology assessment, economics, policy studies – the book lays out the range of concerns encompassed by industrial ecology.

Industrial Ecology and Global Change

Office for Interdisciplinary Earth Studies
Global Change Institute Volume 5
Series Editors: Tom M.L. Wigley and Carol Rasmussen

Produced through support from the Committee on Earth and Environmental
Sciences Subcommittee on Global Change.

Industrial Ecology and Global Change

Edited by

R. SOCOLOW, C. ANDREWS, F. BERKHOUT, and
V. THOMAS

Published by the Press Syndicate of the University of Cambridge
The Pitt Building, Trumpington Street, Cambridge CB2 1RP
40 West 20th Street, New York, NY 10011-4211, USA
10 Stamford Road, Oakleigh, Melbourne 3166, Australia

First published 1994

Printed in Great Britain at the University Press, Cambridge

A catalogue record for this book is available from the British Library

Library of Congress cataloguing in publication data
Industrial ecology and global change / edited by Robert H. Socolow, C. Andrews, F. Berkhout, and
V. Thomas.
    p.   cm.
Includes index.
ISBN 0-521-47197-4
1. Environmental sciences.   2. Industry – Environmental aspects.
3. Environmental policy.   4. Social ecology.   5. Human ecology.
I. Socolow, Robert H.
GE105.I53   1994
363.73'1–dc20   94-11814   CIP

ISBN 0 521 47197 4 hardback

Based on the 1992 OIES Global Change Institute organized by T. Graedel, W. Moomaw, and
R. Socolow

Contents

Contents

Foreword

The millions of life-forms on the planet are supported by complex biogeochemical cycles and physical transfers of materials and energy. Global change science is an interdisciplinary effort to understand these systems. This new field is also concerned with how one of these life forms, human beings, is altering processes, changing material and energy flows, transforming ecosystems, eliminating and rearranging species, and introducing artificial chemicals and species into the environment. But even when the physical and biological processes are illuminated by science, there is still a need to know how and why human societies create these changes and what we might do about them.

In 1986 the Office for Interdisciplinary Earth Studies (OIES) was formed at the University Corporation for Atmospheric Research to stimulate research in global change science and to help scientists, government officials and the international community as they shaped this new field. Under its founding director, John A. Eddy, OIES organized a series of two-week summer workshops held at Snowmass, Colorado. Each of the first three Global Change Institutes addressed some aspect of global change science from an interdisciplinary perspective: greenhouse gases, past climate changes, and earth system modeling. The papers from each of these institutes have been published by OIES.

In 1991, a new and expanded approach to the Global Change Institutes was attempted; global change scientists were joined by social scientists to examine both the human causes and global consequences of altered land-use patterns on the planet. The report of that workshop, *Changes in Land Use and Land Cover: A Global Perspective*, is the companion volume to the present work.

Recognizing that industrial activity rivals land use in producing global change, industrial activity became the focus of study for the 1992 institute. The current volume, *Industrial Ecology and Global Change*, provides not only a collection of the papers from the workshop, but, through judicious editing and the addition of coordinating chapters, an introduction to the field of industrial ecology itself.

Approximately 50 people gathered at Snowmass to address the industrial ecology theme. It seemed an unlikely assemblage of expertise and interests. Scientists long involved with climate change, environmental contamination by heavy metals, and biogeochemical cycling joined specialists in technology assessment and industrial policy, as well as economists and social scientists. A major innovation was the inclusion of industrial scientists and environmental managers from the United States and Europe, their counterparts from environmental organizations, and an environmental minister from one of the German states. Participants from China, India, Bangladesh, and Kenya joined colleagues from North America, Western Europe, and Russia to provide a truly global perspective.

The first few days were a struggle in communication. The working sessions revealed the difficulty of developing a consensus among such a diverse group. Fortunately, the industrial representatives kept the discussions grounded in the actual problems they face as their processes and products are implicated as agents of global change, and the environmental advocates prevented us from losing sight of the real political issues. The international composition of the group also assured that the analysis was not confined to the United States or to the industrialized nations.

As is made clear in this book, global issues pose a new set of challenges for the ecological design of industrial processes. Designing industrial processes so that they are compatible with and do not unduly alter global carbon, nitrogen, and sulfur cycles is a major challenge even to formulate, much less to implement. Understanding the implications of wholly new synthetic chemicals and orders-of-magnitude enhancements of the mobilization rates of metals into the biosphere requires a reexamination of the very fundamentals of industrialization.

In order to address these issues, individual facilities and industries must be understood as being embedded in ecosystems at all levels, from local to global. The natural and social scientists who study global change and identify the scientific concerns must communicate better with the government and industry decision-makers who must respond. In particular, to design effective policies and management strategies, the social sciences must be mobilized to transform scientific findings into effective societal options.

It sounds like a tall order, and it is. I hope that this book will stimulate others to lend their disciplines and insights and to join in addressing this intellectually challenging and critical global challenge.

William R. Moomaw
Chair, OIES Steering Committee

Preface

A Novel Dialog

The 1992 Global Change Institute on Industrial Ecology and Global Change was held in Snowmass, Colorado, from July 19 to 31, 1992. Its principal written product is this book.

The institute was designed to encourage some of the first organized discussions between two communities: (1) natural scientists studying global change, and (2) a community that has focused on technological and policy responses to environmental constraints and on the underlying forces driving human activity. In the jargon of global change researchers, but probably nowhere else, this second community is called the "human dimensions" community.

The priority of the natural scientists studying global change is to understand the earth and its ecosystems—to get the science straight. For the past two decades these scientists have been knitting together previously independent disciplines, such as oceanography, atmospheric chemistry, and ecology, while consciously postponing reaching out beyond natural science.

During the same period, a human dimensions community has been forming, which takes as its starting point whatever current level of imperfect understanding of environmental science is available. This community has been trying to anticipate future levels of human impact on the environment—drawing on historical and economic studies of resource use in industry and agriculture, and on studies of contemporary environment–society interactions in all parts of the world. The human dimensions community has also been designing options for environmentally responsive technology and social organization, and they have been exploring the factors that determine whether such options will be successfully pursued.

The U.S. government agencies who have long sponsored the global change research of the natural scientists were the sponsors of the institute. As a member of the human dimensions community, working for many years on energy efficiency as a response to environmental constraints, I was asked to organize the institute. My charge was to create a dialog between the two communities.

There was indeed a gap to be bridged. The two communities scarcely knew one another and were unaware of many of the ways in which they need one another. The institute did much to educate the two communities about the character of their interdependence, and the opportunities for collaboration. Entering the institute, *industrial ecology* was a phrase looking for a mission; emerging from the institute, *industrial ecology* had the mission of blending the complementary kinds of understanding provided by these two communities.

Models of the earth system rooted exclusively in the natural sciences are adequate for capturing the essentials of preindustrial times, when the environmental consequences of human activity were largely local. Such modeling becomes progressively inadequate as one moves to a description of today's earth and then to projections into the future. For a time, many global natural scientists believed that the human dimensions community could provide deterministic submodels of the future course of human activity for direct incorporation into earth-system modeling. While no such naïveté about forecasting was in evidence at Snowmass, the natural scientists attending nonetheless had been unaware of the vigor of the analysis undergirding environment-driven scenarios. They were introduced to a wide-ranging effort on the part of the human dimensions community over the past two decades to explore new technologies, policies, and patterns of economic development. This body of work has direct consequences for the design of robust global modeling and monitoring. It also suggests a new task for global environmental science: to provide a deeper understanding of the environmental consequences of alternative futures.

In turn, the human dimensions community drew its own lessons from the institute. We became more aware of the profound incompleteness of environmental science when posed in the framework of societal problem-solving. In tracking a long-lived organic molecule or metal atom through the environment (say, from atmospheric emission, to soil deposition, to uptake by a plant, to incorporation into the human body, to health effect), there are islands of well-developed science in a sea of partial understanding. Gas exchange at the earth's surface, crucial to the nitrogen cycle and ecosystem function, is determined by processes involving soil bacteria that are poorly understood. An understanding of the impacts of chemicals on the health of almost all organisms other than humans scarcely exists.

Moreover, every ignorance is transient, and no area of certainty is safe from reappraisal. Important discoveries in environmental science are made almost every week.

One message of this incompleteness for the human dimensions community, it seems to me, is to become intimately involved with environmental natural science as a living and breathing enterprise. The better we know our way around, the more we will become aware of new puzzles and new discoveries, and the more we can expect help from the natural scientists in refining our previous prescriptions.

A second message is that whenever asked for advice by political and industrial leaders, the human dimensions community should not recommend strategies that create massive, difficult-to-reverse commitments to specific technologies and policies. Broadly based approaches, and parallel development of preliminary ideas, are preferable.

A Little History
Institute Number Five

The 1992 Global Change Institute was the fifth in a sequence, and its scope is related to its place in that sequence. The first three institutes dealt with issues in which the role of human beings was secondary. Early on, the fourth institute (1991) and ours were planned as a pair, with the idea that between them they would address the principal global environmental impacts arising from human activity.[1]

The global change natural scientists called for two efforts: "land use/cover" (in 1991) and "industrial metabolism" (in 1992). We will say more about the phrase, "industrial metabolism," in a moment. First, what about the dichotomy expressed by two (and only two) "human dimensions" institutes?

From one viewpoint, this pairing was intended to divide human activities into *rural* and *urban*. From another, it was intended to distinguish two classes of effects of human activity on climate: effects on *surface reflectivity (albedo)* and effects on *atmospheric gas concentrations*. From a third, *physical* modifications of the landscape vs. all *chemical* effects. These dichotomies are not congruent: nitrogen fertilizer is rural and chemical, and it affects both albedo and gas concentrations.

The 1991 Global Change Institute, Global Land-Use/Cover Change, focused on agriculture, forestry, and settlement. It dealt with erosion, desertification, and salinization of lands as a result of irrigation. Its core discipline was geography.

Because of the priority of servicing global environmental science, important human relationships to the natural environment were not caught in the nets of either of the two institutes. Neither institute did justice to human health and human sensibilities. Neither institute made room for informed discussion of lung function, or carcinogenesis, or sterility, or epidemics. Nor did either institute consider in much depth the sorts of aesthetic insults that arise from the degraded transparency of mountain air or from the polyethylene bottle discarded years before at the side of a trail.

Industrial Metabolism and Industrial Ecology

Both "industrial ecology" and "industrial metabolism" are phrases associated with important antecedents of our institute. One research program, carried out under the banner of *industrial metabolism*, has viewed the industrial system as a single entity and has explored its system-wide transformations of materials. Its leaders, Robert Ayres and William Stigliani, both participants in our institute, have been conducting seminal studies of materials flows through the world's industrial system. They have focused on the long-term habitability of the planet, with a corresponding emphasis on documenting the routes to toxification of the global environment.

---

[1] The sixth Global Change Institute (1993) returned to a pure-science theme, the global carbon cycle.

A parallel research program, under the banner of *industrial ecology*, launched by Brad Allenby, Robert Frosch, Tom Graedel, and Kumar Patel (Graedel and Patel were institute participants) has emphasized the industrial firm as agent of change, and has located its analysis at the level of specific industries. Of special interest are the relationships among industries, and the opportunities for the wastes of one industry to become useful inputs to a second.

The word "industrial" has had different meanings for the two groups. "Industrial," in the phrase "industrial metabolism," refers to the totality of civilization. In the phrase, "industrial ecology," it refers, more narrowly, to the activities of specific industrial producers. Industrial metabolism addresses transportation; industrial ecology addresses the automobile, aircraft, petroleum, tire, battery, and other transport industries.

Our institute strove to weave together both strands of research and to demonstrate their enhanced value in combination. It was my decision not to invent a third phrase, but rather to choose one of the two and to set out deliberately to expand its meaning. I chose "ecology" over "metabolism." I am gambling that the phrase "industrial ecology" can be transformed by our exercise, and others, to carry the larger meaning given expression in this book. "Industrial ecology" is intended to mean both the interaction of global industrial civilization with the natural environment and the aggregate of opportunities for individual industries to transform their relationships with the natural environment. It is intended to embrace all industrial activity, specifically including agriculture; both production and consumption; and national economies at all levels of industrialization.

The Snowmass Institute

The most demanding rule of the institute was that everyone had to be in attendance for the full two weeks. In combination with the remoteness of the site, this rule made us feel and act like a sequestered jury. The site was beautiful and hikable. We lost track of much of the outside world.

None of the 50 participants knew even one-half of the others before the institute began. More important, none of us was fluent in even one-half of the disciplinary languages. Deliberate confrontation was the organizing principle behind assignments to subgroups and assignments as official "reader" of another's paper.

The Institute and This Book

Interchanges among institute participants during and after the Global Change Institute have shaped all of the chapters of this book. Many of the chapters are based on drafts that were revised at the institute. Several multiauthored chapters are the result of collaborations initiated there. A few chapters were written by participants after the institute, at the urging of the editors, to close some gaps.

The institute was structured around three Working Groups:

I.   Grand Nutrient Cycles
II. Exotic Intrusions
III. Implementation of Industrial Ecology.

The five sections of this book reflect this structure: Section 1 contains five cross-cutting papers; Sections 2 and 3 contain the papers exploring the topics addressed in Working Groups I and II, respectively; and Sections 4 and 5 address, respectively, the implications of the themes of Working Group III for industrial firms and for governments.

An interesting ecological distinction is inherent in the purviews of Working Groups I and II. Working Group I explored human intrusions into the grand bio-geochemical cycles (carbon, nitrogen, phosphorus, sulfur). Working Group II gave its attention to those human intrusions that are largely unfamiliar ("exotic," "xeno-biotic") from the point of view of the ecosystem (such as "toxic" chemicals, metals, and radioactivity). The distinction is analogous to the distinction between visiting a country where you speak the language and where you don't: we and the plants both speak organic carbon, but only we speak polychlorinated biphenyl. The two categories of involvement of human activity with the earth as a whole are qualitatively different.

Working Group III worked off the axis defined by the first two. It adopted an action orientation towards industrial ecology, while searching for general principles. It explicitly considered current industrial practices and productive strategies for change.

Throughout the second week at Snowmass, the Working Groups concentrated on preparing written reports. Little of this writing is found verbatim in this book. The reports were valiant efforts, full of interesting ideas. Many of these ideas are found in this book in the form of supplementary papers, revised versions of papers, and introductions to sections. But some of the efforts have been left on the cutting room floor, when the editors judged that the work remaining to be done was beyond our powers.

Statement of Goals

I conclude with the seven goals in front of us during the institute:

1. To add coherence to the thinking about industrial ecology, by clarifying what is more and less important and what is well and poorly understood. In short, to launch a field.
2. To inform and give guidance to those activities currently under way or being planned, in industry and elsewhere, designed to reduce the adverse environmental impacts of industrial activities.
3. To locate industrial ecology within broader research fields, in particular, within global change research.
4. To help creative individuals find productive research strategies in industrial ecology.

5. To foster interactions that lead to joint research.
6. To lower thresholds in key federal agencies and corporations for the financial support of research in industrial ecology.
7. To produce a written record, in the form of a book, that is worth reading and keeping around.

Robert H. Socolow
Princeton, NJ

Acknowledgments

We would like to thank our fellow participants at Snowmass for their commitment to working with one another and with us in the preparation of this book. We cannot imagine a tutorial experience that could have been more effective and enjoyable in awakening us to new issues and ways of thinking.

Tom Graedel and Bill Moomaw were deeply involved in structuring and conducting the Global Change Institute and in inviting participants. Along with other members of the Organizing Committee, they also gave invaluable advice during the planning period.

At Snowmass six moderators and six rapporteurs brought partial order to creative ferment in the three Working Groups. Their efforts are reflected in much of the reworking of draft manuscripts for publication.

The Office for Interdisciplinary Earth Studies of the University Corporation for Atmospheric Research was the institutional and spiritual home for this institute. Its departing director, Jack Eddy, secured funding, helped recruit the Organizing Committee, and gave the Global Change Institute its bearings. Sarah Danaher and Pamela Witter brought exceptional professionalism to arrangements for the parti-cipants' welfare, as well as to the conduct of the meeting itself. They were joined in teamwork by Paula Robinson, Lisa Butler, and Diane Ehret. Diane also had responsibility for all manuscripts through numerous revisions. Tom Wigley, as incoming director, found our publisher and offered invaluable editorial advice.

The project is indebted to Jean Wiggs and Michele Brown at Princeton for carrying much of the burden of organizing the institute, and to Samantha Kanaga for drawing and redrawing all of the figures in this book.

Top of the Village, at Snowmass, site of our institute, was run beautifully on our behalf by Stephanie Peterson and David Spence.

The Rockies themselves deserve our thanks. The air 9000 feet (2700 meters) above sea level cleared our heads, taxed our lungs, and strengthened our hearts. A few minutes of hiking on managed ski slopes led to trails well suited to the building of friendships.

The institute and the publication of this volume were made possible by continuing and generous support from the Subcommittee on Global Change Research of

Acknowledgments

the interagency Committee on Earth and Environmental Sciences. The Committee is supported by the agency members of the U.S. Global Change Research Program.

Contributors

Below are the names and affiliations of participants in the OIES 1992 Global Change Institute and contributors to this book.

Stefan Anderberg is currently a researcher at the Department of Social and Economic Geography at Lund University in Sweden. He was previously a research scholar in the Chemical Pollution of the Rhine Basin Project at the International Institute for Applied Systems Analysis in Austria. He is interested in regional comparisons of trends and patterns of chemical pollution.

Susan Anderson is staff scientist in the Energy and Environment Division at the Lawrence Berkeley Laboratory. She studies reproductive and genotoxic effects of environmental contamination on ecosystems. She has also worked on water quality policy and management at the San Francisco Bay Water Quality Control Board. In 1992 she was awarded a Pew Scholarship in Conservation and the Environment.

Clinton Andrews is assistant professor of public and international affairs in the Woodrow Wilson School at Princeton University. His interests include decision science, energy and environmental planning, and regulatory policy. His recent focus has been on the electric power sector.

David Angel is associate professor at the Graduate School of Geography and George Perkins Marsh Institute, Clark University. His research addresses issues of innovation, technical change, and the environment. As Abe Fellow he is also conducting research in Japan.

Robert Ayres is Sandoz professor of environment and management and professor of environmental economics at The European Institute of Business Administration (INSEAD), Fontainebleau, France, and adjunct professor of mineral economics at the Pennsylvania State University. At INSEAD he is co-director of the Centre for the Management of Environmental Resources (CMER). His current research interests include flows of materials through the economy (industrial metabolism), measures of environmental disturbance, and the reorganization of industry to respond to environmental constraints (eco-restructuring).

Frans Berkhout is research fellow at the Science Policy Research Unit at the University of Sussex in the United Kingdom. From 1992 to early 1994 he was research associate at the Center for Energy and Environmental Studies at Princeton. His research interests include nuclear non-proliferation, nuclear power economics, materials life-cycle management, and the diffusion of clean technology.

Michael Braungart is the founding president of the Hamburg Environmental Institute, a non-profit organization giving scientific advice and undertaking pilot projects related to environmental problems. He is also director of the EPEA-Environmental Encouragement Agency, an environmental consultancy aiming to develop "environmentally intelligent" products. He previously worked for Greenpeace Germany and has a chemistry Ph.D.

Stephen Bunker is a professor in the Department of Sociology at the University of Wisconsin-Madison. He has conducted environmental sociology research in Africa and Latin American, and recently received a National Science Foundation grant for research on resource extraction in the Brazilian Amazon.

Robin Cantor is the program director for Decision, Risk, and Management Science at the National Science Foundation. Her recent work has included the development of a research program, based on an accounting framework, to evaluate the external costs of fuel cycles used to produce energy services, and an analysis of the possibilities for cost-sharing arrangements between local jurisdictions and other governmental agencies to clean up hazardous waste sites.

Robert Chen is director of Interdisciplinary Research with the Consortium for International Earth Science Information Network in Saginaw, Michigan, and Adjunct Associate Professor of World Hunger at Brown University. He has held research appointments at Brown University, at the International Institute for Applied Systems Analysis, and with the U.S. National Academy of Sciences. His research interests include global environmental change and its human dimensions, hunger and long-term food security, and the integration of natural and social science data.

Daniel Deudney is Bers Assistant Professor of the Social Sciences, Department of Politics, University of Pennsylvania. His areas of research and teaching are international relations, global environmental politics, and political theory.

Elizabeth Economy is currently an Associate Fellow at the Council on Foreign Relations. Her main research interests are environmental and energy issues in China and Russia. She also works on problems of collective action and environmental regime formation.

Steve Fetter is special assistant to the Assistant Secretary of Defense for Nuclear Security and Counter Proliferation at the U.S. Department of Defense. He is on leave from the University of Maryland, College Park, where he is an associate professor in the School of Public Affairs. His research interests include environmental and national-security policy.

Wayne France is head of the Environmental Research Department at the General Motors Research and Development Center. He is also secretary to the GM Science Advisory Committee, which provides outside advice to the GM Strategy Board on technical and scientific matters.

Ashok Gadgil is a staff scientist in the Energy and Environment Division, Indoor Environment Program, Lawrence Berkeley Laboratory, University of California, Berkeley. His research interests include indoor air pollution, radon in buildings, energy-efficient technology, and energy efficiency in developing countries. In 1991 he received the Pew Award for Scholarship in Conservation and the Environment.

James Galloway is professor of environmental sciences at the University of Virginia. His research interests include atmospheric/biospheric interactions and the past, present, and future impacts of humans on biogeochemical cycles.

George Golitsyn is director of the Institute of Atmospheric Physics of the Russian Academy of Sciences in Moscow. He is chairman of the Council of the International Institute for Applied System Analysis (Laxenburg, Austria) and a member of the Joint Scientific Committee for the World Climate Research Program.

Thomas Graedel is a distinguished member of the technical staff of AT&T Bell Laboratories. His interests include trace gases in the atmosphere and industrial ecology. He is the convener of the Global Emission Inventory Project of the International Global Atmospheric Chemistry Program.

Monika Griefahn is minister of the environment of Lower Saxony, Germany. She was one of the founders of Greenpeace Germany. She is a member of the commission of the federal Social Democratic Party to create a government program responsible for the environment.

Olga Gritsai is professor of social and economic geography in the Institute of Geography, Russian Academy of Sciences. Her interests include regional development and regional policy, and industrial systems and global change. She is a member of the executive committee of the Human Dimensions of Global Change Program and also a member of National Committee for the International Geosphere–Biosphere Program.

Arnulf Grübler is with the Environmentally Compatible Energy Strategies Project of the International Institute for Applied Systems Analysis (IIASA), Laxenburg, Austria. His research interests include long-term technology change in energy and transport systems and allocation regimes for greenhouse gases.

Robert Harriss is professor of earth system science at the University of New Hampshire. His current research interests focus on sources of atmospheric greenhouse gases. He teaches on environmental and energy science and policy.

Bette Hileman is senior editor at *Chemical & Engineering News*. She writes primarily on environmental issues of concern to chemists. She did graduate work in physical chemistry at the Massachusetts Institute of Technology.

Inge Horkeby is head of the Environmental Protection Department at AB Volvo in Gothenburg. He is a chemical engineer with previous experience in paints and finishings development.

Saleemul Huq is the executive director of the Bangladesh Center for Advanced Studies. He is an associate of the Royal College of Science, U.K., and a member of the Institute of Biology, U.K. He was awarded the Duggan Fellowship from the Natural Resources Defense Council in 1989 and the Robert McNamara Fellowship from the World Bank, 1986–87.

Peter Jaffé is an associate professor in the Department of Civil Engineering and Operations Research at Princeton University. His research interests include the chemical, physical, and biological fate of conventional and toxic pollutants in surface and groundwater systems; water quality modeling; and water pollution control.

Jiang Zhenping is a senior research fellow at the Energy Research Institute and the deputy of the Energy System Analysis Division of the State Planning Commission in Beijing, China. He is currently working on "The International Energy Data Base" sponsored by the National Natural Sciences Foundation of China.

Yuri Kononov is chief economist with the Siberian Energy Institute at Irkutsk, Russia. His research addresses the consequences of alternative energy strategies for Russian economic development and the environment. He has made detailed studies of Siberian energy resources, the national energy system of Russia, and the global industrial ecology of carbon.

Vladimir Kotlyakov is director of the Institute of Geography, Russian Academy of Sciences and Chairman of the Russian National Committee for the International Geosphere–Biosphere Program and the Human Dimensions of Global Change Program. He is a geographer who has focused on environmental sciences and glaciology. He is a member of the Russian Academy of Sciences and vice president of the International Geographical Union.

Daniel Lashof is a senior scientist with the Natural Resources Defense Council. Before joining NRDC, he served on the staff of the U.S. Environmental Protection Agency where he was lead author of the report to Congress "Policy Options for Stabilizing Global Climate."

James McNeal is deputy assistant chief geologist in the Geologic Division at the U.S. Geological Survey. A geochemist, he has previously conducted research into the occurrence and distribution of elements in the environment.

William Moomaw is a professor of international environmental policy and director of the International Environment and Resource Program at the Fletcher School of Law and Diplomacy, Tufts University. He currently chairs the Steering Committee of the University Corporation for Atmospheric Research's Office for Interdisciplinary Earth Studies.

Kenneth Nelson is the president of KENTEC, Inc. He was previously manager of energy conservation for Dow Chemical USA. He was recognized as 1990 energy manager of the year for outstanding contributions to industrial energy conservation.

Victoria Norberg-Bohm is assistant professor of environmental policy in the Department of Urban Studies and Planning at Massachusetts Institute of Technology. She has recently completed her Ph.D. dissertation on technology transfer in the electric power sector in Mexico at the Kennedy School of Government at Harvard University. Her current research is on the industrial ecology of chlorine.

Jerome Nriagu is professor in the Department of Environmental and Industrial Health at the School of Public Health at the University of Michigan, Ann Arbor. Until 1992 he was research scientist with Environment Canada for over 20 years. His research interests include toxic metals in the environment. He is a fellow of the Royal Society of Canada.

Jackton Boma Ojwang is professor of law at the University of Nairobi and professorial fellow at the African Center for Technology Studies, Nairobi. His current research interests focus on the role of public law in the formulation of policy and in the design of mechanisms for environmental protection.

Theodore Panayotou is a fellow of the Harvard Institute of International Development and a lecturer in the Department of Economics at Harvard University. A specialist in environmental and resource economics, environmental policy analysis, and development economics, he has advised governments and institutes in Asia, Africa, and Eastern Europe. In 1991 he received the Distinguished Achievement Award of the Society for Conservation Biology.

Kumar Patel is vice chancellor of research at the University of California, Los Angeles. Until March 1993, he was executive director, Research, Materials Science, Engineering and Academic Affairs at AT&T Bell Laboratories, Murray Hill, New Jersey. His work on gas lasers in the early 1960s was followed by contributions in non-linear optics, molecular spectroscopy, pollution detection, and laser surgery. He is president of the American Physical Society.

Bruce Paton is a quality consultant at Hewlett-Packard Company. His responsibilities at HP have included roles in environmental management, Total Quality Management, and business process engineering.

V. Ramanathan is the Alderson Professor of Ocean Sciences at Scripps Institution of Oceanography at the University of California, San Diego, and director of the National Science Foundation Center for Clouds, Chemistry, and Climate. He has worked on the role of chlorofluorocarbons in global climate warming, and has published articles dealing with the greenhouse effect of trace gases, the earth's radiation budget, the general circulation of the atmosphere, and climate change.

Steve Rayner is senior program manager for Global Environmental Management at Battelle Pacific Northwest Laboratories where he also leads the Global Climate Change Group. His research interests include global environmental policy and management; science, technology, and public policy; and risk management.

William Schlesinger is a professor in the Departments of Botany and Geology at Duke University. His research and teaching interests include ecosystem analysis, global change, and biogeochemical cycling.

Jerald Schnoor is a professor in the Department of Civil and Environmental Engineering and co-director of the Center for Global and Regional Environmental Research at the University of Iowa. His research interests include surface water and groundwater quality modeling, exposure assessment, hazardous substances remediation, and global climate change.

Lowell Smith is a research manager for the U.S. Environmental Protection Agency with a principal focus on the agency's Global Change Research Program. He participates in a wide range of research and assessment activities supporting U.S. government's positions at the Intergovernmental Panel on Climate Change.

Robert Socolow is the director of the Center for Energy and Environmental Studies, and professor of Mechanical and Aerospace Engineering at Princeton University. His research interests include energy efficiency in buildings, energy–environment interactions, and environmental constraints on global development.

Richard Somerville is professor of meteorology at Scripps Institution for Oceanography, University of California, San Diego. He serves as director of the Climate Research Division, a group of scientists studying variability and predictability of the earth's climate. His major research interest is the greenhouse effect and global climate change.

Richard Sonnenblick is a doctoral candidate at the Department of Engineering and Public Policy at Carnegie-Mellon University. He is also senior research associate at the Energy Analysis Program at the Lawrence Berkeley Laboratory. His current research is into the economics of energy conservation and the performance of demand-side management programs.

Thomas Spiro is Eugene Higgins Professor of Chemistry and former chairman of the Chemistry Department at Princeton University. His research interests include laser spectroscopic studies of molecular structure, the role of metals in biology, metal binding and mobilization in soils, and industrial ecology.

William Stigliani is a senior research scholar and scientific leader of the Project on Regional Material-Balance Approaches to Long-Term Environmental Policy Planning at the International Institute for Applied Systems Analysis (IIASA), Laxenburg, Austria. He studies the production, transport, and fate of chemicals in the biosphere.

Valerie Thomas is a member of the research staff of the Center for Energy and Environmental Studies at Princeton University. Her research interests include industrial ecology and sustainability, and assessment of pollutant transport and exposures.

Mark Tullis is a research analyst at the Global Development and Environment Institute at Tufts University. His research interests include climate change policy, energy conservation, joint implementation, and environmental technology transfer.

Robert Williams is a senior research scientist at the Center for Energy and Environmental Studies, Princeton University. His current research explores advanced technologies for renewable energy and strategies for accelerating their development. He was awarded a MacArthur Fellowship in 1993.

Clifford Zinnes is an institute associate at the Harvard Institute for International Development and a lecturer on economics at Harvard University. His current research addresses sustainable development and institutional governance.

OVERVIEW

1

Six Perspectives from Industrial Ecology

Robert Socolow

Exploring Industrial Ecology

People are great rearrangers of the earth. Metals that have been locked away in the veins of rocks over the eons of prehistory are mined, freed from their oxide or sulfide drabness, and allowed to shine or cut or channel electrons for our pleasure, for a few decades at most, before being dispersed without plan in soils and streams. Porous sediments more than a kilometer below ground, soaked with oil or laden with natural gas, are penetrated by drill holes to release their burnable contents; people are provided mobility or comfort for a brief moment by the energy accompanying the oxidation of these fuels, and in less than a century the global atmosphere registers five molecules of carbon dioxide for every four that were there before. Chemicals that never existed in the history of our planet are synthesized for the killing of weeds or insects, or for the cooling of transformers. Radioactive isotopes that had decayed to oblivion early in our planet's history are recreated, as the fission of uranium provides another source of electricity and heat. *Industrial ecology* is a metaphor for looking at our civilization through such lenses.

The metaphor of industrial ecology also leads us to look at interrelationships. The interrelationships among producers and consumers determine what becomes waste and what is usable, and how the "natural" is combined with the "synthetic." Industrial ecology explores reconfigurations of industrial activity in response to knowledge of environmental consequences. It is intended to stimulate the imagination and enlarge the sense of the possible, with regard to industrial innovation and social organization. It offers a fresh view of environmental management.

Industrial ecology provides six perspectives:

Long-Term Habitability
The dominant perspective for environmental analysis expands in time: from short-term insult to long-term habitability. Priority issues include persistence of toxic chemicals, depletion, and disruption of grand life-supporting cycles.

Global Scope
The dominant perspective expands in space, from local insult to regional and global impact. "Global" is used in both the sense of global spatial scale (climate) and

3

problems found universally on the globe (pesticides, urban air quality). Economic development paths for the less developed countries receive special attention.

The Overwhelming of Natural Systems

Nature is the measure of man. To overwhelm implies a focus on a ratio, changing over time, that compares some human enterprise with, typically, some pre-existing characteristic of the natural environment. Such ratios complement the traditional indices of economic activity.

Vulnerability

Man is the measure of nature. Vulnerability, and its opposite, *resilience*, establish an appropriately comprehensive scale to evaluate the significance of natural hazard, ecosystem disruption, and disease, for countries, communities, institutions, and individuals. Vulnerability addresses the overwhelming of *human* systems.

Mass-Flow Analysis

Materials are tracked relentlessly through time and space. Mass-flow analysis, well known in chemical engineering, builds on the conservation laws of classical physics (mass cannot be destroyed, atoms are stable, energy can be degraded but does not disappear).

Centrality of the Firm and the Farm

The industrial firm and the industrial farm are put on an equal footing with the household/consumer/voter. The view of the firm and farm changes from culprit to agent of change.

These six perspectives will be outlined in this brief essay, with an emphasis on the concepts they highlight and the integrative frameworks they reveal. The chapters that follow will demonstrate these perspectives in action, in industry and in government. In the aggregate, these six perspectives add up to "new thinking." The contrast, here, is with the more narrowly conceived environmental management perspective ("old thinking") that has been dominant for the past two decades.

The new thinking is as radical for the environmental activist as for the industrialist. The legacy of two decades of environmental regulation is a system of environmental management that has accommodated the strategies of many interest groups. The focus of attention has long been on production facilities—conveniently far from the point of consumer involvement. By contrast, industrial ecology emphasizes the management of products throughout their useful life and beyond, and calls attention to dispersed sources of pollution, such as agricultural chemicals, household wastes, and the chaff resulting from the expected degradation of products like outdoor paints, roofing materials, and brake linings. The importance of the consumer is unmistakable. Rage at the industrial producer recedes in significance as a driver of policy.

Any claim to "new thinking" deserves to be treated with skepticism. Is there something new here, or just a repackaging of common sense? How well grounded are these six perspectives? Is an elitist preference system being proposed, masquerading as rationality? How promising is *any* program advocating rationality, in an irrational world? Won't the program envisioned here be undone by a rage directed against environmentalist–industrialist alliances that interfere with ordinary living? Such doubts cannot easily be dispelled. The ultimate value of the these six perspectives will be established by their ability to transform the environmental agenda in productive directions.

Long-Term Habitability

Environmentalism has been fueled by short-term insult: the beach made unusable by sewage, urban air that keeps children from playing in schoolyards, gasoline lines that result when world energy markets go awry. Industrial ecology addresses a different dimension of environmentalism, not short-term distress but long-term habitability. Quantitatively, attention to habitability stretches the time frame of concern to several decades, even a century. While such a time frame is rarely encountered within local and national environmental regulatory regimes, it is the time frame underlying recently proposed international environmental agreements, such as in climate and forestry. In adopting the framework of habitability, industrial ecology is borrowing a central concept of global change research and inserting it into pragmatic realms like industrial development and environmental regulation.

Persistence and Chemical Toxification

Although much less familiar as a global issue than climate modification, toxification of the environment is another chemical path to loss of habitability. Toxification stalks global change research and is gradually achieving greater prominence. Toxification of soils is one concern: Little is known about the extent to which future nutrient imbalances and loadings of metals and chemicals will diminish the yields currently obtained on agricultural land. Toxification high up the food chain is another concern: Many persistent chemicals concentrate in the food chain, and some of these chemicals reach toxic levels in such varied organisms as raptors, large fish, and human beings. Direct toxification of human environments without the intermediary of environmental processes is a third concern: When persistent chemicals find industrial uses in stable settings, as with lead in the paint on old buildings, the innocent practices of one generation become the hazards of the next.

None of these forms of toxification would occur without durable industrial materials. Yet achieving durability used to be an unquestioned objective of industry. We have been re-educated by the story of the chlorofluorocarbons (CFCs, such as Freon), introduced into the modern economy as refrigerants, foaming agents, spray-can propellants, and cleaning agents, largely because they were not combustible or toxic or reactive with other chemicals. The most widely used CFCs are now being phased out because of their capacity to thin the ozone layer in the stratosphere. It turns out that the same inertness that makes CFCs desirable industrial products allows them to rise intact into the stratosphere, instead of succumbing to the chemical assaults in the radical-rich oxidizing environment of the lower atmosphere that transform most other molecules. We now understand that durability can be a two-edged sword.

Depletion and Physical Degradation

The inhomogeneities of the physical world are the natural endowment of the human species. For many elements, the accessible sites of unusual excesses over average crustal concentrations have become commercial mines. Unusually capacious aquifers have become water supplies. Extraordinary accumulations of fossil fuels have become active drilling provinces. With industrial activity, our natural endowment is dissipated. In the language of thermodynamics, industrial activity is increasing the earth's entropy, making it more similar from place to place. A small exception is where human activity creates a new kind of mine for some future society by reconcentrating an element in a dump or landfill.

Human activity produces not only chemical, but also physical degradation of habitability. Topsoil is blown away. Riverbanks are eroded. Land subsides at sites where natural gas is extracted or water is removed from aquifers. Water is lost from river valleys by evaporation at lakes behind dams. Lakes and inland seas shrink in size (the Aral Sea is a notorious example) when their water sources are diverted for irrigation.

Extinction and Biological Simplification

Loss of habitability is above all a biological issue. Here, more than with chemical or physical degradation, one confronts irreversibility, in the form of loss of species. Loss of species diminishes the robustness of ecosystems. It also removes from the human future a source of enjoyment, education, and possible direct material benefit. In principle (but rarely in practice), one could use some energy source in very large amounts to reconcentrate a mineral resource or to reconstitute an aquifer. Not even in principle can one recover lost species.

Unlike most chemical and physical degradation of the environment, ecological degradation can be abrupt. Fish populations can plunge suddenly when potent industrial maritime technologies for "harvesting" the oceans are introduced. Like physical degradation, biological degradation is, in part, a matter of lost variability and increased entropy. When an ecological system rich with endemic species loses its isolation by breaches at its boundary or by the introduction from afar of common, hardy species, it soon loses its distinctiveness.

When can loss of habitability be reversed? This question is central to the rehabilitation of formerly productive and biologically diverse ecosystems that are now degraded and simplified. In the United States managers of rangelands, forests, and

6

fisheries have been gradually changing their objectives, away from new production and toward rehabilitation and repair. The new objectives are unfamiliar. Highway builders long resisted shifting their attention from the building of new roads to repair and maintenance. But shift they did. So, now, with land and sea.

Global Scope

Most of the impetus to restructure industrial processes continues to come from local impacts: to make a factory into a better neighbor, to comply with local regulations on emissions, to facilitate the siting of facilities in the face of the "Not in My Back Yard" syndrome (NIMBY). Industrial ecology adds an unfamiliar, yet refreshing, challenge by asking which of the strategies of environmental control that emerge in response to local and regional concerns are further justified by global concerns. The dominant perspective expands not only in time but also in space.

Global change research has traditionally addressed two kinds of global systems: (1) systems intrinsically as large in size as the earth, and (2) small but ubiquitous systems. However, research into small, ubiquitous systems, such as rice fields, has focused nearly exclusively on their capacity to induce changes in systems of the first kind, especially climate. Global changes brought about by changes in small but ubiquitous systems not mediated by climate have generally been neglected, such as changes in agricultural practices that could be affecting soil productivity everywhere. A broader framework for global change research would include all disruptions that are ubiquitous, that result from human activities that are themselves ubiquitous, and that have cumulative effects of worldwide significance. Admitted into the canon, for example, would be investigations of the extent to which the various metals mobilized by industrial activity have worldwide deleterious impacts on ecosystems.

Increasingly, local-scale disruptions share common characteristics at sites spatially far apart, because industrialization varies so little in its details from country to country. New automobiles and fuels, new agricultural practices, and new tax structures spread quickly around the world. Everyone is imitating everyone else. Accordingly, studies of the planetary impacts of ubiquitous disruptions of local-scale systems become increasingly important.

Inextricably embedded in a global perspective is an egalitarian morality: everyone, everywhere counts. Industrial ecology integrates over all individuals and all nations. The world's resources are everyone's entitlement and everyone's responsibility. Everyone has a stake in the world's future population and its distribution, as well as in how each of us chooses to live.

With few exceptions, all individuals today wish to live like individuals materially richer than themselves. And all nations aspire to resemble the wealthiest nations. These central facts add portent to the choices among paths of industrialization and population growth made in the less developed countries, and among paths of industrial intensification made in the already industrialized countries. Departures from business as usual are necessary in all countries, at every level of industrialization.

The global perspective of industrial ecology calls attention to the need to combine research on physical systems of global significance with related social science. Behavioral science can be asked to explore how individuals integrate their self-centeredness and tribalism with global concerns. Policy science can be asked for critical assessments of international public and private institutions, with particular attention to their effectiveness in fostering innovation. There is an urgent need to identify, for all countries, effective strategies to build human capital and institutional capacity and to acquire and disseminate information and technique. Lessons can be sought from the international institutions currently at the service of agriculture and public health. As explained in the chapter by Golitsyn (this volume), military conversion offers opportunities, such as augmented capabilities in instrumentation and analysis; these merit attention.

The Overwhelming of Natural Systems

Industrial ecology puts "people" and the "nonhuman environment" into the same picture. Thereby, it gives rise to a set of useful hybrid concepts that capture the interactions of one with the other. Two hybrid concepts are particularly useful: the concept of the overwhelming of natural systems by human activity and the concept of the vulnerability of human systems to natural processes.

In the present period, human beings are perturbing the planet's natural processes significantly on a global scale. We are overwhelming both regional and global environmental systems: lakes, airsheds, fisheries, forests, the ozone layer in the stratosphere, global climate. Our planet has become uncomfortably small.

A central line of inquiry within industrial ecology is directed toward understanding in detail which natural systems are particularly sensitive and therefore likely to be overwhelmed, and how they are affected by particular human activities that appear likely to grow substantially as a result of industrial development. For the first undertaking—identifying sensitive components of the natural environment—environmental science has made some educated guesses: three examples are stratospheric ozone, the Arctic ice cap, and the soil of tropical forests. Today's list is surely incomplete, and it will be improved by a deeper understanding of nonlinear systems. For the second undertaking—identifying fast-growing impacts of human activity—history is one guide, demography a second, and cross-cultural studies of economic development are a third.

Remarkable long-term data sets, such as those related to energy intensity and fuel use, have been teased out of historical records. As the chapter by Grübler in this volume attests, these data sets give helpful insights regarding the engines of industrial growth and the patterns of spread of industrial practices.

Complementing historical studies are the efforts of demographers to relate population, migration, and urbanization to agricultural and industrial activity. Of greatest importance is the total number of people. The need to feed growing numbers of people, for example, drives the intensification of the use of fertilizers and other agricultural chemicals. Indeed, the prevailing methodology for describing environmental consequences of human activity first analyzes per capita activity (diet, habitat, mobility) and then multiplies by total numbers of people. With the greater involvement of demographers, a more refined analysis could be developed that takes into account age structure, family size, population density (both sprawl and crowding), migration within countries and across borders, and other demographic variables.

Cultural studies are a further important component of the analysis of levels of human activity and their consequences. A small set of resource-intensive strategies dominates economic development today, with no rivals, in agriculture, construction, transportation, communications, and other economic sectors. If alternative, less resource-intensive strategies were to be adopted widely, the consequences could include significant reductions in the rates at which, globally and locally, human activity overwhelms natural systems. Cultural studies can identify circumstances where such alternative strategies might be invented by societies blending modernization with preindustrial traditions.

The task of comparing present and potential levels of human activity to thresholds, absorptive capacities, and other quantitative measures of stress on the natural environment is one of the frontiers of industrial ecology. To make progress requires modifying the modus operandi that separates research in natural science from studies of economic development. Natural scientists would surely be giving greater priority to studies of the nitrogen cycle today, for example, if they better appreciated the relentless growth in the production of nitrogen fertilizer that accompanies the intensification of agriculture.

The task of clarifying, case-by-case, the meanings of "overwhelm" abuts another task, that similarly requires a new kind of collaboration, and to which we turn next: What determines how much strain the stresses on a natural system will cause individual human beings, their communities, and their institutions?

Vulnerability

The historical records of natural disasters and, to a limited extent, mathematical models of the weather system, document the frequency of occurrence of acute events such as floods, hurricanes, and typhoons, as well as more chronic departures from average environmental conditions such as prolonged droughts, the accumulation of salts and metals in soil, and a rise in sea level. Elaborate technical apparatus is available to the natural scientist to summarize the implications of this information in probabilistic statements of risk: the height of a river in a "hundred-year" flood, for example.

Such risk analysis for communities is seriously incomplete as a description of the human suffering and economic cost that accompanies nature's excesses. It leaves out any analysis of the capacity of one community to recover from a catastrophe that would devastate another community. One has only to compare the human impact resulting from the Mississippi floods in the summer of 1993 with those from the episodic monsoon floods over the Ganges delta in Bangladesh.

Vulnerability analysis complements risk analysis: It seeks to understand the capacity of a country or community to protect against and cope with both acute disasters and continual incremental environmental changes.

Vulnerability culminates a chain of concepts of increasing inclusivity describing the indirect consequences for human society of the direct environmental manifestations of day-to-day human activities. First, there are measures of *emissions*, rates of introduction of pollutants into the environment; emissions are the target of most environmental regulatory activity and industrial response, as in automobiles and electric power plants. Next, with the addition of an understanding of dispersal and concentration mechanisms in air, water, soil, and industrial practices, emissions are linked to *bioavailability*, a measure of potential for impact on ecosystems and human beings. Linked to bioavailability are *exposure* and *dose*, measures of actual impact that reflect where plants actually grow and what chemicals they mobilize, what people and animals eat and drink, and where they breathe or swim.

Attached to the link from bioavailability to dose is the link from dose to *incidence of disease* and *incidence of loss of ecosystem function*. To develop an understanding of this link requires toxicology and ecotoxicology. Ecotoxicology today is extending knowledge of chemical effects beyond our own species (and a few others studied in medical research) to other species, and beyond individual species to ecosystems. Toxicology is extending our knowledge of human impacts beyond cancer to other biomedical systems.

Thomas and Spiro take us along this part of the chain, in their chapter in this volume, through an analysis of lead that begins with emissions and gets as far as impacts on human health and ecosystem function. Small and particular uses of lead are responsible for most of the environmental damage due to lead moving through the economy. Two examples are the use of lead in gasoline, which becomes particulate lead in urban dust, in turn eaten by young children; and the use of lead in hunters' shot, that becomes lead embedded in the flesh of birds, in turn eaten by other birds. Damage from such uses dominates the damage from other uses of lead in the industrial economy, such as in batteries, that are larger in magnitude but less well linked to ingestion and inhalation.

An analysis of air pollution that considers total exposure and its health impacts, rather than terminating with emissions analysis, leads to a similar shift in attention, in this case from emissions of pollutants into outdoor air to inhalation of pollutants in indoor air. Emissions from cooking, radon drawn into buildings from the soil below, "passive" cigarette smoke, and formaldehyde from furniture and carpets are identified as significant sources for human beings, when a 24-hour-average daily dose is constructed.

When we are determined to confront consequences of damage, rather than only levels of damage, we are brought to the final link in the chain, a link that connects incidence of human disease and ecosystem disruption to *vulnerability*. To make progress exploring vulnerability, and its opposite, *resilience*, requires the further incorporation of social science. Will an ecosystem malfunction be recognized, and are there resources available for restoration? In the case of human illness, are resources available for medical care? Will the illness lead to stigmatization or be socially accepted? The insights from environmental and medical sciences that link emissions to incidence of disruption and illness are made complete by an understanding of how communities and individuals handle stress.

To understand vulnerability requires cultural understanding, including an understanding of wealth and poverty. It also requires an understanding of dysfunctions within communities at all levels of industrialization—dysfunctions such as the domination of waste management by organized crime, and the domination of decisions about the use of chemicals by the purveyors of chemicals.

The better we understand vulnerability, the more effectively we will be able to intervene at an early link in the chain. There is a role, for example, for warning systems and for buffering capacity. What kind of warning system can alert society to vulnerability that is created incrementally, by cumulative effects that extend over several human generations? How can a robust buffering capacity be designed into the management of a natural system, like a river and/or airshed, so that society has room to maneuver and to accommodate environmental surprises and unforeseen future demands?

Mass-Flow Analysis

Industrial ecology seeks a unifying analysis, based on *total flows of materials*, that treats on a common footing all sources, all transport media, and all receptors. For this purpose, mass-flow analysis is proving to be a productive integrative tool.

Mass-flow analysis necessarily focuses on forms of matter that are either indestructible (chemical elements) or persistent on a scale of months to decades or longer (nonchlorinated organics such as methane, and chlorinated organics such as the chlorofluorocarbons and the polychlorinated biphenyls). An application of mass-flow analysis to cadmium in the Rhine Basin is found in the chapter by Stigliani, Jaffé, and Anderberg in this volume.

The analysis of mass flows builds on a previous success in the analysis of energy flows. But mass-flow analysis is more complicated. Generally, there is no interest in an energy flow beyond the point of use, where energy is usually degraded into heat. But mass-flow analysis continues beyond the point of use, because of chemistry-dependent behavior in the environment that affects toxicity. Mercury can be remobilized, as methyl mercury, after apparent discard on a lake bottom; chromium in the environment can become more toxic as a result of a change in valence.

Inevitably, scholars will extend and refine mass-flow analysis in the next few years, at least to the level of comprehensiveness already attained for energy flows: from the mine to the discarded product; at all geographic scales (from facilities to towns to islands and river basins to countries) and in all regions of the world; preindustrial baselines; scenarios of future flows. Ultimately, mass-flow analysis may develop enough downstream capability to be able to track materials through air, water, and soil and to take into account chemical changes that affect solubility, bioavailability, and toxicity. Even then, it will be crucial to combine mass-flow analysis with exposure analysis to avoid missing small flows that happen to be environmentally significant.

The framework of mass-flow analysis has powerful, even subversive, implications for pollution policy. It treats with indifference both what is easy to regulate and what is hard to regulate. In cases where mass-flow analysis reveals the dominance of nonpoint sources over point sources, the focus of pollution regulation may shift to the farm, with its dispersed use of chemicals, and away from the factory, where pollution enters the environment through pipes and stacks. Even more unsettling are the implications of mass-flow analysis in cases where the wear or disposal of goods dominates other releases into the natural environment. This may shift regulatory attention from *production* to *product*.

In part, these shifts of emphasis are a consequence of the success of the first round of environmental control. Point sources are being significantly reduced by initiatives in industry and by regulation, making nonpoint and dissipative sources relatively more important.

Braungart's chapter in this volume takes some first steps in imagining a policy regime sensitive to the insights of mass-flow analysis. He proposes that consumer durable goods—like automobiles and television sets—be leased, not owned, so as to increase the likelihood of recycle and reuse of the many embodied chemicals.

Centrality of the Industrial Firm and the Industrial Farm

Economics and sociology have shed much light on industrial firms and farms as economic actors and as human institutions. Yet very little of this insight has informed the analysis of environmental problems and the design of environmental policy. *Environment* has been framed as a struggle between good and evil, where fault lies entirely with "industry" and not at all with John and Jane Public.

Industry, for the most part, has colluded in this drama, casting itself as helpless victim of misguided public outrage. Industry has rarely stepped forward to organize the debate—failing, for example, to add its abundant expertise to the first stages of discussion of environmental goals. Its posture vis-à-vis environmental policy has generally been: Tell us what you want us to do, assure us that our competitors will be required to do the same thing, and we will do the job.

The result has been an underutilization of technological capability in environmental strategy, with unfortunate consequences. To give a single example, a requirement of "best available control technology," which the layman would reasonably expect to induce continuous competitive innovation to control pollution at ever lower costs, turns into a recipe for freezing technological innovation, because the certification of "best" is so encumbered that whatever was declared "best" years ago is "best" for a very long time.

In industrial ecology, industry becomes a policy-maker, not a policy-taker. Industry demonstrates that environmental objectives are no longer alien, to be resisted and then accommodated reluctantly. Rather, these objectives are part of the fabric of production, like worker safety and consumer satisfaction.

Detailed reports from the industrial frontlines, including several in this volume, offer important insights and challenge simple theories of the firm. At Volvo, industry is inventing design tools to prioritize among environmentally responsive investments, without waiting for precise and transparent methodology. At Dow, industry is achieving savings with short payback periods from environmental investments, with no sign of diminishing returns. At Hewlett–Packard, industry is designing for recyclability and ease of disposal. Reports such as these are invaluable for the crafting of policies that elicit environmental initiatives from industry.

Industrial ecology identifies roles for new industries, including (1) service industries offering efficiency and cost-savings in resource management, and (2) industries fostering renewable energy.

Service Industries Offering Efficiency and Cost-Savings in Resource Management

A service industry has already emerged to provide efficient energy use. Until very recently, around the world, electric utilities were concerned only with electricity supply and saw no role for themselves on the customer's side of the electric meter. Today, the relationship between electric utilities and the owners of commercial and residential buildings is often mediated by an "energy service company" which brings expertise in energy-conserving investments, like low-energy lighting and motor controls. Typically, the service company is either owned by or compensated by the utility, on the grounds that these investments help the utility to avoid the costs of new capital facilities.

Underlying the economic viability of the energy service company are policies embedding the novel concept of "least-cost" provision of amenities, like lighting or well-preserved food. Investments in energy-efficient light bulbs and refrigerators are traded against investments in power plants. Frequently, the investments made at the behest of energy service companies turn out to be justified even without taking into account their energy savings: a more controllable motor permits cost savings through more finely tuned operations. Yet it takes an external stimulus to get expertise directed toward a secondary area of business.

Analogous service industries are beginning to emerge to manage persistent chemicals. An intermediary between a chemical company and a farmer will offer the amenity of a certain crop yield. It will then, internally, trade the provision of fertilizers and pesticides against chemical-conserving, knowledge-intensive investments like genetic selection, biological pest control, and techniques to reduce crop spoilage. In a pollution control regime based on tradable permits for the discharge of chemicals, with caps on total rates of use, a service company might prosper as a dealer in pollution rights. Even in the absence of such a regime, the service company's expertise may produce profits, which the farmer and company can share, from reduced purchases of chemicals.

Overview

There should also be a place in the near future for another class of service companies, managing the discard of goods. Consider, for example, our earlier example of the leasing of television sets: a new kind of service company might be an intermediary between the manufacturer of television sets and the watcher of television, adding the service of final disposal to the service of maintenance. Such a service company would have an economic interest in the producer minimizing the incorporation of low-value, high-volume wastes. In one possible configuration, the intermediary, at the time of retrieval from households of sets that have broken or become obsolete, might obtain pollution credits for avoiding discard of hazardous chemicals. Such specialized industries at the interface between the consumer and the manufacturer are analogous to the decomposers in an ecosystem.

Renewable Energy

Research and development are bringing into view a rich menu of clever ways to collect and transform the energy in sunlight cheaply, efficiently, and in environmentally responsible ways (see Johansson *et al.*, 1993). If renewable energy is to dominate the global energy scene, it will have to be embedded in the world's industrial energy system, centrally managed and grid-connected. In the terms of Amory Lovins's dichotomy of "hard" and "soft" paths (the hard path characterized by large spatial scale, centralization, and management by elites), we will have "hard solar" (Lovins, 1976).

Technologically, "hard solar" could appear in several forms. It could take the form of large arrays of solar thermal collectors, focusing sunlight onto "power towers," where the high temperatures permit efficient thermal cycles based on gas or steam turbines. Or, it could take the form of arrays of large windmills on tall towers. Or, biomass plantations and associated facilities (much like oil refineries) that convert biomass to high-value gases and liquids, as described in the chapter by Williams (this volume). Or, large arrays of photovoltaic collectors—plantations of glass—converting sunlight, via semiconductors such as amorphous silicon, directly to electricity. Commercialization is expanding rapidly on all of these fronts.

One of the most open frontiers of industrial ecology lies where renewable energy production can be combined with land rehabilitation. A project designed to meet both objectives should provide both energy and land repair at lower costs than projects designed for either objective alone. The economic viability of renewable energy projects located on degraded land will depend, in large part, on the extent to which land repair can be documented and monetized. The implications for the employment of specialists are clear: whether a project's renewable energy objective is to harness the wind or to collect sunlight directly or to process biofuels, its success will depend not only on those who understand and can manage high technology, but also on those who understand soil formation and retention, watershed management, and habitat for wildlife.

14

Implications of the Six Perspectives

The Need for Effective Collaborations Beyond the Natural Sciences

In every science, there is a tension between pursuing the most intriguing questions and the apparently most socially useful questions. The global change enterprise is currently experiencing such tension, following a period of self-determination. Government policy-makers are demanding help in drafting and implementing international environmental agreements well rooted in environmental science. Private-sector decision-makers with long time horizons—involved in such activities as natural resource extraction, coastal development, insurance, land conservancy—are demanding guidance in assessing the riskiness of investments. Natural scientists can respond by anticipating the questions that will be asked in ten years that will be answerable *then* if work is started *now*.

To give a single example, the first international agreements bearing on the management of global carbon will extend beyond fossil fuel accounting, to embrace an accounting for natural sources and sinks, only if there has been substantial progress in understanding these sources and sinks by the time these agreements are concluded. A larger accounting framework requires deeper science than is now available.

Basic environmental science is already interacting creatively with technology and policy in the clarification of the impact on stratospheric ozone of hypothetical fleets of long-range civilian aircraft flying in the stratosphere. There is a two-way flow of insights concerning stratospheric chemistry and fluid dynamics on the one hand and altitude–drag tradeoffs and nitrogen oxide formation rates in engines on the other. Each group finds its research questions in the interim results of the other.

Environmental science acquires a human face, and a more forceful mandate, when coordinated with an enterprise with human dimensions like the design of effective treaties or the evaluation of a new technology. The relationships are similar to those between biology and medicine.

To be sure, there are risks in broad collaborations and in an orientation to problem-solving. Science normally proceeds by isolating small pieces of large problems. Ecotoxicology, atmospheric chemistry, cultural studies of economic development—any of these could become paralyzed, if each subproblem were always first embedded in a system that manifested the complexity of the full problem. Or, research could suffer from becoming politicized, if stakes in particular outcomes were too deeply embedded in nominally dispassionate research. No one will be well served if the potential of good science to serve the agenda of industrial ecology is undermined by such modes of failure.

The Illusion of the Plateau

No one sees today's world as a steady state. Our focus is on change. Yet when we think a century ahead, we tend to imagine a world that has gone through a series of transitions and has come out at a place where the pace of change has slowed—where global population, though larger, holds steady, for example. The logistic growth curve is commonly invoked—rising slowly, then rapidly, then slowly again. Those alive today are assumed to be experiencing the period of rapid change, passing through a brief phase of growth, like an adolescent on the way to maturity. Although we exult in the dynamism of our own society, how frequently our debates about the future are about the merits of alternative steady-state societies.

Will human society a century from now really resemble a steady state—with an energy mix, for example, that has not changed in the half century from 2050 to 2100 and that is not expected to change? Will people in 2100 really think of themselves as part of a world in equilibrium and like it that way, guarding against change? The economist, Herman Daly has written insightfully of steady-state economics (Daly, 1991); will he be cited as one of the visionaries of our era?

Or, might the focus in 2100 still be on change, and might public argument be closer to our own? If in 2100 nuclear fusion or renewable biomass is an important energy source, might people be arguing about whether to expand it further or to phase it out? And if in 2100 either is not playing a central role, might people be looking back wistfully at its glorious past, or be looking forward to an imminent breakthrough?

My guess is that the future in 2100 will be a no less restless time than our own. The earth system, although much better understood, will be perceived, even more than now, as a principal shaper of global human activity. A century from now, global environmental constraints will be a much stronger source of stimuli for technology and policy, and the process of accommodation will still be evolving.

I am of the transition generation, who learned the environmental facts of life only after formal education was over. Experiencing the shock of transformation of an already developed world view, my generation is particularly impatient with arguments from neoclassical economics that stress the inevitable costs of moving from one equilibrium to another and that associate every environmental improvement with a trade-off. Society is far from equilibrium. The fulcrum has shifted under the seesaw.

Industrial ecology presents an endless source of fresh questions, never before confronted, that complicate the landscape in interesting ways. Viewed from the six perspectives presented in this chapter, the near horizon provides many peaks that can be climbed, and glimpses of peaks beyond. There are no plateaus.

References

Daly, H. E. 1991. *Steady-State Economics*, 2nd edition. Island Press, Washington, D.C.

Johansson, T. B., H. Kelly, A. K. N. Reddy, and R. H. Williams (eds.). 1993. *Renewable Energy: Sources for Fuels and Electricity*. Island Press, Washington, D.C.

Lovins, A. 1976. The road not taken. *Foreign Affairs 55(1)*.

PART 1

VULNERABILITY AND ADAPTATION

2

Introduction

The Editors

Just as new branches of industry spring up to absorb the wastes of other industries, so new fields of study develop out of established disciplines. In this first part of the book the basic themes of industrial ecology are laid out, together with reviews of these issues from the perspectives of economic history, anthropology, sociology, and development studies. Each of these disciplines has traditionally approached the human dimensions of global change quite differently—asking different questions and using different analytical approaches. It is too much to suppose that we can, in one book, do full justice to each of these streams of research, or resolve the mutual incomprehension and suspicion that exist among them. It is sufficient to show, as these five chapters show, that each separate stream perceives a need to extend its own horizons.

But industrial ecology must be more than an agglomeration of established disciplines. There is an urgent need for a multidisciplinary approach to global change which is itself able to recast the questions to be answered. Industrial ecology has one advantage over other attempts to bridge the disciplines: it has a persuasive metaphor at its heart. The argument that much can be gained by viewing industrial systems, like biological ecosystems, as consumers, digesters, and excreters of energy and materials was first articulated by Frosch and Gallopoulos (1989). In "Industrial Ecology: Definition and Implementation," Graedel develops the biological metaphor in opening a discussion of the definition of industrial ecology. Graedel has been one of the pioneers in thinking about how such apparently esoteric perspectives can be applied in industry. His chapter continues with a description of industrial materials cycles, and how these are linked to environmental concerns.

What is innovative in this presentation is the attempt at a qualitative ranking of key environmental impacts (for example, UV absorption, greenhouse warming, precipitation acidity) across key sources of pollutants (for example, crop production, petroleum combustion, manufacturing). The essential point, and one that is further developed in a later chapter (Graedel, Horkeby, and Norberg-Bohm, this volume), is that priorities must be set in the environmental redesign of industrial and agricultural activities. Pervasive technological and institutional change can occur only if the ground-rules are simple and self-evident. Without widely agreed-upon tools for guiding choices, decisions with far-reaching economic and social effects will not be made. Investment in clean technology and cleaner practices will be sustained only if the incentives are well defined. Graedel ends with a discussion of the short- and longer-term strategies whereby firms might implement industrial ecology in the management of technology.

Grübler, in "Industrialization as a Historical Phenomenon," presents a Schumpeterian analysis of industrialization and technical change and ties this to a discussion of broad historical trends in industrial materials and energy use. For Schumpeter and his followers, industrialization has taken place through the pervasive adoption of "clusters" of technical, organizational, and institutional innovations (Perez, 1983). Phases of industrial development can be characterized by the technological clusters which dominate the most dynamic economies at any given time: textiles (1750–1820), steam (1800–70), heavy engineering (1850–1940), mass production and consumption (1920–2000), and, more speculatively, "total quality" (1980–). By implication, different phases of industrialization have different environmental effects. Grübler suggests that the new phase just beginning may contain industrial and organizational characteristics more respectful of the environment than those of previous phases. In particular he discusses dematerialization and decarbonization as tendencies in more advanced societies. He ends with the important point that in postindustrial societies, industry itself is responsible for a falling proportion of environmental harm, and consumption and leisure activities for a rising proportion.

Cantor and Rayner, in "Changing Perceptions of Vulnerability," review a literature about psychological and cultural attitudes to risk as it applies to global change. Social attitudes toward risk have provided a rich vein of research since the late 1960s, and a number of distinct perspectives have emerged. For example, in explaining why people fear nuclear power more than many other technologies, the cognitive approach considers the basic attributes of risk: whether it is voluntary, whether it is linked to catastrophic effects, and so on. At the other end of the spectrum, cultural analysis is interested in how far individuals' core beliefs about order and freedom may color their attitudes to certain risks, or, more deeply, how prevalent myths about nature affect choices for responding to environmental problems. Cantor and Rayner also consider how the notion of vulnerability can be applied in social risk-perception research.

Chen, in "The Human Dimension of Vulnerability," is concerned with the unequal distribution of the impacts of environmental change across countries and across social strata. His focus is on the most vulnerable populations—the marginal and the poor, especially in the developing world—and he argues that environmental change is a major cause of health and welfare loss for an increasing proportion of the world's poor. Moving beyond effects on individuals, Chen then embarks on a discussion of the vulnerability of entire social systems. He concludes that there are grounds for both optimism and pessimism regarding the vulnerability of future societies to global environmental change.

Huq, in "Global Industrialization: A Developing Country Perspective," writes from the viewpoint of one of the poorest and most vulnerable nations on earth,

Bangladesh. His chapter argues that the fate of the poor in the developing countries must be included in industrial ecology. He further argues that the developing world cannot replicate the paths of development followed by the industrialized world, and must pursue alternative paths.

References

Frosch, R. A., and N. E. Gallopoulos. 1989. Strategies for manufacturing. *Scientific American* *261(3)*, 144–152.

Perez, C. 1983. Structural change and the assimilation of new technologies in the economic and social system. *Futures 15(4)*, 357–375.

3

Industrial Ecology: Definition and Implementation

Thomas Graedel

Abstract

Industrial ecology (IE) is a new ensemble concept in which the interactions between human activities and the environment are systematically analyzed. As applied to industry, IE seeks to optimize the total industrial materials cycle from virgin material to finished product to ultimate disposal of wastes. This chapter provides a discussion of the main ideas of industrial ecology. Industrial impacts on the environment and the means by which industrial processes can be adjusted to lessen these impacts through waste minimization and recycling are also discussed.

Introduction

Industrial ecology (IE) is a new ensemble concept, although some of its individual elements have been recognized for some years. IE arises from the perception that human economic activity is causing unacceptable changes in basic environmental support systems. As applied to manufacturing, this systems-oriented concept suggests that industrial design and manufacturing processes are not performed in isolation from their surroundings, but rather are influenced by them and, in turn, influence them.

In industrial ecology, economic systems are viewed not in isolation from their surrounding systems, but in concert with them. That is, it is the study of all interactions between industrial systems and the environment. As applied to industrial operations, it requires a systems view in which one seeks to optimize the total industrial materials cycle from virgin material to finished material to component to product to waste product and to ultimate disposal. Factors to be optimized include resources, energy, and capital.

A different form of definition arises by analogy to biological ecology. Traditional biological ecology is defined as *the scientific study of the interactions that determine the distribution and abundance of organisms*. The relationship between this concept and that of industrial activities has been discussed by Frosch and Gallopoulos (1990):

In a biological ecosystem, some of the organisms use sunlight, water, and minerals to grow, while others consume the first, alive or dead, along with minerals and gases, and produce wastes of their own. These wastes are in turn food for other organisms, some of which may convert the wastes into the minerals used by the primary producers, and some of which consume each other in a complex network of processes in which everything produced is used by some organism for its own metabolism. Similarly, in the industrial ecosystem, each process and network of processes must be viewed as a dependent and interrelated part of a larger whole. The analogy between the industrial ecosystem concept and the biological ecosystem is not perfect, but much could be gained if the industrial system were to mimic the best features of the biological analog.

It is instructive to think of the materials cycles involved with the earliest of Earth's life forms. At that time, the potentially usable resources were so large and the amount of life so small, that the existence of life forms had essentially no impact on available resources. This individual component process can be described as *linear*, that is, the flow of material from one stage to the next is independent of all other flows. This pattern is designated "Type I" ecology; schematically, it takes the form of Figure 1a.

An aspect of biological ecology that is implied in its definition, but not stated, is that the totality of the ecosystem is sustainable over the long term, although individual components of the system may undergo transitory periods of expansion or decay as a consequence of proximal conditions. In the larger picture represented by an ecosystem in which proximal resources are limited, the resulting life forms are strongly interlinked. This evolution under the pressure of external constraints has produced in nature the efficiently operating system with which we are familiar. In this system, the flows of material within the proximal domain may be quite large, but the flows into and out of that domain (i.e., from resources and to waste) are quite small. We designate such a system as Type II (Figure 1b).

A Type II system is much more efficient than a Type I system, but it clearly is not sustainable over the very long term, because the flows are all in one direction, that is, the system is "running down." To be ultimately sustainable, biological ecosystems have evolved to be almost completely cyclical when sufficiently long time scales are considered. "Resources" and "waste" are undefined, since waste to one component of the system represents resources to another. This Type III system, in which complete cyclicity has been achieved, is pictured in Figure 1c. Note that the exception to the cyclicity of the overall system is that energy (in the form of solar radiation) is available as an external resource. It is also important to recognize that the cycles within the system tend to function on widely differing temporal and spatial scales, a behavior that greatly complicates analysis and understanding.

The ideal anthropogenic use of the materials and resources available for industrial processes (broadly defined to include agriculture, the urban infrastructure, etc.) would be one that is similar to the ensemble biological model. Many uses of

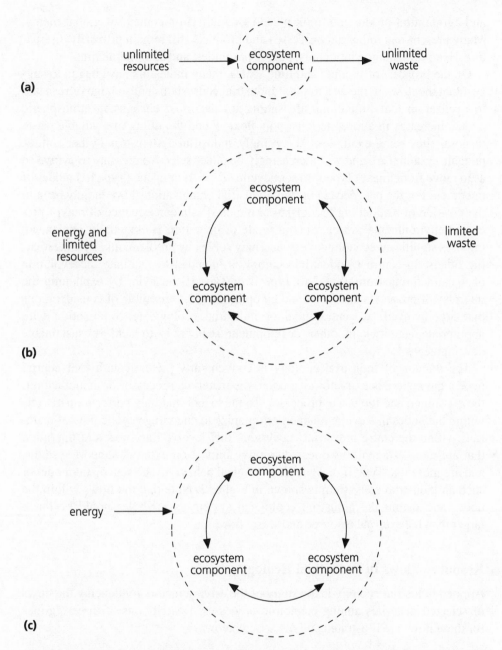

Figure 1. (a) Linear materials flows in "Type I" ecology. (b) Quasicyclic materials flows in "Type II" ecology. (c) Cyclic materials flows in "Type III" ecology.

materials have been and continue to be essentially dissipative, however. That is, the materials are degraded, dispersed, and lost to the economic system in the course of a single normal use (Ayres, 1989), mimicking the Type I pattern above. This pattern can be associated with the maturation of the Industrial Revolution of the 18th century, which, in concert with exponential increases in human population

25

and agricultural production, took place essentially in a context of global plenty. Many present-day industrial processes and products still remain primarily dissipative. Examples include lubricants, paints, pesticides, and automobile tires.

On the broadest of product and time scales, many indicators have begun to suggest that the flows in the ensemble of industrial ecosystems are so large or resistant to cyclization that limitations are setting in: the rapid changes in atmospheric ozone, increases in atmospheric carbon dioxide, and the filling of available waste disposal sites being examples. Accordingly, industrial systems (and other anthropogenic systems) are and will increasingly be under selective pressure to evolve so as to move from linear (Type I) to semicyclic (Type II) or cyclic (Type III) modes of operation. For the past decade or two, industrial organizations have largely been in the position of responding to legislation imposed as a consequence of real or perceived environmental crises. Such a mode of operation is essentially unplanned, imposes significant economic costs, and may solve one problem only by exacerbating others. In contrast, industrial ecology is intended to facilitate the evolution of manufacturing from Type I to Type II or Type III behavior by explaining the interplay of processes and flows and by optimizing the ensemble of considerations that are involved. A central goal of industrial ecology, in combination with appropriate activities in other development sectors, is to achieve sustainable development.

The domain of industrial ecology is conveniently pictured with four central nodes: the materials extractor or grower, the materials processor or manufacturer, the consumer, and the waste processor. To the extent that they perform operations within the nodes in a cyclic manner, or organize to encourage cyclic flow of materials within the entire industrial ecosystem, they evolve into modes of operation that are more efficient, have less disruptive impact on external support systems, and are more like Type II or Type III ecological behavior. The schematic model of such an industrial ecosystem is shown in Figure 2. Note that the flows within the nodes and within the industrial ecological system as a whole should be much larger than the external resource and waste flows.

Resource Flows in Industrial Ecology

Applied industrial ecology is the study of the driving factors influencing the flows of selected materials among economic processes. Descriptions of three regimes for these flows are illustrated and discussed below.

Industrial Production of Materials

The resource flows involved in the production of materials are shown in Figure 4. Beginning with virgin materials, the flows proceed through cycles of extraction, separation and/or refining, and physical and chemical preparation to produce finished materials. A typical example might be the extraction of copper ore from the ground and the eventual production of copper wire.

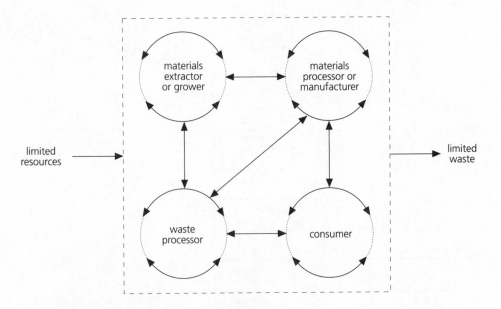

Figure 2. The "Type III" model of the industrial ecosystem.

Figure 3 shows, in addition to the central flows, the materials flows that occur away from the central spine. To the right are the wastes consigned to disposal, typically larger fractions in the early stages of the process and smaller fractions at later stages. To the left are flows of recycled material, such as copper wire recovered from obsolete power distribution systems. Near the center are recycled flows that occur during the production process itself, as when scrap from one process stage is reused in the preceding stage. A complementary flow occurs for any chemicals used in processing the material.

The flows of materials and processes as shown in Figure 3 occur within what has traditionally been termed the "heavy industry" community, and the flows are generally under the sole control of the materials supplier.

Industrial Manufacture of Products

The resource flows involved in the manufacture of products from the finished materials produced by the materials supplier are shown in Figure 4. An important distinction between this figure and Figure 3 is that several finished materials are generally involved, rather than a single one. A typical example is the production of a cable connector from selected metal and plastic. As in Figure 3, this figure indicates waste flows, recycled material flows, and in-process recycle flows. It is in the processes of Figure 4 that materials flows are presently addressed at all in the design process.

It is instructive to examine Figure 4 from the perspective of constraints to optimizing the industrial ecology of the process. At the forming step, material may enter from three types of streams: the virgin materials streams (P1a, b, or c), the

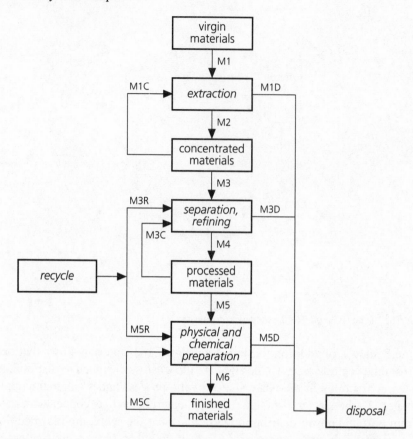

Figure 3. Cyclic materials flows in raw materials processing. Mx, central flows; MxD, wastes consigned to disposal; MxR, flows of recycled materials from outside the production process; MxC, flows of recycled materials during production.

process recycle streams (P1C), and the recycled material streams (P1R). Unrecycled waste exits in disposal stream P2D. Optimization involves decreasing or eliminating P2D, increasing P1C, and using whatever external recycled material is available (P1R), within the constraints of customer preference and existing price structures. The use of any recycled material thus involves a tradeoff between its purity and suitability and the cost of virgin finished materials. A similar analysis applies to each step of the process.

The flows of materials and processes as shown in Figure 4 are within the "manufacturing" community, and the materials flows are generally under the sole control of the manufacturer.

The Customer Product Cycle

The resource flows involved in the customer portion of the industrial ecology cycle are shown in Figure 5. Optimization opportunities here include the need to

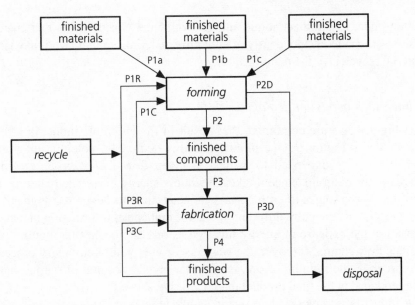

Figure 4. Cyclic materials flows in manufacturing.

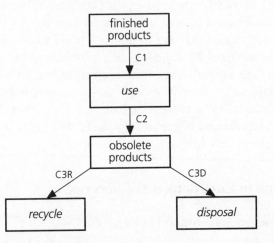

Figure 5. Materials flows in customer use.

avoid resource dissipation, especially in the handling of obsolete products. To the extent that customers favor the recycle stream C3R at the expense of the waste disposal stream C3D, industrial ecology is optimized.

In a free market economy, flows of materials within the industrial system are driven by the purchases of customers. Motivated by their personal preferences and by the information furnished to them concerning the impacts of the use and disposal of specific materials or products, their decisions constitute the ultimate driving force for industrial materials flow. The return flows of materials for reuse are strongly influenced by consumers as well. Although exceptions exist, the situation under existing economic and governmental constraints is that customer decisions are made independently of either the material supplier (who may also be a material recycler) or the manufacturer.

The Integrated Industrial Ecology Cycle

When Figures 3–5 are combined, they result in the ensemble industrial ecology cycle shown in Figure 6. One might view this figure as a potential diagram, in which the energy expended to achieve a given flow increases as one moves upwards in the diagram, since it takes much less energy to recycle materials from one of the lower stages to an intermediate stage than to begin at a higher stage. Ayres (1989) puts the point differently; that the embedded information in materials increases at the expense of energy invested as one approaches the bottom of the materials flow chain. Moreover, the waste associated with preliminary processing stages, such as the refining of ore, is often far greater per unit of output than the waste produced in the final product manufacturing stages.

The industrial ecology cycle is one in which, at present, three essentially decoupled participants are acting in their own interests, relatively independent of the others. Just as the relationships between finished materials suppliers and manufacturers have been markedly changed by just-in-time (JIT) manufacturing techniques, so the relationships indicated in Figure 6 have the potential to be changed by industrial ecology manufacturing techniques. In JIT manufacturing, inventories of components or subassemblies are kept at a minimum at the assembly plant, but delivered as they are needed. This eliminates an important source of waste. As with JIT, implementing IE will require close cooperation among equipment designers, process designers, process engineers, and suppliers. This process may come about as corporations begin to implement design for environment (DFE) procedures (see Paton, this volume).

Industry's Link to Environmental Concerns

Industrial Influences on Atmospheric Concerns

Having described the flows of materials in the industrial process, it is now appropriate to turn to the ways in which those materials relate to some of the major environmental problems that have been identified. To do so, it is helpful to utilize a matrix of the interactions between specific anthropogenic activities and their environmental impacts (Graedel and Crutzen, 1993), as shown in Figure 7. To construct this figure, a critical atmospheric effect like precipitation acidification and its direct and indirect chemical causes are linked with the sources responsible for initiating those interactions. The result shows the impact of each potential source of atmospheric change on each critical atmospheric component. The assessment is qualitative, reflecting the present state of knowledge. It also includes estimates of the reliability of that knowledge, an important component of such an assessment.

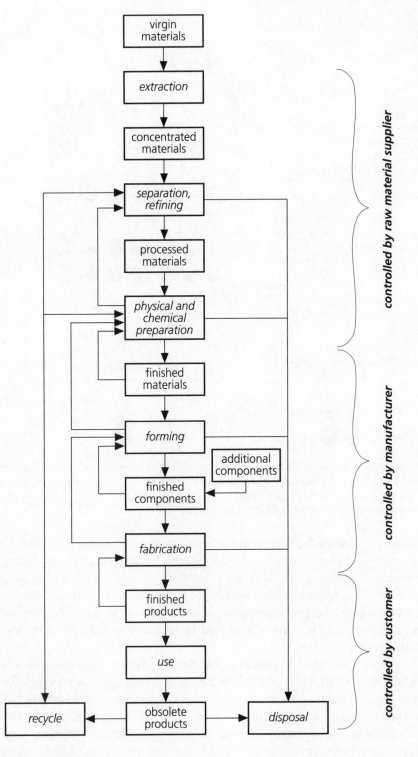

Figure 6. The total industrial ecology cycle.

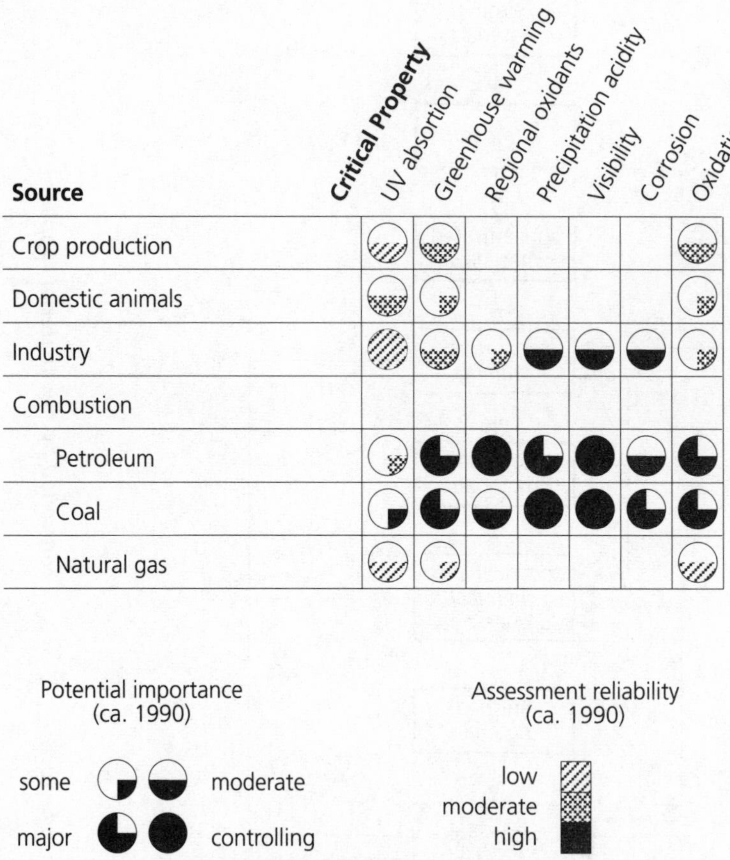

Critical Property

Column headings (diagonal): UV absortion · Greenhouse warming · Regional oxidants · Precipitation acidity · Visibility · Corrosion · Oxidation efficiency

Source	UV absortion	Greenhouse warming	Regional oxidants	Precipitation acidity	Visibility	Corrosion	Oxidation efficiency
Crop production	◐	◐					◐
Domestic animals	◐	◐					◐
Industry	◐	◐	◐	◐	◐	◐	◐
Combustion							
Petroleum	◐	●	●	●	●	◐	●
Coal	◐	●	●	●	●	●	◐
Natural gas	◐	◐					◐

Potential importance
(ca. 1990)

some ◖ ◗ moderate

major ◖ ● controlling

Assessment reliability
(ca. 1990)

low
moderate
high

Figure 7. An initial ensemble assessment of impacts on the global atmosphere. Critical atmospheric properties are listed as the column headings of the matrix. The sources of disturbances to these properties are listed as row headings. Cell entries assess the relative impact of each source on each component and the relative scientific certainty of the assessment.

The matrix is constructed with separate rows for industrial operations and for the major energy-generating processes. This approach makes it easier to distinguish the impacts of specific types of sources, but does not address the allocation question, i.e., whether emissions from energy production should be allocated to the end user rather than to the generating process itself. As will be seen, these secondary links of industry and environment should be regarded as natural components of industrial ecology studies.

The simplest atmospheric impact assessments of Figure 7 involve only a single cell of the matrix. A typical example is the study of the impacts of a single source, such as a new coal-fired power station, on a single critical atmospheric component, such as precipitation acidification. More complex atmospheric assessments have addressed the question of aggregate impacts across different kinds of sources. A contemporary example is the study of the net impact on Earth's thermal radiation budget caused by chemical perturbations due to fossil fuel combustion, biomass burning, land-use changes, and industrialization. An alternative approach, especially useful for the purposes of policy and management, is to assess the impacts of a single source on several critical atmospheric properties. The coal combustion study noted above would fall into this category if the impacts were assessed not only on acidification, but also on photochemical oxidant production, materials corrosion, visibility degradation, etc. If desired, the columns could be summed in some way to give the net impact of the ensemble of sources on each critical property. Similarly, the rows could be summed in some way to give the net effect of each source on the ensemble of properties.

Figure 7 shows that the sources of most general concern for atmospheric impacts are fossil fuel combustion (especially coal and petroleum) and industrial processes. Emissions from crop production, especially methane (CH_4) from rice paddies, also have significant effects on climate and atmospheric chemistry. When contemplating these source–impact assessments it is useful also to consider some of the differing attributes of the sources. Food production, through the growing of crops and the raising of animals, is potentially sustainable, but current operations require very large resource inputs, and sustainability would require substantial and far-reaching changes in approach. The combustion of fossil fuels, necessarily preceded by extraction and purification, involves little transformation and is purely dissipative in nature. In contrast, industrial operations exist to make major transformations to raw materials, and are potentially nondissipative.

Among the most troublesome interactions between development and environment are those that involve cumulative impacts. In general, cumulative impacts become important when sources of perturbation to the environment are grouped sufficiently closely in space or time that they exceed the natural system's ability to remove or dissipate the resultant disturbance. The basic data required to structure such assessments are the characteristic time and space scales of the atmospheric constituents and development activities. For example, perturbations to gases with very long lifetimes accumulate over decades to centuries around the world as a whole. Today's perturbations to those gases will still be affecting the atmosphere decades or centuries hence, and perturbations occurring anywhere in the world will affect the atmosphere everywhere in the world. Long-lived emittants tend to be radiatively active, thus giving the greenhouse effect its long-term, global-scale character. At the other extreme, heavy hydrocarbons and coarse particles, being short-lived, drop out of the atmosphere in a matter of hours, normally traveling a few hundred kilometers or less from their sources. The atmospheric properties of visibility reduction and photochemical oxidant formation associated with these chemicals thus take on their acute, relatively local or regional character. Species with moderate atmospheric lifetimes include gases associated with the acidification of precipitation and fine particles, all with characteristic scales of a few days and a couple of thousand kilometers.

Industrial Influences on Water and Soil Concerns

An approach similar to that for atmospheric impacts can be used to put the impacts of anthropogenic activity on water and soil into perspective. This is done in Figure 8, where the sources of Figure 7 are retained. The figure contains several messages. One is that industrial activities, unlike the other sources, play at least a small role in all of the impacts. For toxicity, industry plays the major role. For impacts on species diversity, industrial activities are minor compared with crop production (and consequent forest clearing). The same is true of soil productivity loss. In the case of groundwater quality, industrial activity and food production can each be important, the dominant influence varying from location to location. Industry (defined in a very broad sense) can be held predominantly responsible for the depletion of landfill capacity.

In the case of materials extraction, industry is a major factor, but not as large as the removal of fossil fuel resources for the production of energy. The profligate use of energy thus bears two burdens: one as a producer of carbon dioxide, the principal anthropogenic greenhouse gas, the other as a major negative factor on the environment through the impacts that accompany resource extraction.

In the context of global change, product and process designers need to consider, at a minimum, stratospheric ozone depletion (hence chlorofluorocarbons [CFCs] and halons), greenhouse warming (hence CO_2, especially in the context of energy use; CH_4, sulfur dioxide or SO_2, and nitrous oxide or N_2O) and materials resources (hence use of abundant and easily obtained materials, minimization of solid waste, and maximization of recycling potential). Regional change issues (often hard to separate from global change issues and, in any case, also the responsibility of the design engineer to consider) include tropospheric acidity (hence SO_2 and oxides of nitrogen or NO_x), photochemical smog (hence NO_x and volatile organic carbon compounds), aquifer depletion (hence rate of water use), and groundwater chemical changes (hence heavy metals, nitrate, pesticide residues, and selected organics).

Aspects of Industrial Ecology

Manufacturing industrial products involves physical and chemical changes of materials and cannot be accomplished without interactions with the external world. These interactions have the potential to be relatively benign or to be relatively injurious; the choice between the two is often within the province of the informed product or process designer. Design engineers have a whole host of goals to optimize as they design their products: high performance, high reliability, low cost, attractive appearance, safety, and (more recently) modest and carefully chosen environmental impacts. The following environmental interactions are among those that designers need to consider as they develop their products and processes.

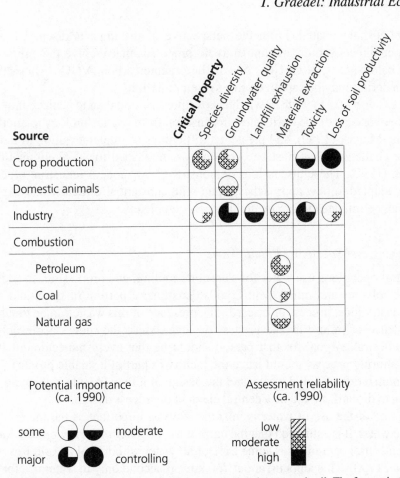

Figure 8. An initial ensemble assessment of impacts on global water and soil. The format is the same as that of Figure 7.

Choosing Raw Materials

In the long term, society cannot be sustainable if it uses up important resources that are in short supply. Despite uncertainties in defining *important* and *short supply*, efforts to assess the depletion of resources have been and will continue to be made; for example, those of Goeller and Zucker (1984) and Frosch and Gallopoulos (1989). Although one should place limited confidence in the accuracy of the results of such efforts, for resources that are now crucial and are known to be in increasingly short supply, such as copper and petroleum, prudent design engineers should make special efforts to develop substitute materials. Industrial product and process designs using any of the potentially limited materials should, if possible, be avoided.

Minimizing and Specifying Air Emissions

Most industrial processes involve the emission of materials to the air. The most common emittants are solvents or cleaning agents; these are usually organic materials, often halogenated. From the perspective of the impacts discussed above, design engineers should attempt to avoid processes that involve the emission of CFCs, halons, CH_4, N_2O, NO_x, or volatile organic carbon (VOC). Alternatively, these materials may be captured prior to their emission.

It is easier by far to write down the above prescription than to achieve it in practice. An overview of the intricacies of transition from one technology to another is given by Manzer (1990), and the actual difficulties of implementation in industry and the challenges that must be overcome are described for CFC elimination by Boyhan (1992). These discussions demonstrate how modern industrial processes can be adapted to new approaches, often with substantial difficulty and expense, but cannot eliminate environmental impacts completely.

Minimizing and Specifying Liquid Waste

Industrial operations use water for transport, cooling, and processing. In addition, organic solvents and other spent liquids constitute a portion of the liquid waste that industry must process or discard. The presence of this waste can be thought of as evidence of failure by the process designer, since the (never achievable but always desirable) goal for that person should be that every molecule entering a manufacturing process should leave the facility as part of a salable product. Thus, the minimization of liquid waste, and the design of it to make it less costly to recycle than to discard, should be a central element in process design.

The rate of the use of water is in some ways as important as the materials put into the water. It is estimated that the current use of water in energy production and other industrial operations is some 21% of all water withdrawal (World Resources Institute, 1990). This amount, about 760 km^3, is second only to water use for agricultural irrigation, and is much higher than the flows for domestic and municipal water withdrawal. About 13% of industrial use is consumptive; the remainder is returned as wastewater, often with contaminants added.

Minimizing and Specifying Solid Waste

The disposal of solid waste has become one of the major problems of developed societies. In Europe, landfill capacity is so exhausted that only several years' dumping at the present rate is possible; in the United States, some two-thirds of all landfills have closed within the past decade. Solid waste disposal has thus become increasingly expensive. Industry, which generates a substantial fraction of all solid waste, is increasingly under pressure to minimize its rate of solid waste production and disposal.

A major key to the minimization and recycling of solid waste is close working relationships with suppliers, recyclers, and marketers. These relationships can often result in the return of wooden pallets for reuse, reductions in packaging volume, minimization of packaging materials diversity for products leaving the facility, and the return of process sludge to chemical suppliers for regeneration and purification. A simple example of what can be achieved is a recent AT&T effort to change packaging of a computer terminal keyboard. From an initial packaging involving a large cardboard exterior box and a styrofoam insert, a revised approach involving a smaller external box and a folded cardboard insert was developed. The result was a package that was 30% smaller, easier to recycle, and cost effective, since the smaller, lighter package was less costly to ship and store.

Designing for Energy Efficiency

Ross (1989) has discussed the impact of industrial operations on the use of energy. Manufacturing activities consume 25–30% of all energy use in the United States and similar fractions elsewhere. Much of this consumption can be minimized by attention to process design and by the reuse of energy expended by manufacturing processes. Gibbons and Blair (1991) suggest that major reductions in energy use are possible through the use of more efficient motors, cogeneration, and the substitution of more energy-efficient manufacturing processes. The payback for the investment involved should be calculated not only on the basis of energy costs saved but also on the avoidance of emissions of carbon dioxide and other deleterious gases and particles as a consequence of energy generation.

A second aspect of design for energy efficiency is the minimization of energy consumption by products once they are in service. Such design often carries with it little cost penalty; it may involve a bit more insulation for a refrigerator or the alteration of an electrical design to minimize energy use. Recent legislation encouraging the labeling of the energy performance of products is a step in the right direction but provides little incentive for customers unless energy costs are perceived to be high. An obvious example of the potential energy efficiency gains to be made by customer purchase decisions is the tendency of consumers to purchase energy efficient automobiles when gasoline prices are high.

Recycling during Manufacture

The concept of recycling during manufacture was shown in Figure 5. The capability for doing so is a strong function of the design of manufacturing processes with the retention and reuse of materials in mind. In the case of plastics, regrinding and reuse of sprues, runners, and other waste fragments is often possible. Metal components can sometimes be remelted, or an unsatisfactory surface stripped and replated. It is not uncommon for a manufacturer to recycle factory waste successfully only by forming a close working relationship with a raw materials processor or a recycler.

An aspect of solid waste generation in manufacturing that has received too little attention in many places is the minimization and recycling of packaging waste from components entering the manufacturing facility. Here again, close relationships with suppliers can aid in the minimization of component packaging destined for disposal.

Recycling after Use

The materials cycle in industrial ecology is effective only if materials are efficiently returned to the system for reuse. This process is greatly aided by product designs that aid disassembly, avoid the use of a multiplicity of materials or of irreversible materials bonding techniques (Stein, 1992), identify the materials used in the product, and avoid or minimize the inclusion of materials difficult or dangerous to recycle (Levenson, 1990). Designing durable goods for disassembly may seem like an oxymoron, but it is being done. An example is a teapot recently brought to market by Polymer Solutions, a joint venture of GE and Fitch Richardson Smith, which uses injection molded parts from Vermont and British heating elements and switches. The product is satisfactory to customers and to the manufacturer as well. Because the parts snap together, engineers found that the tolerance requirements were much more severe than for older assembly methods; more accurate tooling was needed. In this way, the implementation of industrial ecology encourages more precise manufacturing techniques, which will be put to good use in designs across the corporations' product lines.

Closing the Industrial Ecology Cycle

One factor that is overlooked by many is that manufacturing often involves trade-offs among alternative materials or process chemicals rather than complete elimination, as in the removal of a reactive chemical from a surface following a process like etching or soldering. This removal inherently produces a residue containing the reactive chemical and the removal agent. Unless the process can be completely redesigned to eliminate the necessity for cleaning, the best one can do may be to choose the most environmentally benign removal agent and the least sensitive medium that is impacted. For example, it may be possible to clean etchants from a metal with either chlorofluorocarbons or organic solvents. Which is preferable on a per-molecule basis? If it takes ten times as much organic solvent to do as effective a cleaning job as CFCs, does the answer change? Does it change if the organic solvent is volatile (thus affecting either local photochemical smog generation) or nonvolatile (thus affecting either local ground or surface water or requiring well-regulated landfill disposal)? Designers must make such choices, and they are aware that there are no simple answers. Even if there were, the designers of products and processes probably would not have the perspective to implement the state of knowledge properly.

Since it is impossible for environmentally responsible corporations to avoid having some level of impact on the external world, if industrial ecology is to be truly responsive to global human health and ecological concerns those concerns must be ranked. This can only be done by the experts: the environmental scientists. Better yet, the scientists should be willing to assign some sort of severity index, to indicate how much worse one impact is than another. This sort of action is antithetical to the "pollutant of the month" approach that has been so prevalent, but it is not impossible; several possible approaches are discussed in this volume by Graedel, Horkeby, and Norberg-Bohm. To set the industrial impacts of this chapter into perspective, I will use the following somewhat arbitrary ranking of 12 major environmental concerns: (1) minimizing or eliminating toxic effluents; (2) protecting species diversity; (3) greenhouse warming; (4) the ozone layer; (5) ground and surface water quality; (6) landfill exhaustion; (7) photochemical smog; (8) resource depletion through materials extraction; (9) loss of the atmosphere's oxidizing capacity; (10) acid rain; (11) corrosion; (12) visibility.

To bring the whole picture together, the aspects of industrial operations discussed earlier are coordinated with their impacts in Table 1. The table suggests that product and process designers should assess their environmental interactions, with the following priorities being of most importance:

- Toxic species in wastewater and elsewhere should be eliminated.
- Air emissions should be minimized to the fullest possible extent, especially emissions of CFCs, VOCs, and SO_2.
- The use of energy in manufacturing and during the life of products should be minimized.
- Strong efforts should be made to reduce solid waste from the manufacturing process.
- Designing for the disassembly and reuse of materials following product obsolescence should be actively pursued.

21st Century Engineering

How should engineers move toward implementing industrial ecology? They need to realize a few important differences from traditional engineering practice. One is that environmentally responsible choices are sometimes not obvious. A second is that some choices require a very long time perspective on the problems. For exam-

Table 1: *Relating aspects of industrial activity to impacts*

Aspect	Impact Designation (see text)											
	1	2	3	4	5	6	7	8	9	10	11	12
Raw materials extraction		L						M				
Air waste	L		L	M			M		M	M	M	M
Water waste	M				M							
Solid waste	L					M						
Energy use			M						L			
Recycling		L				L	L					
Nonmanufacturing[1]		M				M			L			

M: major impact.
L: lesser impact.
[1] Land-use change due to population and agriculture increase.

ple, the increase in greenhouse gases suggests that energy-efficient designs will become more and more important. Third, since environmental science will continue indefinitely to be a field in flux, one cannot depend on regulations to be a guide, but must try to anticipate the science to some extent. This requires that physical designers have environmental consultants whom they trust and who understand their needs and constraints. For example, designers are now working with manufacturing engineers to eliminate CFCs as a consequence of the Antarctic ozone hole and related data. Many are taking the standard public advice to change processes in minor ways and substitute hydrochlorofluorocarbons (HCFCs). Anyone familiar with the science can see, however, that HCFCs will be regulated sooner rather than later, and such specialists should advise designers to eliminate manufacturing using any halogenated solvents.

Finally, environmentally-responsible design will require great ingenuity. An example is the recent effort by BMW, whose Z1 sports car has an all-plastic skin designed to be completely disassembled from its metal chassis in 20 minutes. The skin components are of recyclable thermoplastic. An unexpected side benefit of this design is that it has proven much easier to repair (Nussbaum and Templeman, 1990).

What is the role of the nonindustrialist in these activities? Such a person can help most by collaborating in the prioritization of environmental concerns. Given a choice of energy efficient or chemically efficient cleaning, for example, is emission of a molecule of methyl chloroform better or worse than one of carbon dioxide? Does reduction of sulfur dioxide emissions from materials extraction processing take precedence over the use of organic solvents for the cleaning and preparation of recycled material? What materials resources are the most crucial, considering their abundances, difficulty of extraction, and potential for reuse? Product and process designers make such choices every day, often implicitly. Since some global change as a consequence of industrial activity is part of the modern societal process, global change scientists need to work with industrial product and process designers to choose those impacts that are both most realizable and most benign.

Industrial ecology as a concept is an idea whose time has come (Woolard, 1992). Its implementation will not be easy, and considerable codification of its techniques remains to be done. Notwithstanding its complexities and difficulties, it is clear that industrial ecology is the way 21st century engineering will be done—the *only* way.

References

Ayres, R. 1989. Industrial metabolism. In *Technology and Environment* (J. H. Ausubel and H. E. Sladovich, eds.), National Academy Press, Washington, D.C., 23–49.

Boyhan, W. S. 1992. Approaches to eliminating chlorofluorocarbon use in manufacturing. *Proceedings of the National Academy of Sciences of the USA, 89*, 812–814.

Frosch, R. A., and N. Gallopoulos. 1989. Strategies for manufacturing. *Scientific American 261(3)*, 144–152.

Frosch, R. A., and N. Gallopoulos. 1990. Toward an industrial ecology. Paper presented to the Royal Society, London, 21 February.

Gibbons, J. H., and P. D. Blair. 1991. US energy transition: On getting there from here. *Physics Today 44(7)*, 21-30.

Goeller, H. E., and A. Zucker. 1984. Infinite resources: The ultimate strategy. *Science 223*, 456–462.

Graedel, T. E., and P. J. Crutzen. 1993. *Atmospheric Change: An Earth System Perspective*, W. H. Freeman, New York, 480 pp.

Levenson, H. 1990. Wasting away: Policies to reduce trash toxicity and quantity. *Environment 32(3)*, 10–15, 31–36.

Manzer, L. E. 1990. The CFC–ozone issue: Progress on the development of alternatives to CFCs. *Science 249*, 31–35.

Nussbaum, B., and J. Templeman. 1990. Built to last—Until it's time to take it apart. *Business Week*, 17 September, 102–104.

Ross, M. 1989. Improving the efficiency of electricity use in manufacturing. *Science 244*, 311–317.

Stein, R. S. 1992. Polymer recycling: Opportunities and limitations. *Proceedings of the National Academy of Sciences of the USA, 89*, 835–838.

Woolard, E. S., Jr. 1992. An industry approach to sustainable development. *Issues in Science and Technology 8(3)*, 29–33.

World Resources Institute. 1990. *World Resources, 1990–91*, Oxford University Press, New York, 378 pp.

4

Industrialization as a Historical Phenomenon

Arnulf Grübler

Abstract

Industrialization is described as a historical succession of periods of pervasive adoption of clusters of technological and organizational innovations. Combined they have enabled vastly rising industrial output, productivity, and incomes, as well as reductions in the amount of time worked. The resource and environmental intensiveness of different industrialization paths is illustrated with quantitative data on energy consumption and carbon emissions. It is concluded that industry in principle moves in the right direction of dematerialization and decarbonization; however, to date not fast enough to compensate for increasing output volumes. Continued structural change from industry to services and from work to pleasure will require a redefinition of the scope of industrial activities from artifacts to integrated solutions to satisfy consumer service demands in an environmentally compatible manner.

Introduction

Industrialization is a process of structural change. Sources of productivity and output growth as well as of employment move away from agriculture toward industrial activities, manufacturing in particular (Figures 1 and 2). Rising productivity and output in industry have been main drivers for economic growth and increased national and per capita incomes, which in turn provide an ever enlarging market for industrial products.

Like any pervasive process of economic or social change, industrialization is driven by the diffusion of many individual (but interrelated) *innovations*. These are not only technical, but also organizational and institutional, thus also transforming the social fabric. In fact, the term "industrial society" has come to describe a particular type of economic and social organization, from science and industrial management to the fine arts. An industrial society is based pervasively on the economics of standardization and specialization of human activities to produce, not only ever more, but, paradoxically, an ever larger *variety* of final products.

Industry is an important part of human activities and a powerful agent of global change. It accounts for about 20% of employment and 40% of value added, final energy consumption, and carbon emissions (Table 1). However, the relative

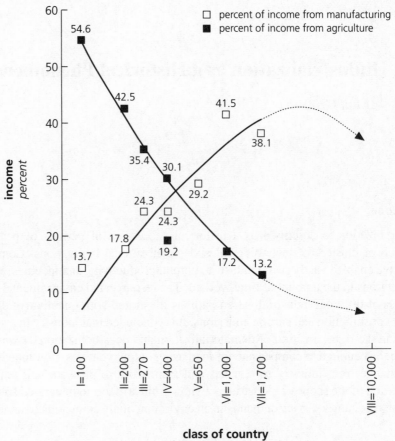

Figure 1. Industrialization as a process of structural change. Value generation and employment (cf. Figure 2 opposite) shift away from agriculture to industrial activities, manufacturing in particular. Source: Kuznets, 1958.

weight of industry varies widely in time and space primarily as a function of the degree of industrialization (or postindustrialization) and the overall level of economic development.

Since the middle of the 18th century, global industrial output and productivity have risen unimaginably. Based on updated estimates of Bairoch (1982), global industrial output has risen by about a factor of 100 since around 1750. Over the last hundred years, industry has grown by a factor of 40, or at an annual growth rate of about 3.5%. Per capita industrial production increased over the same period by a factor of about 11, or at a rate of 2.3% per year. This suggests that rising per capita activity levels were a more powerful agent of change than was human population growth. The growth in industrial labor productivity has been even more spectacular than output growth. Again the data are uncertain, but recent quantitative evidence does not change the impressive account of industrial productivity

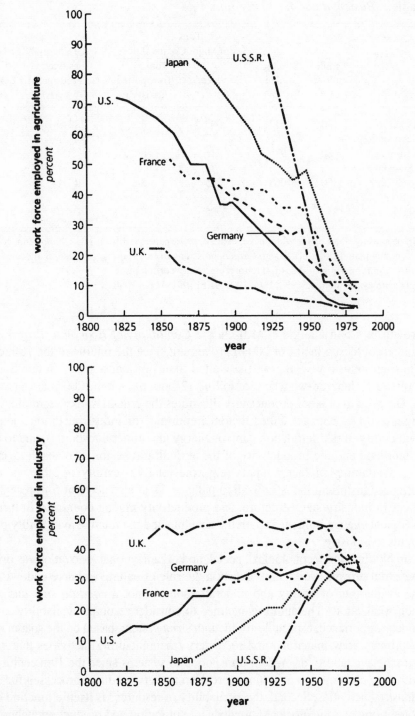

(a)

(b)

Figure 2. Industrialization as a process of change in occupational structure: (a) percentage of work force employed in agriculture vs. (b) percentage of workforce in industry. Note that industry now performs many activities previously residing in agriculture (from Nakićenović *et al.*, 1990).

Table 1: *Basic industrial activity data, 1990*

	People Employed ($\times 10^6$)	$ Value Added ($\times 10^9$)	Tons of Seven Major Commodities[1] Produced ($\times 10^6$)	Tons / km Transported ($\times 10^{12}$)	Final Energy Consumed (w/o Feedstocks) (GW/ yr/s)	Tons of Carbon Emissions[2] ($\times 10^6$)
Market economies	130	4632	1095	6	1164	766
Reforming economies	80	975	515	8	851	584
Developing economies	300	1068	895	8	1116	733
World	510	6675	2505	22	3131	2083

[1] In decreasing order of global tonnage: cement, steel, paper, fertilizer, glass, aluminum, copper.
[2] Including manufacture of cement. Carbon emissions from electricity production allocated to industry in proportion of industrial to total electricity consumption.
Data sources: Economist, 1990; ILO, 1991; IRF, 1991; U.N., 1990.

growth in Colin Clark's (1940) classic *Conditions of Economic Progress*. Data indicate at least a factor of 200 improvement since the middle of the 18th century. Thus, an industrial worker in the United States produces today in one hour what took a U.K. laborer two weeks of toiling 12-hour days some 200 years ago.

The growth in labor productivity illustrates the crucial role of technological and organizational change. Other factors important for industrialization include the availability of (skilled) labor, capital, energy and mineral resources, and to a lesser extent land and the productivity of the agricultural sector. The absolute and relative availability of factor inputs (e.g., the relative scarcity of labor vs. capital) influence historical industrialization paths and can also account for present differences in industry size, structure, and productivity among countries. In turn, technological change influences both the absolute and the relative availability of factor inputs to industry.

In highlighting technological, social, and organizational innovations as drivers of industrial growth, we seem to question the role of natural resource endowments. Is the availability of energy and mineral resources not a *conditio sine qua non* for industrialization? The author is inclined to consider resource availability as of secondary importance, especially for the industrial system, based on the spatial division of primary (raw materials) and secondary (manufacturing) activities that emerged with the availability of modern transportation systems since the 19th century. First, without technology no natural resource can be harvested and processed for input to industrial activities. Second, the availability of resources is itself a function of technology (via, e.g., geological knowledge, exploration and production technologies). Thirdly, technology development can provide for *substitutes* such as the replacement of natural nitrate by manmade fertilizers, or natural by synthetic rubber.

From this perspective, different degrees of development and industrialization are *technology gaps* resulting from differences in accumulation and innovativeness, and not so much from resource endowments or scarcity. Innovative capacity (and thus production, income and growth possibilities) is *created* (among others by human capital and an appropriate socio-institutional framework), and not given. Historical analysis indicates a number of cases where successful industrialization was achieved even with only modest national natural resource endowments (e.g., France, Scandinavia, Austria, Japan). Considering the resource and environmental intensiveness of different industrialization paths (discussed below), the abundance of resources even could be a mixed blessing. One might wonder if coal-rich China will develop along the energy-intensive development path of the United States, or alternatively along more energy-efficient pathways of industrialization as in the French or Japanese experience.

The Spread of Industrialization: Technology Clusters, Sources of Growth, and Spatial Heterogeneity

Below we illustrate that industrialization, embedded within a broader framework of economic growth, proceeded through a succession of development periods based on the pervasive adoption of various "technology clusters," i.e., a set of (interrelated) technological, organizational and institutional innovations driving industrial output and productivity growth. Such a succession is, however, not a rigid temporal sequence as various clusters coexist (with changing weights) at any given time. Older technological and infrastructural combinations coexist with the dominant technology cluster, and in some cases previous clusters (compared to the dominant technology base in the leading industrialized countries) are perpetuated, as was largely the case in the post-World War II industrial policy of the former USSR.

At any time most industrial and economic growth is, however, driven by the dominant technology cluster, frequently associated with the most visible technological artifact or infrastructural system (or "leading sector") of the time (e.g., the "railways era" [Schumpeter, 1939] or the "age of steel and electricity" [Freeman, 1989]). We emphasize the concept of technology clusters because studies under the leading sector hypothesis (e.g., Fishlow, 1965; O'Brien, 1983; Tunzelmann, 1982) have shown that these can explain only a fraction of economic and industrial growth. Only the combination of a whole host of innovations in many sectors and technological fields can account for sustained industrial and economic growth.

Table 2 is an attempt to categorize various phases of industrial and economic development through the concept of technology clusters. It lists the dominant cluster in the top row and the emerging cluster (dominating in the successive phase) below. Examples of key technologies in the areas of energy and transport systems, materials and industry, and the final consumer sphere are listed. Finally, we categorize the dominant "organizational style" (Perez, 1983), i.e., the predominant mode regulating industrial, economic, and social relations, and give a geographical

47

Table 2: Important technology clusters for economic growth and industrialization

Cluster	1750–1820: Textile	1800–1870: Steam	1850–1940: Heavy Engineering	1920–2000: Mass Production/Consumption	1980– Total Quality
Dominant					
Energy	Water, wind, feed, wood	Wood, feed, coal	Coal	Oil, electricity	Gas, electricity
Transport and communication	Turnpikes	Canals	Railways, steamships, telegraph	Roads, telephone, radio, and TV	Roads, air transport, multimedia communications
Materials	Iron	Iron, puddling steel	Steel	Petrochemicals, plastics, steel, aluminum	Alloys, specialty materials
Industry	Castings	Stationary steam, mechanization	Heavy machinery, chemicals, structural materials	Process plants, numerically-controlled machinery, consumer goods, drugs	Environmental technologies, disassembly and recycling, consumer services
Consumer products	Textiles (wool, cotton), pottery	Textiles, chinaware	Product diversification (imports)	Durables, food industry, tourism	Leisure and vacation, custom-made products
Emerging					
Energy	Coal, coke	City gas	Oil, electricity	Gas, nuclear	Hydrogen?
Transport and communication	Canals	Mobile steam, telegraph	Roads and cars, telephone, radio	Air transport, telecommunication, computers	Hypersonic? high-speed trains
Materials	Puddling steel	Mass produced steel	Synthetics, aluminum	"Custom-made" materials, composites	Recyclables and degradables
Industry	Stationary steam, mechanical equipment	Coal chemicals, dyes, structural materials	Fine chemicals, drugs, durables	Electronics, information technologies	Services (software), biotechnologies
Consumer products	Chinaware	Illuminants	Consumer durables, refrigeration	Leisure and recreation products, arts	Integrated "packages" (products + services)

Organizational Style				
Plant/company level	Individual entrepreneurs, local capital, small-scale manufacture	Small firms, joint stock companies	"Giants," cartels, trusts, pervasive standardization	Fordism/Taylorism, multinationals, vertical integration	"Just-in-time," Total quality control (TQC), horizontal integration
Economy and society	Breakdown of feudal and medieval economic structures	"Laissez-faire," Manchester liberalism	Imperialism, colonies, monopoly and oligopoly regulation, unionization	Social welfare state, Keynsianism, "open" society	Economic deregulation, environmental regulation, networks of actors
Industrial Geography					
Core	England	England, Belgium	Germany, USA, Benelux, France, England	USA, Canada, JANZ, European Community, England	OECD
Rim	Belgium, France	France, Germany, USA	Central Europe, Italy, Scandinavia, Canada, JANZ[1], Russia	USSR, Central and Eastern Europe, Southern Europe	4 Tigers,[2] Russia, Central and Eastern Europe, ??

[1] JANZ = Japan, Australia, New Zealand
[2] 4 Tigers describes the dynamic rapidly industrializing economies of Asia: Hong Kong, Singapore, South Korea, and Taiwan.

taxonomy[1] of centers of industrialization ("core") and regions catching up (newly industrializing or "rim" countries). All regions/countries not listed separately in Table 2 are classified as "industrial periphery" for the purposes of this discussion.

Four historical and a prospective fifth future cluster are identified, named after their most important carrier branches or functioning principles. These are: the *textile* industrialization cluster, extending to the 1820s; the *steam* cluster until about the 1870s; heavy engineering, lasting until the eve of World War II; and *mass production/consumption* until the 1970s and 1980s. Currently we appear in the transition to a new age of industrialization. Both its characterization as a "total quality" cluster (i.e., with control of both the internal and external, or environmental, quality of industrial production) and the technological examples given are necessarily speculative.

It has to be emphasized that the classification presented in Table 2 is a crude one and the examples are illustrative, not exhaustive. Also the timing of the various clusters in Table 2 is only approximate. In view of space limitations, the following qualitative[2] discussion of Table 2 will be brief and (over)simplified.

1750–1820: Textiles

Industrialization as a process of structural change began in 18th-century England. Technological innovations transformed the manufacture of textiles and gave rise to what later became a new mode of production: the factory system. Important bottlenecks for industrialization and its concomitant spatial concentration of population and economic activities began to be overcome. Coal and Darby's coke combined with the stationary steam engine (particularly important for coal mine dewatering) put an end to fuelwood and charcoal shortages and provided for spatial power densities previously found only in exceptional locations of abundant hydropower. The improvements in parish roads and turnpikes and especially the "canal mania" around the turn of the 19th century enabled the supply of rapidly rising urban and industrial centers with food, energy, and raw materials. Charcoal and the puddling furnace produced the first industrial commodity and structural material: wrought iron. Innovations in spinning (and after the 1820s also in weaving) enabled falling costs and rising output, particularly in the manufacture of cotton textiles. The introduction of fine porcelain from China gave rise to an expanding chinaware industry.

---

[1] This taxonomy is introduced to account for persistent spatial disparities in levels of industrialization, technology base, and degree of interconnectedness (exchange of information and goods) between countries/regions. Note that this is a functional categorization and not necessarily one based on geographical proximity. For similar concepts discussed within the framework of sustainability, see Brooks, 1988.

[2] For a quantification using principal component analysis see Glaziev, 1991; for an analysis based on innovation diffusion cf. Grübler (1990). The rise (and fall) of particular "technology clusters" has also been described using particular sectors or representative technologies (e.g., energy and transport infrastructures) as "metaphors." In view of abundant literature (e.g., Hoffmann, 1931, 1958; Woytinsky and Woytinsky, 1953; Landes, 1969; Rostow, 1978; Mokyr, 1990) containing valuable historical data and easily available output statistics of principal industrial commodities (e.g., Mitchell, 1980, 1982, 1983), this information is not further discussed here.

The nexus of innovations involving cotton textiles, the coal and iron industries, and the introduction of steam power constitute the heart of England's Industrial Revolution. However, in order for these developments to take place, important preconditions must be mentioned. More complex crop rotation patterns, abandonment of fallow lands, field enclosures, new crops, and improved animal husbandry allowed fewer people to grow more food (cf. Grigg, 1987; Grübler, 1992). Freed from agriculture, people sought urban residence and industrial employment. In the institutional sphere, the separation of political and economic power, new institutions for scientific research and dissemination of its results, organization of market relations, etc., all mark the breakdown of feudal and medieval economic structures with their associated monopolies, guilds, tolls, and restrictions on trade. Perhaps the intellectual and institutional/organizational changes were indeed the most fundamental (Rosenberg and Birdzell, 1986) as enabling and encouraging changes in the fields of industrial technology, products, markets, infrastructures, etc. Under a general laissez faire attitude, no provision was made to socially smooth the disruptive process of structural change in employment, rural–urban residence, value generation, and distribution of income, leading to violent manifestations of social and class conflict (e.g., Luddists, or the Captain Swing movement; cf. Hobsbawn and Rudé, 1968).

1820–1870: Steam

In this period, lasting to the recession in the 1870s, industrialization emerges from a spatially and sectorally confined phenomenon to a pervasive principle of economic organization. Industrialization continues to be dominated by England, which reaches its apogee as the world's leading industrial power by the 1870s, accounting for nearly one-quarter of the global industrial output. Industrialization spreads to the continent (Belgium, and the Lorraine and the Ruhr in France and Germany, respectively) and to the eastern United States much along the lines of the successful English model (textiles, coal and iron industry).

Coal (fuelwood in the United States) provides the principal energy form for industry, whereas transportation and household energy needs continue to be supplied mostly by renewable energy sources (animal feed and wood). The steam period is characterized by the emergence of mobile steam power (locomotives and boats), but transport infrastructures are still dominated by inland navigation and canals, reaching their maximum network size by the 1870s (in England, France and the United States). Important innovations emerge in the fields of materials (Bessemer steel production), transport and communications (railways and telegraphs), energy (city gas), and the (coal-based) chemical industry. These were to become the dominant technological cluster of the second half of the 19th century until the Great Depression of the 1930s.

1870–1930: Heavy Engineering

Fueled by coal, this industrialization phase is dominated by railways, steam, and steel: it is the most smokestack-intensive period of industrialization. Dominated by the output of primary commodities and capital equipment, the industrial infrastructure spreads on a global scale. Enlarging the industrial and infrastructural base becomes almost a self-fulfilling purpose, driven by economies of scale at all levels of industrial production and organization. Standardization of mass-produced components and structural materials, perhaps best symbolized by the Eiffel Tower, is another characteristic of heavy engineering.

England loses its position as industrial leader (in terms of production and innovations) to Germany and the United States. The latter emerges as the world's largest industrial power by the 1920s, accounting for 40% of global manufacturing output (Bairoch, 1982), 60% of world steel production (Grübler, 1987), and 80% of cars registered worldwide (MVMA, 1991).

Railway networks and ocean steamships draw distant continents into the vortex of international trade, dominated by the industrialized core countries. Free world trade, greatly facilitated by the universal adoption of the Gold Standard, grows exponentially, but its political counterparts are imperialism and colonialism. Trade flows are dominated by trade between the industrialized core countries (see Table 2) and the rapidly industrializing rim (Russia and Japan). The industrial periphery (regions with the weakest industrial base) provides ever-enlarging markets for the products of the industrialized core and supplies raw materials and food (long-distance trade being made possible after the invention of canned food and refrigeration).

The pace of technological change accelerates with the emergence of oil, petrochemicals, synthetics, radio, telephone, and, above all, electricity, but the institutional and regulative picture is less progressive. Emerging industrial giants, monopolies, and oligopolies, perhaps best symbolized by Rockefeller's Standard Oil Company, are at the focus of government regulatory efforts, while social issues are only beginning to be tackled. Legislation to limit child labor, provide for elementary health care, and reduce long working days (up to 16 hours per day) is introduced at a slow pace and implemented at an even slower one. Dissatisfaction with the prevailing capitalistic accumulation regime stimulates the development of alternative theoretical expositions (Marxism) and the emergence of new social movements (the labor movement, trade unionization), aiming at a more equitable distribution of productivity gains. Workers reap some of the benefits in the form of increasing employment, falling real-term prices of food and manufactured products, and (to a smaller extent) rising wages. But the inability of the social/institutional framework to provide for a more equitable distribution of productivity gains causes increased social conflicts. These begin to be resolved only by progressively internalizing labor costs into the economics of industrial growth, as symbolized by the social welfare state emerging in the 1920s.

1930–1980: Mass Production/Consumption

The post-World War II economic boom was based on a cluster of interrelated technical and managerial innovations, leading to productivity levels clearly superior to those of the heavy engineering paradigm. The extension of the continuous flow concept of the chemical industry to the mass production of identical units enabled unprecedented real-term cost and price decreases and thus mass consumption. Typical products include the internal combustion engine and the automobile, petrochemicals and plastics, farm machinery and fertilizers, consumer durables, etc. The prototype of the associated production organization was the Fordist assembly line, complemented on the organizational level by a separation of management and administration from production along the ideas of Taylor's scientific management. Additional economies of scale were realized by the increasing vertical integration of industrial activities and the emergence of enterprises operating on a global scale (multinationals).

New energy, transport, and communication infrastructures proved vital. Petroleum was available at low (real-term) costs and became the principle energy carrier and feedstock. Roads and vehicles powered by internal combustion engines (cars in market economies and buses in formerly planned and in developing economies) replaced railways as dominant transport systems. Air transportation and global communication networks (telephone, radio, and TV) have not only reduced physical distances but also enhanced cultural and informational interchanges. Science has grown "big" (de Solla-Price, 1963) and has been integrated systematically into industrial activities, from industrial R&D laboratories to product quality control and even consumer research.

Although industrialization has become a global phenomenon, an analysis reveals only a few examples of successful catching up (notably Japan). Instead, catching up happens more within given geographical regions or between regions with not too different degrees of industrial development. In terms of the spatial taxonomy adopted here, this implies that some former members of the industrial rim (Canada, Japan, Scandinavia, Austria, Switzerland, Italy) have joined the core, but the dominance of the core is as great as ever. The members of the Organization for Economic Cooperation and Development (OECD) countries account for 70% of the world's industrial output and for 75% of the world merchandise trade (World Bank, 1992). Over 80% of OECD's imports of manufactured goods is imported from other OECD members, another 9% from the industrial rim (Eastern Europe and 4-Tigers), and only about 10% from the rest of the world (World Bank, 1992).

Examples of the social-institutional framework associated with the mass production/consumption regime include Keynsian policies leading to various forms of demand management (enabling mass consumption) via public infrastructure, defense, and public service spending, and via income redistribution (the welfare state). Other examples include socio-institutional innovations such as large-scale consumer credits, publicity, development of mass communication, institutional embedding of labor unions, or the development of various forms of *Sozialpartnerschaft* as the institutional framework of a social consensus on the general growth trajectory. However, it appears that we are witnessing a widening mismatch (Perez, 1983) between this socio-institutional framework and the attainment of (market, environmental, and social acceptance) limits to the further expansion of its production/consumption paradigm.

Industrialization: Output and Productivity Growth

Estimates of global industrial output growth (Rostow, 1978; Bairoch, 1982; Haustein and Neuwirth's 1982 update of Hoffmann, 1958) indicate a growth over the last 100 years of some 3.5% annually. Although estimates differ on global industrial growth during the early industrialization phase, they agree on an exponential growth pattern since the latter half of the 19th century. This, however, applies only to estimates of the monetary value of industrial output and not to its physical equivalent. The material intensiveness of industrial output varies over time and, especially in the OECD countries, has been declining for decades (cf. Williams *et al.*, 1987).

Table 3 summarizes the geographical distribution of industrial output growth, following the spatial taxonomy adopted here. Based on Bairoch's estimate, the industrial output of England in 1900 is used as normalizing index. Thus, Table 3 indicates that the industrial output of England in 1900 approximated that of the entire globe 150 years earlier. Conversely, global industrial output in 1980 was a factor over 100 larger than in England 80 years earlier. Since the mid-19th century, the industrialized core countries account persistently for up to two-thirds of global industrial output. Table 3 indicates that the industrial core has persistently higher growth rates than the rim and periphery. Only in 1930–80 does the rim show higher growth rates than the core, i.e., it is catching up. But the absolute and relative gap between the industrial core and the periphery widens. It is beyond the scope of this chapter to discuss reasons (or possible remedies) for persistent, even widening, disparities in levels of industrial development. One frequent argument points to falling real-term primary resource prices and resulting deteriorating terms of trade. However, one has also to keep in mind the constant change in the industrial structure of the core, and especially its falling materials intensity. Thus, deteriorating terms of trade can partly explain why industrial growth rates in the periphery were smaller than to be expected from their factor endowments. However, they are an insufficient explanation for the persistently higher growth rates in the core. Instead, the success of the core appears more related to its dynamics of industrial innovation and the resulting rise in factor productivity.

Figure 3 presents estimates of the improvement in labor productivity in manufacturing for a number of industrialized countries. The international comparison of industrial and manufacturing labor productivity is far from easy. Differences in industrial output mix, relative price structure, exchange rates, labor qualification, industrial relations, hours worked, etc., still await definitive methodological and empirical resolutions. Therefore, the data primarily illustrate the evolution of labor productivity over time within a given country, rather than serving as a yardstick for international comparisons.

Table 3: *The global geography of industrialization (level of industrialization in the U.K. in 1900 = 100)*

	1750	1830s	1870s	1920s	1980	1980 / 1750
Level in:						
Core	2	20	180	950	7400	3080
Rim	5	20	40	190	2300	430
Periphery	120	145	100	220	1300	11
World	127	185	320	1360	11000	87
Growth rates, %/yr:						
Core		2.6	4.6	3.6	4.0	3.6
Rim		1.7	1.3	3.3	5.0	2.7
Periphery		0.2	–0.7	1.7	3.5	1.1
World		0.5	1.1	3.1	4.1	2.0
Regional shares, %:						
Core	2	10	56	70	67	
Rim	4	11	12	14	21	
Periphery	94	79	31	16	12	

All figures rounded. Regional shares and factor increases calculated from original data may differ from rounded figures.
Data source: Bairoch, 1982.

Persistent differences in levels of labor productivity in manufacturing among the industrialized countries (not to mention the developing ones) emerge from Figure 3. Apparently, distinct national industrial systems (in terms of sectoral structure, technology base, etc.) with associated institutional settings (working time regulation, wage negotiation, etc.) have evolved. The cumulativeness of such industrialization paths is responsible for persistent differences in productivity despite intense international trade and competition. Some of the historical differences can also be related to the relative availability of various factor inputs. For example, labor was comparatively scarce for U.S. industry. Consequently, compared to that of England, U.S. industrial labor productivity was already higher when the U.S. was still a newly industrializing country.

Industrialization and Environment

The environmental implications of industrialization can perhaps best be described by Gray's (1989) paradox of technological development. Industrialization has brought unprecedented levels of environmental impacts stemming from effluents whose impacts are fairly well understood. It has also introduced new materials and substances (e.g., chlorofluorocarbons) with hitherto unknown impacts on the envi-

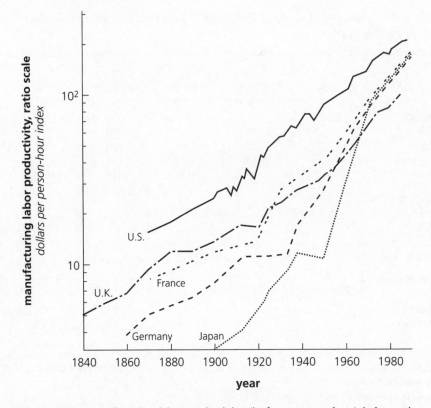

Figure 3. Growth in manufacturing labor productivity (in $ per person–hour) index, ratio scale. Comparative productivity levels are only approximate; therefore weight should be given only to the relative evolution of productivity in a given country over time. Industrial labor productivity gains have been extraordinary and have allowed rising incomes (wages) and shortening of working hours. Industrial output and employment data are from Liesner, 1985, and Mitchell, 1980, 1983; working hours from Maddison, 1991. Productivity figures between 1840 and 1930 have been harmonized with the estimates of Clark, 1940.

ronment. But at the same time, the technological change that goes along with industrialization and the growing incomes generated by rising productivity have also enhanced our technological and economic capacities for remedies.

Industry has built in an *inherent incentive structure* to minimize factor inputs. This is primarily driven by economics and by continuous technological change. Therefore, industry moves in the right direction, and the real issue is how to accelerate this desirable trend. "The right direction" means, in principle, two things: (1) minimizing resource inputs per unit of economic activity, i.e., *dematerialization*, and (2) improving the environmental compatibility of the materials used, processed, and delivered by industry, i.e., with respect to industrial energy use, *decarbonization*. Energy-related carbon emissions are the largest global expression of industry's metabolism, hence they are used as an illustration below.

Toward Industrial Dematerialization and Decarbonization

An analysis of industrial energy intensity per unit value added over time shows two important trends: decreasing energy intensities in the industrialized countries, and increasing intensities in newly industrializing ones. The much higher energy input per unit value added in the latter is frequently interpreted as *potential* for short- to medium-term energy efficiency improvements. However, higher energy intensities are in most cases the result of differences in degrees of industrialization and resulting differences in the structure and technology base of industry. This is illustrated in Figure 4, where the industrial energy intensity per unit value added is plotted against per capita levels of industrial value added. From such a perspective, the energy intensity of the Brazilian industry is in fact quite similar to that of the Japanese *at similar levels of industrial per capita output.* Conversely, the Nigerian example gives rise to concerns: increasing intensities of factor input use, but no significant growth in per capita levels of industrial output. The most spectacular improvements in industrial energy intensity were achieved in South Korea, illustrating that rapid industrial development and vigorous efficiency improvements are not mutually exclusive. Again, we observe only conditional convergence between countries and persistent differences between intensity "trajectories" of industrial development (e.g., United States vs. Japan).

The existence of specific industrialization trajectories illustrated above is also consistent with comparative macroeconomic studies of industrial development. Chenery *et al.* (1986) developed a typology of industrialization paths based on a differentiation of three classes of variables: size of the economy (small vs. large), sector orientation (primary vs. manufacturing), and trade orientation (inward vs. outward orientation). Over the post-World War II period, the highest industrial growth rates in semi-industrialized countries were achieved in small, manufacturing, and outward-oriented economies. Convergence is confined to countries belonging to a particular typological group rather than existing between groups. Chenery's typology constitutes an important differentiation of Rostow's (1978) stage theory of economic development. Instead of a single linear development model, a number of distinct development trajectories exist. Success appears also to be contingent on developing at least part of the industrial base on the technological productivity frontier. Perhaps the former USSR, or China's experience with rapid industrialization during the Great Leap Forward, can provide lessons on the feasibility of industrialization based on outdated technological vintages and industrial structures.

Figure 5 illustrates industrial carbon emissions as an environmental indicator of industrialization. An analog to Figure 4, it shows the industrial carbon intensity vs. per capita levels of industrialization. Carbon emissions from electricity generation are attributed to industry in proportion to industry's share in total electricity consumption and based on the (changing) average fuel mix in electricity generation. Overall, the decreasing carbon intensity of industrial activities is dominated by improvements in energy efficiency (cf. Figure 4).

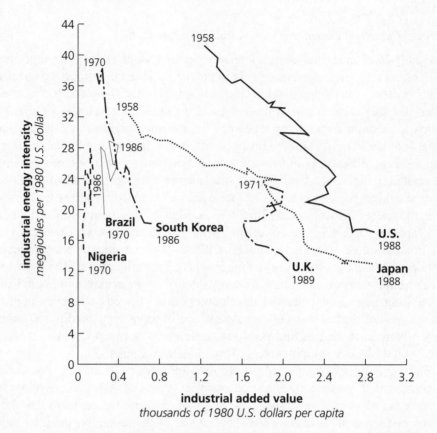

Figure 4. Industrial energy intensity vs. degree of industrialization as a more functional scale to assess the evolution of industrial energy intensity. Data from: Lawrence Berkeley Laboratory (LBL) data base and IEA, 1991.

Another factor explaining differences in industrial carbon intensity and its changes over time are changes in the structure of the industrial output. For instance, about 50% (some 230 million tons C) of U.S. industrial carbon emissions result from products contributing only to 15% (some $200 billion) of the industrial value added, whereas 50% ($780 billion) of the industrial value added is produced with only 13% (60 million tons C) of the sector's carbon emissions (Marland and Pippin, 1990). The skewed distribution function of industrial carbon emissions (Figure 6) indicates the importance of changes in the output mix, albeit these are difficult to model, yet to predict.

A Case Study of Carbon Emissions in the U.S. Steel Industry

This section uses the U.S. steel industry to illustrate the importance of structural shifts in process technologies and energy supply mix in moving in the direction of industrial dematerialization and decarbonization. Figure 7 illustrates specific and total sector carbon emissions since the middle of the second half of the 19th

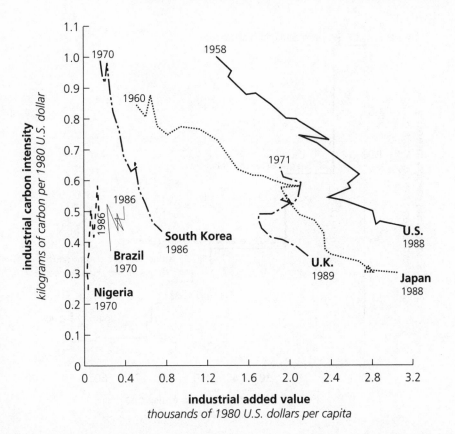

Figure 5. Industrial carbon intensity (kg C per U.S. $ 1980 value added) versus per capita level of industrialization (1000 U.S. $ 1980 per capita), cf. Figure 4 above. Data source: energy and value added: LBL data base, carbon emissions: emission factors based on Ausubel *et al.*, 1988; electricity production structure from IEA, 1991.

century. Although minimizing carbon emissions has not yet been on the agenda of the industry, it is interesting to note the significant improvement (factor of 20) in the carbon emissions per ton (pig iron) produced. The secular trend follows a typical industrial learning curve when plotted against the cumulative output as done in Figure 7. Thus, specific carbon emissions decrease by 17% for each doubling of cumulative output. As significant as these improvements have been, their rate has fallen short of output growth. Consequently, total sector emissions (including emissions from the generation of the electricity consumed by industry) have increased over time, but apparently have already passed through their historical maximum. However, the important point here is to compare actual emissions with what they would have been if growth had been achieved by simply intensifying existing production methods. (In reality, the tremendous output increases could only be sustained precisely because of technological change.) The historical role of technology change has been, therefore, two-fold: first, enabling significant output growth (and emissions) and, second, at

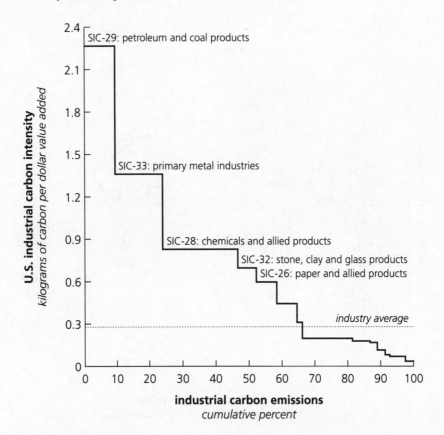

Figure 6. 1987 distribution of U.S. industrial carbon emissions by carbon intensiveness (kg C per $ value added) for 2-digit SIC-code level product categories. The heterogeneity of the emission intensiveness between different industrial products indicates the importance of changes in industrial output mix for lowering overall specific industry emissions. Data source: Marland and Pippin, 1990.

the same time averting even worse impacts, due to significant efficiency improvements.

Improvements in the carbon intensity of steel manufacture were achieved by a combination of gradual, incremental, and radical changes in both process technology (Figure 8) and the energy supply mix (Figure 9). These two sets of changes operating in tandem are yet another illustration of the importance of interlinkages among different technological systems. Changes in the fuel mix are closely tied to changes in industrial process technologies and both are instrumental for achieving energy efficiency improvements. They also point to the holistic nature of measures needed to accelerate desirable rates of industrial dematerialization and decarbonization.

Impacts of Industrialization on Consumption and Leisure

Industrialization had and continues to have far-reaching social impacts. Changes in

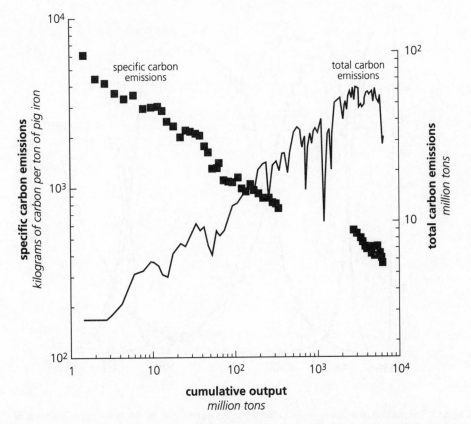

Figure 7. U.S. steel industry: specific and total sector carbon emissions vs. cumulative output. Carbon emissions from electricity production included in proportion of industrial to total final electricity consumption. Specific carbon emissions decrease by 17% for each doubling of cumulative output. Data source: Grübler, 1987; IEA, 1991.

employment structure, urbanization, increased life expectancy, rising incomes, and reductions in working time are examples of social changes directly and indirectly resulting from industrial output and productivity growth. Contingent on a social consensus, productivity gains have been distributed among rising wages and incomes (cf. Phelps Brown, 1973) and reductions of working time (Figure 10).

Perhaps the changes in time allocation patterns are among the least known of the social impacts of industrialization. Some 100 years ago, a U.K. laborer had an average life expectancy at the age of 10 of about 48 years and at age 20 of about 40 years, i.e., a total life span of less than 60 years. Before education became mandatory, labor began young, and essentially men who were healthy enough worked until they died (average length of a work career: about 47 years). Over his lifetime a male worker worked about 150,000 hours, or 60% of his available lifetime after subtracting necessary "physiological" time (i.e., the time required to eat and sleep). Today a typical male worker in the U.K. works some 88,000 hours during his lifetime. Due to reduced working time and increased life expectancy he spends

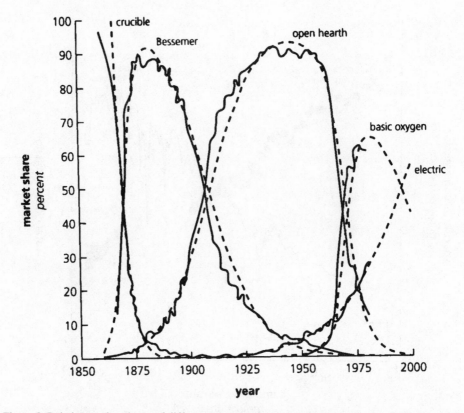

Figure 8. Relative market shares of different processes in raw steel production. Jagged lines are historical data and dashed lines model estimates by a set of coupled logistic equations. The dynamic pattern of technological change over time shown here is almost invariant across sectors and countries, but the timing and regularity of such technological substitution processes can vary considerably (Nakićenović, 1990).

only about 25% of his available lifetime at the work place. Trends in working time reductions (at paid work) for women have been less pronounced, but nevertheless noteworthy (cf. Ausubel and Grübler, 1990). International and intertemporal time-budget studies report on a broadly converging change in the structure of time allocation of the population (Figure 11).

More free time, coupled with higher incomes, has led to the development of lifestyles centered around private consumption and demand for services (cf. Gershuny, 1983). The structure of employment, industry, and production has followed suit. It is important to note to what extent resource consumption in post-industrial societies has become dominated by private consumption and leisure activities. Schipper *et al.* (1989) present data on final energy consumption for the FRG, indicating a dramatic shift in the relative share of energy consumption between productive (i.e., industrial) and consumptive (i.e., services and private households) uses of energy. Industry accounted in 1950 for two-thirds of final energy consumption, whereas today it accounts for only one-third. In future, it will become increasingly

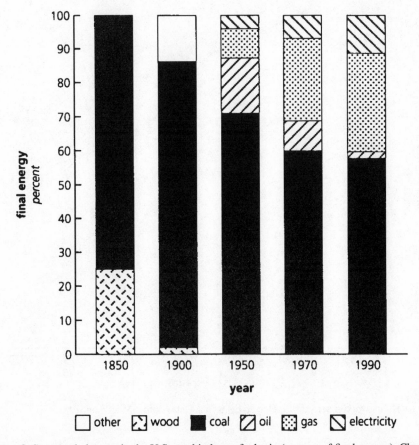

Figure 9. Structural changes in the U.S. steel industry fuel mix (percent of final energy). Changes in energy supply structures have accompanied changes in industrial process technologies (cf. Figure 8). Data source: 1850–1970: Grübler, 1987; 1990: IEA, 1991.

important for industry to take up the challenge to assist consumers in more environmentally compatible lifestyle choices—in providing not only new ("green") products, but also ways to ensure that environmentally friendly products are adopted, used and dispensed appropriately. All this implies redefining traditional markets for industrial products and services in the direction of *integrated packages,* focusing on the delivery of end-use *services* rather than on artifacts.

Conclusion

Industrialization as a historical phenomenon is conceptualized as a succession of phases, characterized by the pervasive adoption of "technology clusters." The introduction of a host of technological, institutional, and organizational innovations leads to productivity gains, impossible by a mere intensification of traditional solutions. From this perspective, industrialization is a time-specific phenomenon,

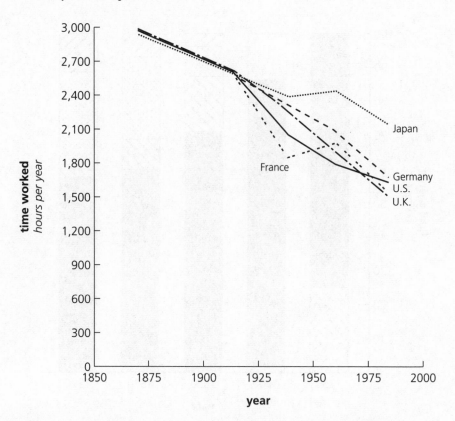

Figure 10. Hours worked per year in selected countries. Data source: Maddison, 1991.

characterized by (discontinuous) processes of change in the areas of economic structure, technological base, and social relations. History matters because of the cumulativeness of socio-institutional and technological change. This results in distinct development trajectories, spanning the extremes of high-intensity and high-efficiency industrialization paths, clearly discernible from historical data.

With respect to environmental impacts, minimizing factor inputs is an inherent part of the incentive structure of industry. Improved factor productivity and lowered resource intensiveness of industrial production have historically accompanied structural changes in industry. In principle, industry is moving in the right direction, referred to here as dematerialization and decarbonization. This gives reasons to be cautiously optimistic, albeit historical trends will have to be accelerated significantly to reduce the absolute levels of emissions and environmental impacts. As in the past, changes in technology, energy, and transport infrastructures, and in social and institutional regulatory mechanisms, will be instrumental.

If environmental compatibility indeed could become a new dominant paradigm of industrial development, future sources of industrial growth will be primarily in this area. Such tendencies will be first discernible in the most advanced postindustrial economies (i.e., in the industrial core). It is our contention that (as in the past)

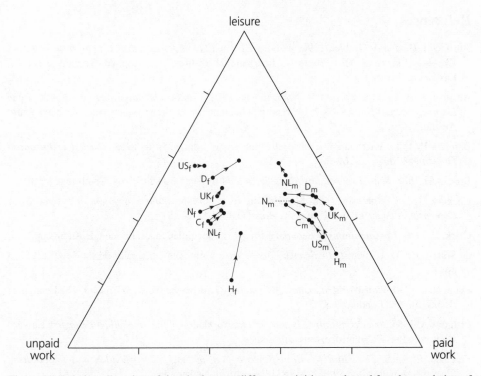

Figure 11. Relative allocation of time budgets to different activities, male and female population of seven countries, 1960s to 1980s. The figure illustrates an international and gender convergence away from formal, contracted work to unpaid work (e.g., family care) and leisure activities. This transition in activity patterns can also be clearly discerned in energy demand statistics. In industrialized countries today about two-thirds of final energy is consumed outside the productive sphere (i.e., industry) for services and leisure uses of energy (source: Gershuny, 1992).

successful catching up will only be possible if based on technological and institutional solutions not in conflict with the dominant industrial paradigm of the core.

Industrialization has brought tremendous productivity gains and resulting rising incomes and reduced working time—in short, *affluence* and *leisure*. From an environmental perspective activities outside the productive sphere are increasingly the determinants of resource consumption and environmental impacts. Furthermore, private and leisure activities are more difficult to steer with traditional policy instruments, such as price signals to which industry adheres. The decision-making criteria of consumers are complex and far from the rationality concepts underlying most economic models. Perhaps this will provide the largest future challenge to industry: providing consumers not with products, but with environmentally friendly *integrated solutions* to satisfy a particular *service demand*.

Acknowledgments

Comments by Bob Chen and Jerry Schnoor, and assistance of Andreas Schäfer in the preparation of the tables and figures, are gratefully acknowledged.

References

Ausubel, J. H., and A. Grübler. 1990. *Working Less and Living Longer. Part I: Long-term Trends in Working Time and Time Budgets*. International Institute for Applied Systems Analysis, Laxenburg, Austria.

Ausubel, J. H., A. Grübler, and N. Nakicenovic. 1988. *Carbon Dioxide Emissions in a Methane Economy*. Report No. RR-88-7, International Institute for Applied Systems Analysis, Laxenburg, Austria.

Bairoch, P. 1982. International industrialization levels from 1750 to 1980. *Journal of European Economic History 11*, 269–333.

Brooks, H. 1988. *Some Propositions About Sustainability*. Harvard University, Cambridge (draft).

Chenery, H., S. Robinson, and M. Syrquin (eds.). 1986. *Industrialization and Growth: A Comparative Study*. Oxford University Press, Oxford, U.K., 387 pp.

Clark, C. 1940. *The Conditions of Economic Progress*. Macmillan Press, London, U.K., 589 pp.

de Solla-Price, D. J. 1963. *Little Science, Big Science*. Columbia University Press, New York, 118 pp.

Economist. 1990. *Vital World Statistics: A Complete Guide to the World in Figures*. The Economist Books Limited, London, U.K.

Fishlow, A. 1965. *American Railroads and the Transformation of the Antebellum Economy*. Harvard University Press, Cambridge, Massachusetts.

Freeman, C. 1989. *The Third Kondratieff Wave: Age of Steel, Electrification and Imperialism*. Research Memorandum 89-032, MERIT, Maastricht, Netherlands.

Gershuny, J. I. 1983. *Social Innovation and the Division of Labour*. Oxford University Press, Oxford, U.K.

Gershuny, J. I. 1992. La répartition du temps dans les societés post-industrielles. *Futuribles 165–166*, 215–226.

Glaziev, S. 1991. *Economic Theory and Technological Change*. Nauka, Moscow, Russia, 232 pp. (In Russian)

Gray, P. E. 1989. The paradox of technological development. In *Technology and Environment* (J.H. Ausubel *et al.*, eds.), National Academy Press, Washington, D.C., 192–204.

Grigg, D. B. 1987. The industrial revolution and land transformation. In *Land Transformation in Agriculture* (M.G. Wolman and F.G.A. Fournier, eds.), John Wiley and Sons, Chichester, U.K., 79–109.

Grübler, A. 1987. Technology diffusion in a long wave context: The case of the steel and coal industries. In *Proceedings of the International Workshop on Life Cycles and Long Waves*, Montpellier, France, July 8–10, 1987. International Institute for Applied Systems Analysis, Laxenburg, Austria.

Grübler, A. 1990. *The Rise and Fall of Infrastructures, Dynamics of Evolution and Technological Change in Transport*. Physica Verlag, Heidelberg, Germany, 305 pp.

Grübler, A. 1992. *Technology and Global Change: Land-use, Past and Present*. Report No. WP-92-2, International Institute for Applied Systems Analysis, Laxenburg, Austria.

Haustein, H.-D., and E. Neuwirth. 1982. *Long Waves in World Industrial Production, Energy Consumption, Innovations, Inventions, and Patents and Their Identification by Spectral Analysis*. Report No. WP-82-9, International Institute for Applied Systems Analysis, Laxenburg, Austria.

Hobsbawn, E. J., and G. Rudé. 1968. *Captain Swing*. Pantheon Books, New York.

Hoffmann, W. G. 1931. *Studien und Typen der Industrialisierung. Ein Beitrag zur quantitativen Analyse historischer Wirtschaftsprozesse.* Revised English translation: Hoffmann, W.G. 1958. *The Growth of Industrial Economies.* Manchester University Press, Manchester, U.K.

IEA (International Energy Agency). 1991. *Energy Statistics of OECD Countries 1960–1989.* Vols. I and II, IEA, Paris, France.

ILO (International Labor Office). 1991. *Yearbook of Labour Statistics.* ILO, Geneva, Switzerland.

IRF (International Road Federation). 1991. *World Road Statistics.* IRF, Washington, D.C.

Kuznets, S. S. 1958. *Six Lectures on Economic Growth.* Free Press, New York.

Landes, D. S. 1969. *The Unbound Prometheus: Technological Change and Industrial Development in Western Europe From 1750 to the Present.* Cambridge University Press, Cambridge, U.K.

Liesner, T. (ed.). 1985. *Economic Statistics 1900–1983.* The Economist Publications Limited, London, U.K.

Maddison, A. 1991. *Dynamic Forces in Capitalist Development, a Long-run Comparative View.* Oxford University Press, Oxford, U.K., 333 pp.

Marland, G., and A. Pippin. 1990. United States emissions of carbon dioxide to the earth's atmosphere by economic activity. *Energy Systems and Policy 14*, 319–336.

Mitchell, B. R. 1980. *European Historical Statistics: 1750–1975.* Macmillan Press, London, U.K.

Mitchell, B. R. 1982. *International Historical Statistics: Africa and Asia.* MacMillan Press, London, U.K.

Mitchell, B. R. 1983. *International Historical Statistics: The Americas and Australia.* Macmillan Press, London, U.K.

Mokyr, J. 1990. *The Lever of Riches, Technological Creativity and Economic Progress.* Oxford University Press, Oxford, U.K.

MVMA (Motor Vehicle Manufacturers Association of the United States, Inc.). 1991. *World Motor Vehicle Data.* MVMA, Detroit, Michigan.

Nakićenović, N. 1990. Dynamics of change and long waves. In *Life Cycles and Long Waves* (T. Vasko, R. Ayres and L. Fontvielle, eds.), Lecture Notes in Economics and Mathematical Systems, Springer Verlag, Berlin, Germany, 147–192.

Nakićenović, N., A. Grübler, L. Bodda, and P.-V. Gilli. 1990. *Technological Progress, Structural Change and Efficient Energy Use: Trends Worldwide and in Austria.* International part of a study supported by the Österreichische Elektrizitätswirtschaft AG, International Institute for Applied Systems Analysis, Laxenburg, Austria.

O'Brien, P. (ed.) 1983. *Railways and the Economic Development of Western Europe 1830–1914.* Macmillan Press, London, U.K.

Perez, C. 1983. Structural change and the assimilation of new technologies in the economic and social system. *Futures 15(4)*, 357–375.

Phelps Brown, E. H. 1973. Levels and movements of industrial productivity and real wages internationally compared, 1860–1970. *The Economic Journal 83(329)*, 58–71.

Rosenberg, N., and L. E. Birdzell. 1986. *How the West Grew Rich: The Economic Transformation of the Industrial World.* I.B. Tauris and Co., London, U.K.

Rostow, W. W. 1978. *The World Economy: History and Prospect.* University of Texas Press, Austin, Texas, 833 pp.

Schipper, L., S. Bartlett, D. Hawk, E. Vine. 1989. Linking life-styles and energy: A matter of time? *Annual Review of Energy 14*, 273–320.

Schumpeter, J. 1939. *Business Cycles: A Theoretical, Historical and Statistical Analysis of the Capitalist Process.* Vols. I and II, Mc-Graw Hill, New York.

Tunzelmann, G. N. 1982. Structural change and leading sectors in British manufacturing 1907–1968. In *Economics in the Long View, Vol. 3 Part 2: Applications and Cases* (C.P. Kindleberger and G. di Tella, eds.), New York University Press, New York, 1–49.

U.N. (United Nations). 1990. *1987 Statistical Yearbook.* U.N., New York.

Williams, R. H., E. D. Larson, and M. H. Ross. 1987. Materials, affluence, and industrial energy use. *Annual Review of Energy 12*, 99–144.

World Bank. 1992. *World Development Report 1992: Development and the Environment.* Oxford University Press, Oxford, U.K.

Woytinsky, W. L., and E. S. Woytinsky. 1953. *World Population and Production, Trends and Outlook.* The Twentieth Century Fund, New York, 1268 pp.

5

Changing Perceptions of Vulnerability

Robin Cantor and Steve Rayner

Abstract

Although the scientific definition of human impacts is important to our under-standing, human perception and cultural attitudes to these problems are primary determinants of political action. The psychological and sociological literature of human attitudes to risk are reviewed. In a synthesis of the literature the importance of cultural bias and organizational interests are emphasized over the individual's psychological preferences in our responses to risk.

The Perception of Vulnerability and the Industrial System

It has been widely suggested that satellite pictures of the earth from space may have fundamentally changed human perceptions of the vulnerability of life on the planet (Clark, 1988). For example, the principal architect of the Montreal Protocol on Protection of the Ozone Layer writes:

> Perhaps the most poignant image of our time is that of earth as seen by the space voyagers: a blue sphere, shimmering with life and light, alone and unique in the cosmos. From this perspective, the maps of geopolitics vanish, and the underlying interconnectedness of all the components of this extraor-dinary living system—animal, plant, water, land, and atmosphere—becomes strikingly evident (Benedick, 1991).

This vision of the earth as a fragile system of natural interdependence endan-gered by human activity is currently a popular one, particularly in the developed world. However, it has arisen as a result of satellite photographs. Early maps of a world surrounded by dragons remind us that the idea of the earth encapsulated in a hostile milieu dates back many centuries at least. Our response to the satellite images may be conditioned by our own accepted myths, perhaps including our memories of these early maps.

When we begin to explore these issues, we discover that the myth of the fragile earth is by no means universally accepted, even within our own culture. Earlier in western history, a vision of the earth as a threatening and untamed wilderness to be conquered and exploited by humanity (for example, in Thomas Hobbes' *Leviathan*) seems to have coexisted and alternated in popularity with myths of noble savagery (for example, in the works of Jean-Jacques Rousseau) and the Garden of Eden (Nash, 1967). In recent years, fundamentalist preachers in the U.S. have attacked the "neo-paganism" of the environmental movement, and the myth of global fragility has little influence in the logging communities of the Pacific Northwest, where jobs and the viability of communities are the objects of concern over vulnerability (Gerlach, in press). In addition, we can observe a broadening array of corporate and governmental responses to changing environmental and regulatory concerns that are linked to global change.

As we extend the geographical reach of our inquiry, the myth of the fragile earth may prove to be quite at odds with equally popular myths of resilience and of cyclical destruction and renewal in non-western cultures. Is it merely narrowly conceived self-interest that leads nations such as China and India to emphasize development in responding to environmental concerns? Or could it be that different perceptions of urgency prevail in a Confucian culture that has weathered the changes of centuries or that different perceptions of fragility predominate in a culture firmly rooted in a cyclical perception of death and renewal?

These are not merely concerns of academic or scholarly interest. We focus this chapter on human perceptions of vulnerability because so much of the global environmental change issue depends on the recognition and understanding of scientific problems, creation and communication of the technical information, human judgments of facts and values, management preferences, and acceptability thresholds. All of these activities are informed by deeply held, often unexamined perceptions and beliefs about the nature of our world and of our role within it.

A significant issue in studying the perception of vulnerability is deciding exactly what is perceived to be vulnerable and what is threatening. We are faced at global, national, and institutional levels with a plurality of perceptions. Environmentalists clearly regard nature as vulnerable and the threat to nature as arising from human industry. President Bush, loggers in the Pacific Northwest, the emerging counterenvironmentalist movement, and significant portions of U.S. industry view the economy as vulnerable. The threat to human welfare arises from misplaced human concern with spotted owls and snail darters. A third view, espoused by advocates of sustainable development, sees human populations as vulnerable to an imbalance between their desire for development and the capacity of natural systems to support that development. Additionally, when we deal with such a wide variety of decision-making entities as we encounter in the issue of global environmental change, we are constantly challenged by shifting units of analysis. What can be said about the roles of perceptions at the levels of individuals, institutions, nations, or the world?

Consideration of the appropriate unit of analysis provokes us to examine more carefully what we mean by perception. The field of study that seems to be closest to our concern with perception of vulnerability to global change is that of risk perception. Strictly speaking, the term refers to individual acts of cognition.

However, the early literature on risk closely identified the issue of technology acceptance with risk perception or misperception. This practice has led to a situation in which the term "perception" has been expanded beyond cognitive response to the probability and magnitude of events. In the pages of the journal *Risk Analysis* and at meetings of the Society for Risk Analysis, the term risk perception encompasses issues of risk judgment, preference, opinion, attitude, communication, decision, choice, acceptance, action, and management, as well as of cognition. If customary usage is to be respected in defining our terminology, risk perception must be regarded as a broad term encompassing issues of technology choices arising from preferences for management procedures involving trust, liability, and consent as well as stricter issues of the cognition of probability and the magnitude of consequences.

Different social science disciplines, however, highlight different perceptual biases on environmental vulnerability. Our discussion is organized in the spirit of Bradbury (1989) and Wildavsky and Dake (1990). We marry the prior work on rival theories of risk perception with the environmental vulnerability issue by emphasizing the dominant bias applied by analysts to model or frame their subjects. Cognitive bias approaches include the psychometric, knowledge theory, and personality theory approaches because they rely on a paradigm of the mental processing of information by individual subjects. The self-interest bias approaches subsume the economic and political approaches since they model differences in individual risk perceptions as differences in payoffs to the perceivers. Cultural bias approaches combine emphases on the risk perceiver and world views.

Empirical and experimental studies indicate some evidence about the dominance of different paradigms and the factors that drive perception formation and change. Rival approaches and the fact that existing evidence is too limited, unfocused, and segmented make it impossible to provide reliable conclusions about the distribution and intensity of any particular perception of environmental vulnerability at this time. Instead, we suggest some lines of integration and social science research to elucidate the perceptions of vulnerability and their relevance to the development of a sustainable industrial future.

Cognitive Bias

The cognitive paradigm of risk perception asserts that perception is an outcome of the mental processing of information. The individual, or a sum of individuals, is the unit of analysis. There is a large literature on the cognitive underpinnings of risk perception phenomena. Several components of this literature are worth distinguishing here. The heuristics and biases component emphasizes the rules of thumb and consistent errors in judgment that characterize decision-making under uncertainty. The psychometric component emphasizes the clash between deeply held social-psychological values and attributes of activities that impose risks. The mental model component emphasizes the link between experience and risk responses.

Heuristics and Biases

Observed psychological and behavioral consequences of decision-making under uncertainty provide the stimuli for investigations into heuristics and biases. Of primary concern is how people interpret alternative, probable conditions and make selections across uncertain outcomes.

Experimental research on actual decisions under uncertainty (Tversky, 1972; Tversky and Kahneman, 1974) and experience on risk-taking behavior (Kunreuther *et al.*, 1978; Fischhoff, 1989) suggest that individuals make judgments about risk that consistently oppose the laws of probability. Cognitive psychologists have identified a number of heuristics and biases that are characteristic of the decision-making process under uncertainty and are easily elicited in experimental settings (Tversky and Kahneman, 1974; Kahneman *et al.*, 1982). Fischhoff (1981) provides a discussion and a research agenda that address the significance of the heuristics and biases for climate change. He suggests that the most important heuristics for climate change judgments are overconfidence, editing, availability, and anchoring.

Calibration and Overconfidence

People are often over- or underconfident regarding their judgments depending on the difficulty of the assessment. For moderate or extremely difficult assessments, people with general knowledge are often overconfident (Lichtenstein *et al.*, 1982). Overconfidence means that people believe their judgments to be more accurate than they actually are about assessing risks, even if this confidence cannot be supported by statistical evidence.

Editing and Isolation

In low-probability events this tendency is expressed by either careful selection (editing) of parameters from which to make the decision or setting aside (isolation) of the less favored among mutually exclusive parameters.

Availability

This tendency is expressed by overweighing events that can be easily imagined or recalled. Availability is common when information about an event is readily retrieved but the sources of information have little to do with the attributes of the risk. This heuristic is consistent with an excessive reaction to current information and an underestimation of future uncertainties.

Adjustment and Anchoring

People may anchor risk judgment to some starting point in the assessment (e.g., the last point on a time series), and adjust from the initial position while ignoring all previous evidence. This heuristic plays an important role in correctly eliciting judgments from people.

Fischhoff (1989) also articulates some general guides for understanding public responses to highly complex risks such as might be implied by global change:

- People simplify.
- It is difficult to change people's minds once they are made up.
- People remember what they see.
- People cannot readily detect omissions in statistical evidence.
- People disagree more about the definition of risk than about the magnitude.
- People have difficulty sorting through risk disputes.
- People have difficulty evaluating expertise.

Psychometric Links

Psychometric studies, derived from the seminal work of Fischhoff, Lichtenstein, and Slovic, investigate subjective risk judgments to understand the cognitive structure and the social factors that combine to influence the perceived magnitude and acceptance of risks (Slovic, 1987). While the focus is very much on human cognition, this literature provides the empirical base and evidence to extend the scope of risk analysis beyond its historically technical and physical domains to a multidimensional concept.

Psychometric factors have been linked to discussions of environmental vulnerability, but few psychometric studies have been focused on global change. In fact, most of the studies further explore some subset of the 90 risk sources first investigated by Slovic *et al.* (1980), who did not address global change risks explicitly. Fischer *et al.* (1991) find people perceive less personal control relative to social control over large, ecological risks and less control overall relative to other risks mentioned by respondents as concerns. This finding is significant for the evolution of perceptions since controllability has consistently emerged as an important factor in technology-hazard studies, although its influence is mixed across studies (Rohrmann, 1991). Schmidt and Gifford (1989) also address global environmental impacts and distinguish between environmental risks for humans and risks for the state of the environment.

Mental Models

The objective of the mental model approach "is to learn how people think about their particular situation, what they know, and what they misperceive about the facts and processes" (Lave and Lave, 1991, p. 260). Mental modeling follows the tradition of the cognitive bias paradigm in that it extends the notion of judgmental errors to misconceptions about risks. However, being knowledge based, this approach allows the subjects to define the salient information about risk perceptions, which is then contrasted with an "expert" representation of the issues, i.e., a composite representation based on facts and expert judgment. Additional information on sociodemographics and experiences is also collected from subjects to interpret mental model results.

Vulnerability and Adaptation

Although in an early stage, mental model research has produced some results for global climate change. First, there is an overall tendency to use the terms and concepts "weather" and "climate" interchangeably (Kempton, 1991; Bostrom *et al.*, in preparation). Extreme weather events may then have a significant effect on perceptions of climate change. Second, one of the more salient concepts linked to global warming is stratospheric ozone depletion (Kempton, 1991; Lofstedt, 1991; Bostrom *et al.*, in preparation). Finally, a concern for future generations seems to be an important reference for environmental values.

Self-Interest Bias

The most widely accepted paradigm in social science for the explanation of human behavior is that of self-interest (Lasswell, 1958; March and Simon, 1958; Easton, 1965). This concept dominates the economists' account of the search for profit and the political scientists' analysis of the quest for power. Risk perception approaches characterized by a self-interest bias emphasize the influences of perceived personal and social payoffs from risk sources. Health risk and costs for one's descendants are a significant part of what people fear in the absence of environmental protection (Hays, 1987). Kempton (1991) finds that his subjects define most environmental value in economic and anthropocentric terms.

Economics

The notion that individuals reveal their preferences about risks and benefits through their choices among goods and services is well established in the economics literature (Smith, 1986). Accordingly, if concerns about risks are linked to perceived personal costs, individuals should be willing to pay some amount to reduce their risk exposure. However, this assumption about risk behavior has not performed well as an indicator of perceived risk where risks are complex. Gardner and Gould (1989) show that broader definitions of perceived benefits and costs do little to help explain risk acceptability.

To explore risk concerns and individual priorities, Fischer *et al.*, (1991) use a phased procedure to elicit important risk concerns and willingness-to-pay (WTP) amounts from respondents. Although environment is mentioned more often by respondents than health, safety, and social risks, environmental risk evokes the smallest WTP amount from respondents. On the other hand, health risks, which are very directly tied to self-interest, evoke the highest WTP value.

Politics

Dunlap and Scarce (1991) argue that recent trends show public concern for environmental quality is increasing, as is support for government actions to improve it. Furthermore, there is growing support for increased environmental regulations on business and industry even at the expense of economic growth. Self-identification as an "environmentalist" has also increased in the United States.

These trends are reinforced in the larger, more recent Health of the Planet Survey (Dunlap *et al.*, 1992). Environmental problems increasingly are viewed as risks to human health and well-being. Furthermore, differences between rich and poor nations regarding the perceived seriousness of environmental threats, at least on the global level, seem to be lessening. Differences over responsibility are also less apparent, as indicated by responses that both industrialized and developing countries are equally responsible for environmental problems.

Many authors distinguish the politics of risk by recognizing different stakeholders in risk decisions. The identified stakeholder groups vary from one study to another, but all share a common characterization: the magnitudes of perceived risks and benefits from a risk source are stakeholder-dependent. Internationally, the North–South, industrialized–developing, and rich–poor distinctions are used to classify nations. Intranational distinctions follow more traditional political categories: Democrat and Republican parties, or left-wing and right-wing ideology. There is some evidence that for environmental risks, ideological distinctions are more correlated with risk perceptions than political party affiliations (Dake and Wildavsky, 1991).

Lewis (1990) distinguishes between the anti-technology elite (upper middle class in the United States and Europe) and the guardians of risky technology (complacent bureaucracies) as the central actors in risk controversies. Edwards and von Winterfeldt (1985) find it more complete to distinguish among three classes of stakeholders: proponents, opponents, and regulators. The technical elite is commonly identified as a proponent of risk-taking in studies of risk controversies, and a substantial literature has emerged that highlights the differences between expert and lay judgments.

Similarly, Krimsky and Plough (1988) focus more on the distinction between the regulators and "the public" to explore stakeholder interests. "Environmental risks in a community can evoke concerns about equity, the moral responsibility of government, and participatory democracy" (p. 301). They point out that concepts of rationality (often underlying the technical notion of risk perception) and concepts of democracy are antagonistic to one another. Political constituencies may use risk concerns to place other nonrisk demands on public authorities. The growing democratization of nations and escalating environmental concerns may exacerbate tensions over these hidden objectives (Jasanoff, 1986).

The interest bias paradigm is difficult to refute in a conceptual exercise since almost any hypothesized difference in perceived risks and benefits may be rationalized as calculated individual or social selfishness. In application, the external validity of self-interest approaches is continually challenged, and there is no robust set of benefit categories that dominates our understanding of risk behavior. However, the attention to individual and social payoffs as determinants of risk perceptions moves the self-interest bias paradigm away from the quest to reveal errors in judgment and closer to a search for the nontechnical features of risk behavior. The concept of risk is no longer multidimensional; it is polythetic. A polythetic approach treats technical and social definitions as equally important foundations of responses to risk. The critical distinction is that the self-interest bias preserves the technical definition of risk as its core concept with other dimensions entering in a subordinate way into the analysis.

Cultural Bias

Cultural analysis of risk looks behind the perception of physical risks to the social norms or policies that are being attacked or defended. Of all the things that people can worry about, they will be inclined to select for particular attention those risks that can be used to reinforce the solidarity of their institutions. Cultural theory differs from behavioral approaches to risk perception in that it focuses not on the individual, but on an institution that is driven by organizational imperatives to select risks for management attention or to suppress them from view.

Cultural theorists have proposed that contemporary environmental debates can be understood in the context of people invoking differing mythologies of the workings of nature to support their various political and moral beliefs (Douglas, 1985; Rayner, 1984). The climate change issue provides ample evidence that there are abiding and sometimes contradictory views of nature and philosophies of risk management—in short, plural rationalities.

Anthropologists have traced the ways in which perceptions of nature are constructed to reinforce choices about social organization (Douglas, 1966; 1970; 1978). An appeal to nature is the trump card that closes any argument. Invoking danger from the environment is a universal rhetorical technique of persuasion in moral and political debate in contexts as diverse as biblical prophesies of plague, floods, fire, or famine and contemporary concerns about deforestation, desertification, sea-level rise, or sunburn. From the anthropological perspective, the attribution of blame is an important key to understanding perceptions of vulnerability. The case of climate change clearly demonstrates that these perceptions are diverse and pertinent to decision-makers. The South blames the North for its energy profligacy; the North blames the South for soaring population growth and deforestation. Fossil fuel producers point the finger at land use while producers of tropical hardwoods blame industry.

Cultural theory suggests that there are four primary "nature myths" (Holling, 1986; Schwarz and Thompson, 1990). These describe nature as either fragile, robust, resilient, or capricious. As shown in Figure 1, where nature is represented as a ball on a differently shaped surface, each myth leads to a particular moral imperative, preference in response strategy, and type of social organization.

Key factors emphasized by the cultural bias paradigm are constituency agreement and conflict, and the preferred principles for trust, liability, and consent. In practice, cultural models are difficult to test systematically, and there are few examples of rigorous empirical validation of the propositions. Trust emerges as a significant factor in several studies of risk where other cultural variables are not investigated (Slovic *et al.*, 1991; Bord and O'Connor, 1990). Wildavsky and Dake (1990) show that cultural distinctions matter but they are highly correlated with

Social Organization	Egalitarians	Markets	Bureaucracies	NIMBYs Not In My Backyard
Image of Nature	Fragile 	Robust 	Resilient 	Capricious 
Moral Imperative	Don't mess with nature	Don't curb growth	Preserve choice	Don't tread on me
Response Strategy	Prevention	Adaptation	Sustainable development	Fatalism Denial

Figure 1. Nature myths.

political attributes. Tentatively, we may say that egalitarians are more concerned with technological and environmental risk, while bureaucratic and market organizations are more worried about social deviance and economic troubles.

Integration and a Proposed Framework

Our review of the social science approaches to perceptions of risk and vulnerability suggests multiple perspectives and insights. Using the cognitive bias paradigm, the literature has emphasized psychometric factors, but the link of these factors to acceptability is not established. Feelings of controllability as an indicator for risk concerns, may be important for climate change, but the evidence is unclear. So, too, is the importance or meaning of the weather/climate and ozone depletion/climate change misconceptions.

Social science is ambivalent about whether or not the subjectivist paradigm should dominate the technical approach, as reflected by the following two statements:

> Our attitudes toward risk vary according to what has happened to us, what we expect, what we feel, what we know, and what we care about. We ignore some risks, overestimate others. Our perceptions are selective and change as social life changes (Teuber, 1990, p. 235).

> Establishing the social character of the scientific perceptions of the climate change problem does not alter the physical realities of climate phenomena, or diminish the role of the physical sciences and technology in dealing with them (Boulding, 1983, p. 178).

Thus, a persistent criticism of the analysis of risk behavior is that it takes risk, not the risk perceivers, as the starting point for analysis. Neither risk nor information are things; they are complex relationships, human and nonhuman, that cumulatively alter their bearers. This is the crux of the debate between the subjectivists and the objectivists. The behavioral scientists urge risk managers to listen to the noise—the sociopsychologically shaped risk concerns arising within the system—while the technical risk analysts seek to devise risk communication schemes to eliminate noise so that everyone will understand probabilities and choose among them rationally. The cognitive bias approach may condemn the intentional manipulation of perceptions, but in seeking misconceptions instead of genuine disagreements about risk, it reinforces the technical view that the core dimensions of risk are independent of the social context.

The self-interest perspective directs us to health risks and concern for one's descendants as keys to climate change responses. However, while taking a more polythetic definition of risk and beginning with subjective impressions, the paradigm continues to seek numerical assessments of key characteristics. Perhaps even more disconcerting is the evidence challenging the links between risk concerns and personal willingness to bear the costs of actions to remedy risk threats. Here we see economic and poll data in direct conflict, and much more work remains to be done to sort out the messages. Furthermore, since climate change is characterized by diverse victims and diffuse agents, it does not mesh neatly with prior insights on stakeholders and technological risk. The increasing pressure from democratization of national and international decision processes is an additional complication that limits our confidence in predicting responses to climate change risks.

The cultural bias paradigm, by emphasizing world views and ideology, may be better suited to unravel constituency responses to climate change risks than other approaches. Our understanding from the production of scientific knowledge also supports a highly subjectivist perspective for an issue like climate change with high decision stakes and high systems uncertainty (Funtowicz and Ravetz, 1985; Rayner, 1987). However, as suggested by Kasperson *et al.* (1988), we may also expect climate change risks to be highly amplified, given the lack of personal experience with the changes and high reliance on networks for information and response guidance.

Social science has not yet reconciled the individual and constituency levels of analysis. However, poll data suggesting growing demands for public intervention and management of environmental risks may also be an indication that an institutional approach is appropriate to understand pivotal decisions and perceptions of environmental vulnerability at the level of organizations.

The cultural bias paradigm can accommodate the self-interest bias but not the cognitive bias. The diversity of cultural types embracing environmental issues indicates that self-interest may be critical for some. Because the functions of organizations do not differ systematically on a single dimension, and because organizational interests tend to be specific to the issues being debated, a general framework to predict institutional responses to risk may prove elusive. Table 1 is a preliminary effort to provide such a framework.

The rows of Table 1 consist of the cultural types in declining order of organizational formality from hierarchy to equality. The columns indicate broadly defined organizational functions ordered in increasing levels of generality from the specific goal of environmental protection to the broadest issues of economic and societal development. The table is completed by the inclusion of examples of the type of organization that corresponds to each of the nine cells, a prediction of its reasoning style (reductionist, pragmatic, or holistic), and a prediction of its interpretation of prudence (its choice of the precautionary principle or the proof-first principle).

The prediction of reasoning style varies consistently with culture and is unaffected by function. Since we know that hierarchies like to routinize decision making, often combining incompatible agendas without acknowledging differences, we anticipate a tendency to reductionist reasoning that renders parts of a problem susceptible to a decision rule, and is unlikely to expose contradictions in the institution's utterances or actions; hence the attractiveness of benefit–cost analysis or probabilistic risk analysis to these institutions. By way of contrast, success in the market depends on judgment skills and flexibility to recog-

Table 1: *Interaction of organizational function and structure in framing uncertainty*

Structure	Function		
	Environmental Protection	Social/Political Regulation	Economic Development/Growth
Hierarchy	e.g., federal and state environmental regulators	e.g., courts, public utility commissions	e.g., federal and state energy and commerce departments
	• Reductionist style • Precautionary principle	• Reductionist style • Proof-first principle	• Reductionist style • Proof-first principle
Market	e.g., environmental entrepreneurs, energy service companies	e.g., federal and state legislatures	e.g., utilities, manufacturing companies
	• Pragmatic style • Precautionary principle	• Pragmatic style • Mixed prudence	• Pragmatic style • Proof-first principle
Collective	e.g., grassroots environmental groups	e.g., town meetings	e.g., nongovernmental economic think tanks
	• Holistic style • Precautionary	• Holistic style • Precautionary	• Holistic style • Mixed prudence

nize and exploit opportunities. Resort to rules is likely to give way to pragmatic assessments of gain or loss. Finally, egalitarian collectivists seek to marshal the fullest range of arguments in favor of their position. They are likely to be adept at connecting technical, ethical, and socioeconomic arguments in a holistic fashion.

Unlike variation in reasoning styles, the interpretation of prudence seems to depend on the interaction of culture and the extent to which the goals of the organization are dominated by the specific function of environmental protection. The institutions charged with environmental protection consistently advocate precaution. Even the hierarchical environmental protection agencies that might be organizationally disposed to proof-first are aware that the constituency to which they must ultimately answer, and which will lobby most effectively for their funding, is precautionary environmental activists. Institutions charged with nurturing economic growth tend to want proof before agreeing to environmental investment that may constrain economic growth. However, the nongovernmental organization think tanks can afford the luxury of open speculation about the down side of uncertainties, which, for political and economic reasons respectively, the commerce agencies and utilities cannot. Finally, the institutions whose goals require them to mediate between the demands of environmental protection and economic growth tend to find their approach to prudence heavily modified by their institutional structure: courts favoring proof, legislatures favoring some balance of prior proof and precaution, and participatory structures favoring precaution.

The relationship between cultural orientation based on social organization and the strategic attitude of an institution toward uncertainty in debates about environmental and technological risk is not a simple one of risk-loving markets, risk-averse collectives, and risk-managing hierarchies. The constraints upon discourse and the credibility of arguments within various institutions interact with functional preferences based on self-interest to shape the style and content of intervention in the wider debate. Broadening the scope of uncertainty relationships challenges us to consider how the cognitive bias paradigm might be included as well. Certainly we should expect individual heuristics and errors that are found in uncontentious risk problems to be reflected in problems as complex as climate change. A new research agenda might address this integration by considering the relationships between individuals and the institutions they come to select for membership and how these institutions accommodate individual biases over time.

References

Benedick, R. 1991. *Ozone Diplomacy,* Harvard University Press, Cambridge, Massachusetts, p. 199.

Bord, R., and R. O'Conner. 1990. Risk communication, knowledge, and attitudes: Explaining reactions to a technology perceived as risky. *Risk Analysis 10(4),* 499–506.

Bostrom, A., M. G. Morgan, and B. Fischhoff. Lay mental models of risk control: What about hairspray, styrofoam cups and global warming? (in preparation).

Boulding, E. 1983. Introduction to psychological dimensions of climate change. In *Social Science Research and Climate Change: An Interdisciplinary Appraisal* (R. S. Chen, E. Boulding, and S. H. Schneider, eds.), D. Reidel, Dordrecht, Netherlands, 177–179.

Bradbury, J. 1989. The policy implications of differing concepts of risk. *Science, Technology, & Human Values 14(4),* 380–400.

Clark, W. C. 1988. The human dimensions of global environmental change. In *Toward an Understanding of Global Change,* Committee on Global Change, National Academy Press, Washington D.C., p. 187.

Dake, K., and A. Wildavsky. 1991. Individual differences in risk perception and risk-taking preferences. In *The Analysis, Communication, and Perception of Risk* (B. J. Garrick, and W. C. Gekler, eds.), Advances in Risk Analysis Volume No. 9, Plenum Press, New York, 15–24.

Douglas, M. 1966. *Purity and Danger,* Routledge and Kegan Paul, London.

Douglas, M. 1970. *Natural Symbols,* Barrie and Rockliffe, London.

Douglas, M. 1978. *Cultural Bias,* Royal Anthropological Institute Occasional Paper No. 35, London.

Douglas, M. 1985. *Risk Acceptability According to the Social Sciences,* Russell Sage Foundation, New York.

Dunlap, R., and R. Scarce. 1991. Environmental problems and protection. *Public Opinion Quarterly 55,* 651–672.

Dunlap, R., G. Gallup, Jr., and A. Gallup. 1992. *The Health of the Planet Survey.* Preliminary Report, The George H. Gallup International Institute, Princeton, New Jersey, 45 pp.

Easton, D. 1965. *A Framework for Political Analysis,* Prentice Hall, Englewood Cliffs, New Jersey.

Edwards, W., and D. von Winterfeldt. 1985. Public disputes about risky technologies: Stakeholders and Arenas. In *Environmental Impact Assessment, Technology Assessment, and Risk Analysis* (V. Covello, J. Mumpower, P. Stallen, and V. R. Uppuluri, eds.), NATO Advanced Science Institutes Series, Volume G4, Springer-Verlag, Berlin, Germany, 877–915.

Fischer, G., M. G. Morgan, B. Fischhoff, I. Nair, and L. Lave. 1991. What risks are people concerned about? *Risk Analysis 11(2),* 303–314.

Fischhoff, B. 1981. Hot air: The psychology of CO_2-induced climate change. In *Cognition, Social Behavior, and the Environment* (J. H. Harvey, ed.), Lawrence Erlbaum Associates, Hillsdale, New Jersey, 163–184.

Fischhoff, B. 1989. Risk: A guide to controversy. In *Improving Risk Communication,* National Research Council, National Academy Press, Washington, D.C., 211–319.

Funtowicz S. and J. Ravetz. 1985. Three types of risk assessment. In *Risk Analysis in the Private Sector* (C. Whipple, and V. Covello, eds.), Plenum, New York.

Gardner, G., and L. Gould. 1989. Public perceptions of the risks and benefits of technology. *Risk Analysis 9(2),* 225–242.

Gerlach, L. P. Resources, conflict, and change: Negotiating the natural resources future through disputes and debates. *Annual Review of Energy,* in press.

Hays, S. 1987. *Beauty, Health, and Permanence: Environmental Politics in the United States, 1955–1985,* Cambridge University Press, Cambridge, United Kingdom.

Holling, C. S. 1986. The resilience of terrestrial ecosystems: Local surprise and global change. In

Sustainable Development of the Biosphere (W. Clark, and R. Munn, eds.), Cambridge University Press, Cambridge, United Kingdom, 292–316.

Jasanoff, S. 1986. *Risk Management and Political Culture,* Social Research Perspectives Volume 12, Russell Sage Foundation, New York, 93 pp.

Kahneman, D., P. Slovic, and A. Tversky, eds. 1982. *Judgment Under Uncertainty: Heuristics and Biases,* Cambridge University Press, Cambridge, United Kingdom, 555 pp.

Kasperson, R., O. Renn, P. Slovic, H. Brown, J. Emel, R. Goble, J. Kasperson, and S. Ratick. 1988. The social amplification of risk: A conceptual framework. *Risk Analysis 8(2),* 177–187.

Kempton, W. 1991. Lay perspectives on global climate change. *Global Environmental Change 1(3),* 183–208.

Krimsky, S., and A. Plough. 1988. *Environmental Hazards: Communicating Risks as a Social Process,* Auburn House Publishing Company, Dover, Massachusetts, 333 pp.

Kunreuther, H., R. Ginsberg, L. Miller, P. Sage, P. Slovic, B. Borkan, and N. Katz. 1978. *Disaster Insurance Protection,* John Wiley and Sons, New York.

Lasswell, H. 1958. *Politics,* Meridian, Cleveland.

Lave, T., and L. Lave. 1991. Public perception of the risks of floods: Implications for communication. *Risk Analysis 11(2),* 255–267.

Lewis, H. 1990. *Technological Risk,* W. W. Norton and Company, New York, 353 pp.

Lichtenstein, S., B. Fischhoff, and L. Phillips. 1982. Calibration of probabilities: The state of the art to 1980. In *Judgment Under Uncertainty: Heuristics and Biases* (D. Kahneman, P. Slovic, and A. Tversky, eds.), Cambridge University Press, Cambridge, United Kingdom, 305–351.

Lofstedt, R. 1991. Climate change perceptions and energy-use decisions in Northern Sweden. *Global Environmental Change 1(4),* 321–324.

March, J., and H. Simon. 1958. *Organizations,* John Wiley and Sons, New York.

Nash, R. 1967. *Wilderness and the American Mind,* Yale University Press, New Haven, Connecticut.

Rayner, S. 1984. Sickness and social control. *Listening: A Journal of Religion and Culture 19(2),* 143–154.

Rayner, S. 1987. Risk and relativism in science for policy. In *The Social and Cultural Construction of Risk* (B. Johnson, and V. Covello, eds.), D. Reidel, New York.

Rohrmann, B. 1991. *A Survey of Social-Scientific Research on Risk Perception,* Studies on Risk Communication Volume No. 26, Research Center Julich, KFA Julich, Germany, 96 pp.

Schmidt, F., and R. Gifford. 1989. A dispositional approach to hazard perception: Preliminary development of the environmental appraisal inventory. *Journal of Environmental Psychology 9,* 57–67.

Schwarz, M., and M. Thompson. 1990. *Divided We Stand,* Harvester Wheatsheaf, London.

Slovic, P. 1987. Perceptions of risk. *Science 236,* 280–290.

Slovic, P., B. Fischhoff, and S. Lichtenstein. 1980. Facts and fears—understanding risk. In *Societal Risk Assessment* (R.C. Schwing, and W.A. Albers, eds.), Plenum, New York, 181–218.

Slovic, P., M. Layman, and J. Flynn. 1991. Risk perception, trust, and nuclear waste: Lessons from Yucca Mountain. *Environment 33(3),* 6–9, 28–30.

Smith, V. K. 1986. Benefit–cost analysis and risk assessment. In *Advances in Applied Micro-Economics* (V. K. Smith, ed.), Volume 4, Risk, Uncertainty, and the Valuation of Benefits and Costs, JAI Press, Greenwich, Connecticut, 190 pp.

Teuber, A. 1990. Justifying risk. *Daedulus 119(4)*, 235–254.

Tversky, A. 1972. Elimination by aspects: A theory of choice. *Psychological Review 79*, 281–291.

Tversky, A., and D. Kahneman. 1974. Judgment under uncertainty: Heuristics and biases. *Science 185*, 1124–1131.

Wildavsky, A., and K. Dake. 1990. Theories of risk perception: Who fears what and why? *Daedalus 119(4)*, 41–60.

6

The Human Dimension of Vulnerability

Robert S. Chen

Abstract

One of the key ingredients of industrial ecology is the concept of vulnerability, which includes an assessment of individual and social risks associated with anthropogenic environmental change, together with an appreciation of the robustness of people and societies to adapt to or accept such risks. This discussion of human vulnerability focuses on the world's most vulnerable populations: poor, urban, and rural inhabitants in the developing world. Environmental change can lead to significant societal disruption, especially among peoples less able to adapt.

Introduction

Much recent debate about global environmental change has tended to separate discussion of the natural and human *sources* of change from consideration of the *impacts* on human welfare posed by environmental variation and change. On the one hand, human activities such as fossil fuel consumption, land use, and industrial and agricultural production are widely recognized as major contributors to "environmental vulnerability," driven primarily by rapid population growth, economic development, and technological change. On the other hand, many of these same activities, along with other aspects of natural resource management and human health and welfare, are themselves key factors in "human vulnerability" to environmental fluctuation and change, both now and in the future.

Unfortunately, separate consideration of these two aspects of vulnerability has led to poor integration of research and understanding, especially for policy applications. Decisions to modify human activities to reduce perturbations to the environment inevitably require tradeoffs between the costs and benefits of the activities, which in turn depend in part on the sensitivity of human welfare to different environmental variations.[1] For example, developing countries are understandably reluctant to forego the immediate benefits of industrial and agricultural development—which among other things help to reduce vulnerability to present environ-

---

[1] For a mathematical treatment of this point, see Chen and Fiering (1989).

mental variation—in order to help avert the longer term and less certain impacts of a changing climate.

Preventive and mitigative activities may themselves alter the nature of human interactions with the environment beyond their intended objectives. A shift from fossil fuels to renewable energy resources would affect future sensitivity to environmental variation not just with respect to the reliability of energy supply but also in terms of air and water pollution, land-use patterns, and associated resource management. Efforts to reduce methane emissions from livestock and rice production could have significant—and immediate—effects on agriculture, food security, and other related industries. Recognizing such interactions is especially important since there is no guarantee that unilateral or even global preventive efforts will succeed in maintaining the environmental "status quo" for all; some degree of adaptation will still be necessary even after extensive efforts to prevent environmental change.[2]

Other changes in human management of the environment may also have implications for both environmental and human sensitivities. Continuing evolution of irrigation technology is likely to affect not only drought sensitivity but also agricultural energy, water, and fertilizer use, and therefore greenhouse gas emissions. Increased combustion efficiencies and changes in land use could reduce carbon monoxide emissions, leading to higher concentrations of the hydroxyl radical (OH) and in turn accelerating the removal of methane and chlorofluorocarbons from the atmosphere. Failure to deal with issues of poverty and sustainable development in developing countries could lead to accelerating population growth, environmental degradation, and loss of biological diversity and might well inhibit agreement on, and effective implementation of, worldwide measures to reduce trace gas releases.

Potential interactions of this kind highlight the need to reexamine traditional divisions between studies of the human sources of environmental change and the potential impacts of such change on human systems. For instance, many early research efforts emphasized the time delay between sources and impacts: present-day perturbations to the environment were not expected to have significant effects until the "distant" future. Yet as surprises like the Antarctic ozone hole emerge, this assumption is becoming less applicable. Likewise, the tendency to assess sources and impacts independently for different regions has led to neglect of both environmental and human "teleconnections"—effects mediated by long-distance flows of energy, material, people, wealth, and information in an increasingly interconnected world. It is therefore important to consider the *overall* ability of interlinked natural and human systems to adjust in the short term and adapt in the long term to the additional complications posed both by environmental change and by preventive or mitigative actions. It is also important to examine the *distribution* of impacts and vulnerabilities across human populations in the context of concerns about equity and the emergence of global responses to environmental threats.

[2] Curiously, two recent U.S. National Research Council reports (Stern *et al.*, 1992; NRC, 1991; 1992) give very little attention to the interaction between environmental and human vulnerabilities.

In this chapter, I characterize some of the variability and dynamics of human vulnerability to the environment over space and time and highlight some possible feedbacks to environmental vulnerability. Vulnerability is defined as "the capacity to suffer harm or to react adversely" (Kates, 1985).[3] Notably, this definition requires some measure of harm or adversity—clearly a normative judgment based on human perceptions and values. The term "environmental vulnerability" thus includes a presumption of harm or adversity either to human beings or perhaps to some human conceptualization of "nature" or "life."

I focus first on human adaptation to the environment through history and then review key features of the vulnerability of specific "marginal" population groups and of social systems. My primary objective is to give recognition to the rapid global change in natural and human systems now under way that is contributing to both environmental and human vulnerabilities.

Human Adaptation to Environmental Variation and Change

A Historical Perspective

Humanity has depended on natural ecosystems for most of its two million years of existence. Only in the relatively recent past, through the domestication of plants and animals some 10,000 years ago, have humans begun to modify and manage their local environments consciously to improve their own welfare. Although differing theories exist as to the origins of agriculture and the role of the environment, it is evident that agriculture spread widely during a period of climatic and ecological variation characterized by major changes in lake and ocean levels, glacial boundaries, monsoon behavior, and vegetation patterns. Indeed, as early as about 7000 years ago, human settlements had begun to grow rapidly in fertile alluvial floodplains created by rising sea levels, at first taking advantage of seasonal flooding and later constructing canals and other structures for gravity-fed irrigation (Matthews *et al.*, 1990).

Continuing technological and social innovations in agriculture, industry, and transportation clearly permitted higher population densities and provided increased protection from environmental variation. The Roman Empire, for example, grew to depend heavily on long-distance grain shipments; as much as 200,000 metric tons of food crossed the Mediterranean in small ships each year by one estimate (Rickman, 1980). Between 1831–35 and 1909–13, the supply zone for agricultural imports into London more than tripled in extent, and agricultural imports into the United Kingdom grew to nearly half of apparent consumption (Chisholm, 1990).

But early population growth was not consistently upward. As illustrated by four case studies of long-term population change (Figure 1), population densities have fluctuated significantly on time scales of centuries to millennia. In the case of the

---

[3] See Timmerman (1981), Liverman (1989), and Downing (1991) for much more detailed discussions of definitions of vulnerability and their application in different settings.

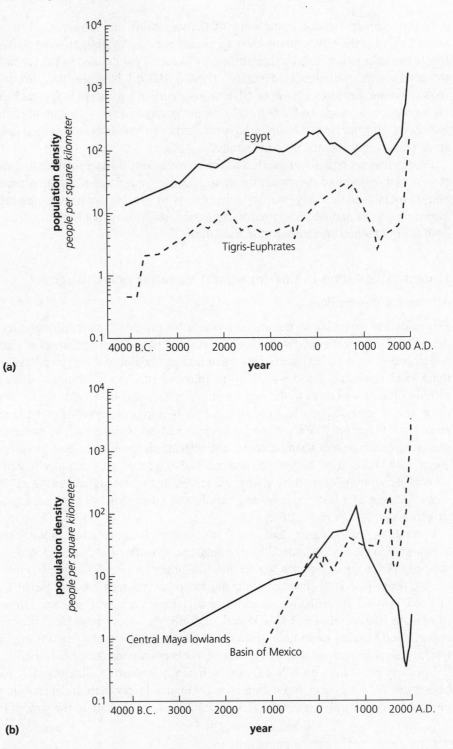

Figure 1. Four case studies of long-term fluctuations in population density: (a) Tigris-Euphrates lowlands and Egyptian Nile Valley; (b) Basin of Mexico and Central Maya lowlands (Whitmore *et al.*, 1990).

Tigris-Euphrates lowlands, evidence of widespread salinization, waterlogging, and siltation suggests the environment's key role in a major population decline beginning around 1900 B.C. (Whitmore *et al.*, 1990). In the Mayan lowlands, most of the original tropical forest had been cleared by 300 A.D. and cultivation intensified to support more than 3 million people by 800 A.D. But over the next millennium, the population collapsed rapidly, dropping to less than 7000 people by 1900 A.D. However, it is not clear that this collapse stemmed from a "Malthusian" scenario of population growth outstripping agricultural production (Turner, 1990). In other cases, links between population change and environmental degradation are not evident. Thus, history suggests that there is no simple or predictable connection between long-term population fluctuations and environmental variation and transformation.

One intervening factor is clearly migration, long a response to climatic (and other) perturbations. Periods of drought, cold, or other extremes have led to large population movements from many regions throughout history, including Mycenaea in 1230–1000 B.C., Iceland and Sweden before and during the so-called Little Ice Age (ca. 1450–1900 A.D.), Ireland in the mid-1800s, and the Great Plains in the 1890s and 1930s (NRC, 1992). In the case of Ireland, warm moist summers encouraged the spread of the potato blight fungus and the subsequent "potato famine," which in turn led to emigration of up to half of the Irish population. In contrast, the Great Plains droughts displaced hundreds of thousands of people but did little to halt the long-term increase in the region's population, which quintupled between the 1890s and 1980s (Riebsame, 1990).

Recent Changes in Vulnerability

During the past two centuries of industrialization, population growth has accelerated dramatically at both regional and global scales along with humanity's ability to influence and control the environment. Beginning with the introduction of the steam engine in the mid-1800s, the dependence on water, wind, animal, and human power has steadily declined (Headrick, 1990). Sensitivity to environmental fluctuation has also diminished. Diverse technological and social innovations have enabled human settlements to grow dramatically in size and density, even where environmental conditions are harsh (e.g., Ausubel, 1991). Refrigerants based on chlorofluorocarbons have permitted radical changes in food storage and distribution, diet, and patterns of agriculture and trade (Stern *et al.*, 1992). Refrigerants also allowed development of air conditioning technologies that facilitated population growth and movement into the U.S. Sunbelt, with consequent impacts on energy demand and supply among other factors.

Whether or not overall vulnerability to environmental variation has declined as the result of technological and social changes of this kind remains controversial.[4]

---

[4] See, for example, NRC, 1991 and the dissent by Jessica Mathews on pp. 45–46.

Indeed, this may be a misleading question given the many dimensions of vulnerability and change evident in both natural and human systems and the value judgments implicit in selecting specific indicators of vulnerability. What is clear is that the patterns and scale of vulnerability have changed over time, as have some of the major sources of vulnerability. Thus, in developed countries natural hazards such as drought, flood, and storms are now much less life-threatening than in the past, though their threat to property in absolute economic terms remains high, due in part to increasing concentrations of population and wealth in vulnerable areas such as coastlines (White and Haas, 1975; Burton *et al.*, 1993). The emergence of a "global food system" provides greater flexibility of response when climate extremes occur at local or regional levels (Millman *et al.*, 1990).

In the developing world, vulnerability to the environment has evolved at several different levels. At the international level, continuing improvements in communications, monitoring systems, and response mechanisms have in many instances increased the ability of the world community to provide timely humanitarian assistance in times of disaster (Downing, 1991). However, this international system is itself threatened by rapid growth in needs, pressing resource and institutional limitations, and political controversy. In the decade since 1981 the number of internationally recognized refugees has doubled to more than 16 million people—an annual growth rate several times larger than the rate of population growth in the developing world. The number of internally displaced persons may be much larger. Recent reports indicate a significant increase in the level of hunger and malnutrition among those refugees and others receiving assistance from international humanitarian agencies (Chen, 1990a, 1992; ACC/SCN, 1992). Violent conflict has become a major cause of famine and famine deaths (Millman *et al.*, 1991). Of course, recent deadly floods in Bangladesh, China, Nicaragua, Europe, and the U.S. midwest illustrate that there is still much reason for concern about natural hazards now and in the future.

Within countries of the developing world, the vulnerability of specific population groups and of specific regions has also changed over time. Certainly the most vulnerable to a wide range of threats are the poorest and hungriest subgroups. In recent decades, the proportion of people who have inadequate access to even minimal diets has decreased by half, but due to population growth the absolute number of people has not fallen (Kates *et al.*, 1989). Recent World Bank projections indicate that poverty rates are expected to decline in all regions except sub-Saharan Africa by the end of the century. In the latter region, deteriorating conditions combined with population growth imply nearly a 50% increase in the number of "poor" people.[5]

In Mexico, losses due to natural hazards have increased greatly since the 1940s, primarily due to increasing population and poverty in hazard-prone areas such as the arid north and the flood-prone coasts. There is no indication of any increase in

---

[5] "Poor" is here defined as no more than $370 annual income per capita in 1985 purchasing parity dollars (World Bank, 1992).

the severity of weather events during this period. Drought losses in the relatively poor communal (*ejido*) agricultural sector have been greater than in the private sector, in part because of differences in land quality and level of technology (e.g., irrigation and fertilizer use) between these two sectors (Liverman, 1989).

An alternative approach is to examine regions that may be especially vulnerable to environmental variation or degradation. Kates and Haarmann (1992) have recently reviewed links between poverty and "environmentally threatened areas"—drylands, highlands, and tropical forests. These areas contain as much as one-quarter of the world's population. Unfortunately, detailed and reliable data on income distribution and demographic trends are lacking for most of these regions. However, it seems likely that many of these people are poor and that the size if not the proportion of population in these areas is increasing.

These examples suggest the importance of discriminating between vulnerability in developed vs. developing countries, between the vulnerability of population subgroups vs. overall social systems, and between past, present, and future sources of vulnerability. I focus first on the vulnerability of specific "marginal" groups of people whose survival and welfare for various reasons may be especially susceptible to environmental variation.

Vulnerability of Marginal Groups

Assessments of the implications of environmental change often consider only "net" impacts, i.e., overall changes in welfare or other measures of impact regardless of the magnitude and distribution of gains and losses in welfare within a population or region. It is assumed either that impacts are distributed "evenly" across socioeconomic groups or that differential effects across such groups are less important than the overall magnitude of change. Unfortunately, even the definition of "evenly" is ambiguous, as it could refer to a fixed absolute change or a fixed percentage change in some measure of welfare, and it does not necessarily take into account existing differences in underlying income or wealth. Thus, reducing a poor person's income by a given amount or a given percentage can certainly be much more devastating for that person than the identical reduction would be for someone with greater wealth. A transfer of wealth from the poor to the rich would be more harmful to general welfare than an equivalent flow in the opposite direction.

This difference in vulnerability reflects a difference in the likelihood of adverse consequences given exposure to a particular risk. Poor people—including the landless urban and rural poor and resource-poor smallholders—are generally more vulnerable to economic disruption than richer people, since they have fewer financial resources or other entitlements. Their health and nutritional status is generally lower, due in part to inadequate access to food and primary health care, lower educational levels, and higher incidence of disease and parasites. They may also live in areas that are less accessible or that have more limited capacities to respond to disaster.

Vulnerability and Adaptation

A second source of variation in vulnerability relates to the likelihood of exposure. Again, it is poor people who most often live in the least productive, most hazardous, and/or ecologically fragile areas: on steep slopes, in flood- or drought-prone regions, near industrial or waste sites. Shelter is usually inadequate and reliable transportation inaccessible. Wealthier groups may also live in hazardous areas for various reasons, but they can usually limit or control their exposure through greater mobility and access to protection. In Mexico City, for example, the rich "can afford air conditioning, bottled water, rural retreats, good nutrition, and health care" (Liverman, 1989:27). Employment and other income-generating activities frequently subject unskilled workers to a variety of health and safety hazards including high accident rates, unsanitary conditions, toxic chemicals, and exposure to pollution and adverse weather and climate conditions.

Poverty and Vulnerability

Taken together, these two sources of vulnerability suggest that poverty is a key indicator of vulnerability to the environment. Poverty forces poor people to overuse or degrade environmental resources and to abandon sustainable practices in favor of short-term survival. Unfortunately, research on the linkages between poverty and the environment has been limited to date. Kates and Haarmann (1991, 1992) found only 30 case studies that examine causal linkages between poverty and "threatened" environments to some degree. In many situations, the poor become trapped in a "spiral" of impoverishment and degradation (Figure 2) in which underlying forces of population growth, poverty, development and commercialization, and hazards combine to displace people and divide their resources, resulting in further environmental degradation and intensification of underlying problems.

A key feature of the spiral in Figure 2 is the potential for feedback between environmental degradation and the four underlying forces. Once environmental degradation begins, it is likely to increase the costs of development, to exacerbate imbalances between population and resources, to impact more heavily on the poor and therefore increase the extent and intensity of poverty, and to heighten the likelihood and severity of losses due to natural hazards. In many highland areas, for example, deforestation stemming from the need for fuelwood and farmland increases soil erosion and the risk of flooding, avalanches, and mudslides. This can in turn reduce agricultural productivity and income from forest resources in both highland and lowland areas and increase pressure on other areas and resources. Possible results include accelerated loss of biomass and biological diversity, increased reliance on fertilizers and irrigated agriculture, and substitution of fossil fuels for fuelwood.

Urbanization

Vulnerability is not limited to rural areas. Leonard (1989) estimates that more than 100 million extremely poor people live in squatter and other urban settlements. In

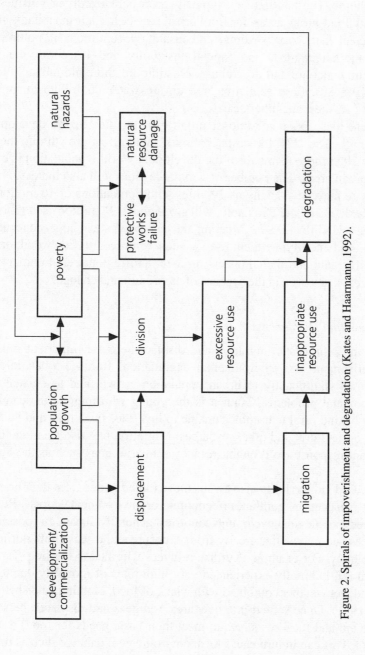

Figure 2. Spirals of impoverishment and degradation (Kates and Haarmann, 1992).

the late 1960s and 1970s, more than one-third of the populations of at least a dozen large cities in the developing world lived in slums or squatter settlements. Calcutta alone had more than 5 million slum-dwellers and squatters in 1971—two-thirds of its population. These settlements typically grow haphazardly on hillsides or in polluted and flood-prone areas, far from urban services. Their inhabitants suffer from "insufficient diet, overcrowding, and insanitary conditions"; "pollution, traffic, stress, and alienation"; and "social instability and insecurity that…includes promiscuity, alcohol and drug abuse, prostitution, and child labour" (Tabibzadeh *et al.*, 1989:30). Poor sanitation and unsafe water contribute to high rates of malaria, parasites, and other diseases.

Despite these poor conditions, urban growth rates in the developing world remain high. The U.N. forecasts continued rapid urban growth into the next century; by 2025 more than half of the developing world's population is expected to live in urban areas. The number of very large cities will also increase: by the year 2000 there may be as many as 19 cities with populations of 10 million or more. Two, Mexico City and São Paulo, will swell above 24 million each (Tabibzadeh *et al.*, 1989; UNFPA, 1991). Meeting the needs of such large concentrations of people—and modifying their use of such resources as fuelwood, fossil fuels, and refrigerants and their impacts on local and regional environments—will be formidable tasks even in the absence of environmental change.

Migration and Displacement

As Figure 2 suggests, migration and displacement are important links between impoverishment and environmental degradation. Rural-to-urban migration has contributed significantly to urban population growth and associated social and environmental problems. Often it is the young who migrate in search of work, leaving behind rural households unable to invest as much time and labor into sustaining agricultural and other resources. Migration has also been one of the most important responses to drought in many semiarid areas such as the Sahel (NRC, 1983).

Increased vulnerability also awaits those forced to flee from their homes for reasons of persecution, conflict, or economic or environmental stress. Refugees and displaced people in developing countries generally have few possessions and resources and may suffer greatly from exposure and disease both during and after their journey. For example, Kurdish refugees who fled to the Iraq–Turkey border in April 1991 initially experienced very high rates of mortality due to unusually cold and wet weather combined with a lack of food, clothing, and shelter (Sandler *et al.*, 1991). Once settled in a new area, refugees and displaced persons rely on outside aid and local resources to meet their basic needs for food, water, shelter, and fuel. This can in turn cause local environmental damage such as deforestation and water pollution.

At the global level the present number of refugees and other displaced peoples is relatively small compared with world population, though growth rates are high.

However, at regional or national levels proportions can be much higher. Africa is estimated to host as many as 20 million refugees and internally displaced people, about 3% of the continent's population. One-fifth or more of the populations of Afghanistan, Mozambique, and Somalia have fled their homes (USCR, 1992). Rapid expansion of population movements has raised concern about the environmental impacts in host countries and about the potential for even larger flows stemming from environmental threats such as sea-level rise, environmental degradation, and recurrent drought and famine (e.g., Refugee Policy Group, 1992; UNFPA, 1991; IPCC, 1990; Hinnawi, 1985).

Vulnerability of Social Systems

The vulnerability of marginal groups such as migrants and the very poor is one manifestation of the potential vulnerability of social systems to environmental variation. Again, it is useful to distinguish between the likelihood of exposure and the likelihood of adverse consequences.

Environmentally Sensitive Activities and Resources

A wide range of human activities are sensitive to environmental variations. The sectors of the economy most often cited as especially sensitive to the environment are agriculture, water resources, forestry, fisheries, and tourism, but human settlements, energy resources, transportation, health, and some industries also have received attention (e.g., IPCC, 1990; NRC, 1992). I focus here primarily on agriculture.

Agriculture's economic importance varies greatly between developed and developing countries. Agriculture generated about 30% of the total 1990 gross domestic product (GDP) in 35 countries classified as low income (gross national product less than $600 per person in 1990 U.S. dollars). This compares with 12% in middle-income countries and less than 5% in high-income countries (World Bank, 1992). The contrast is even greater in agricultural employment: 60% of the economically active population in the developing world was engaged in agricultural activities in 1990, compared with only 8% of those in the developed world (FAO, 1991).

Agriculture is strongly linked with other economic sectors. For example, most food processing and distribution activities and fertilizer production and equipment manufacture are in support of agriculture. Thus, in 1980 only one-fifth of energy use in the U.S. food and fiber sector was attributable to farm production; the remainder was divided among linked economic activities (Duncan and Webb, 1980).

The degree of agriculture's vulnerability to environmental variation depends greatly on the nature and timing of change, the types and areas of agriculture affected, and interconnections with impacts in other sectors such as energy and water resources. What is worth emphasizing here from the viewpoint of the vulnerability of social systems is the diversity of potential threats and the importance of interregional and intersectoral linkages.

Diversity of Threats

Many different environmental factors are likely to affect agriculture. Projections of global climate warming tend to emphasize potential changes in mean temperature and precipitation, and to a lesser degree changes in spatial patterns of these and related variables such as soil moisture. But local changes in climate may have more significant effects on agricultural systems (see Figure 3). Changes in the ozone layer may already be increasing levels of ultraviolet radiation, which could well be damaging to crops and other natural ecosystems. Increasing carbon dioxide (CO_2) levels will benefit some crops, but also will have effects on nutrient and water uptake and on weeds. Changes in air pollution levels and atmospheric OH levels could significantly affect plant productivity. Indirect ecological effects of potential importance to agriculture include changes in plant and animal pests and diseases; wind- and water-induced soil erosion; soil salinization, alkalinization, and leaching; water quality and availability; and soil loadings of metals and persistent organics (see Schnoor and Thomas, this volume; Stigliani, Jaffé, and Anderberg, this volume).

No studies to date have investigated the combined effects of these environmental and ecological changes. Initial studies of yield impacts associated with altered temperature and precipitation patterns in a world of doubled atmospheric CO_2 suggest first-order changes of between +20% and –30% in northern midlatitude nations (Parry, 1990). Increased CO_2 levels might compensate somewhat for yield decreases for the so-called C3 crops like wheat, rice, and soybeans, but would have more limited effects on C4 crops like maize, millet, and sugarcane common in the tropics. Other effects such as plant sensitivity to increased UV-B radiation have been studied in isolation, but little is known, for example, about how UV-B and climatic effects might interact (Anderson, this volume).

Potential implications for other key aspects of agriculture have also been neglected. For example, relatively little attention has been given to rice, which makes up more than one-fourth of all world cereal production but which is almost entirely produced in the developing world.[6] It is known that many rice cultivars are sensitive to levels of solar radiation as well as to temperature. The pattern of solar radiation and temperature over the crop season (i.e., whether it rises, falls, or stays constant from time of sowing to harvest) can make a 20–60% difference in total rice yields (Venkateswarlu, 1989; Seshu *et al.*, 1989).

Interregional and Intersectoral Linkages

Linkages between regions and sectors will be extremely important to the overall sensitivity—and vulnerability—of agriculture to environmental variation. Obviously, yield reductions that affect many agricultural regions concurrently would have more

---

[6] FAO data for 1990 indicate that only 5% of rice was produced in developed countries—half of that in Japan (FAO, 1991).

Possible Global Environmental Changes

Possible Measures to Prevent or Delay Global Environmental Changes

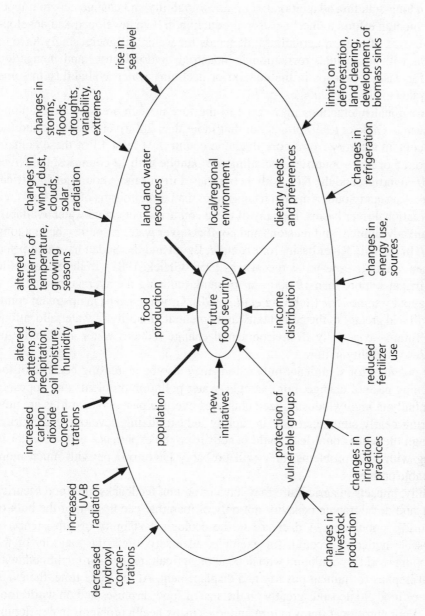

Figure 3. Possible links between future food security and the occurrence or prevention of global environmental change. Thick arrows indicate linkages that have been most often studied; thin arrows indicate those that have received little or no careful examination to date. Reprinted by permission of the publisher from Chen, 1992. © 1990 by Elsevier Science Inc.

severe effects on world food security than those that affect only a few regions (Parry, 1990). But in the long run, interregional effects may be more subtle and dynamic.

Agriculture's links with other sectors such as water and energy are critical to overall socioeconomic sensitivity to environmental change. For example, irrigation is a long-standing adaptation to climatic variability and change and an important element in efforts to increase crop production in both developed and developing countries. However, agricultural demands for water will increasingly have to compete with municipal, recreational, industrial, hydroelectric, and navigation needs for water—perhaps in the context of declining water availability in some regions due to climatic change.

Interregional interactions may come to the fore in such areas as international river basins. Gleick (1989) points out that more than 200 river basins around the world contain territory from more than one country. At least 13 of these contain land from 5 or more countries, including the Danube with 12 countries, the Niger with 10 countries, and the Nile with 9 countries. Fifty or more countries in Africa, Europe, Asia, and South America have greater than three-fourths of their land area in international river basins. In many of these countries, per capita water availability is already limited, and tensions and conflicts over water resources have a long history. In the Nile River basin, for example, Egypt and the Sudan have competed and even fought over water as recently as 1959 (Gleick, 1989). In the Sudan, Nile water irrigates more than 19,000 km^2 of land, reducing the average annual flow into Egypt by some 15% (Schwarz *et al.*, 1990). Growing needs in upstream countries will add greatly to the complexity and potential volatility of water allocations in the future—especially if environmental change reduces water availability or increases variability in flow.

The potential for significant interactions may also be increasing because of the quickening pace of change. For example, water pollution problems that appeared sequentially in highly industrialized countries over the past two centuries are now appearing nearly simultaneously in rapidly industrializing developing countries. Although the latter countries should benefit from the experience of the former in dealing with such problems, they will probably have to cope with much more intractable problems.

Finally, impacts in agriculture may have important feedbacks into food security, health, and demographic behavior not only of marginal peoples but of the bulk of the world's population. A decrease in the ecological viability of subsistence or small-scale agriculture could, for example, adversely affect the majority of the developing world's population who depend on agriculture for their livelihood, with all that implies for human poverty and displacement. At the same time, the availability of food aid depends greatly on the size of food surpluses and on world food prices. Poor nutritional status in turn underlies many health problems in developing countries (e.g., Millman *et al.*, 1991; Chen, 1990a,b; WHO, 1990). Notably, environmental change will likely affect human health not only through agriculture and nutrition but also through the availability of drinking water and sanitation services; alteration in patterns of disease vectors and hosts such as mosquitoes, livestock, and snails; changes in exposure to pollution and environmental extremes; and perhaps through increased violent conflict and population displacement.

Societal Resilience

A social system that has a high degree of reliance on environmentally sensitive activities and resources may still have some "resilience" or capacity to resist harm, e.g., by minimizing or absorbing losses, by distributing or transferring risks, or by rapidly recovering or rebuilding (Kates, 1985; Timmerman, 1981). As with vulnerability, societal resilience is likely to vary to some degree with the specific perturbation or threat, but it also makes sense to think about indicators of general resilience that capture the overall ability of a social system to withstand or cope with adversity. Such an integrated assessment of resilience—or the lack of it— may be especially important in an interconnected world where responses to an environmental threat in one region or country may have large and perhaps even magnified effects elsewhere.

It has become part of the conventional wisdom that developed countries will be more resilient than developing countries in coping with environmental change. This assertion is based not only on the greater wealth and resources of the former,[7] but also on the "lessening hypothesis" that states that over time societies adapt to recurrent environmental stresses through their technological and social organization. Evidence for this hypothesis can be found in the diminishing impact of drought in the Great Plains and the Sahel, in part the result of the growing integration of these areas into regional and global socioeconomic systems (Bowden *et al.*, 1981). A complementary hypothesis is that the vulnerability of societies to rarer and more extreme events remains high or may even increase, in spite of—or perhaps because of—reduced vulnerability to more frequent events. However, evidence for this second hypothesis is less clear (Burton *et al.*, 1993).

Identifying countries or regions with *greater* resilience to environmental change than others is difficult given the many different types of possible change, differing time scales for judging resilience, distributional effects associated with different adjustment and adaptation mechanisms, and controversies over what contributes to resilience or vulnerability. On the other hand, it is easier to identify those countries or regions with the *least* resilience to environmental variation.

Poverty at the national level is clearly one defining characteristic of less resilient countries. Countries poor in resources and wealth relative to their needs have less capability and flexibility in responding to crises or to long-term change. This might, for example, include the 43 countries that the World Bank considers "low income" (GNP per capita <$600 per year in 1990)—countries with more than 3 billion people.

---

[7] For example, the recent NRC report (NRC, 1991) points out that the United States is "a country richly endowed with natural and human resources, and one benefiting from a geography that encompasses many climate zones. Compared to other countries, the United States is well situated to respond to greenhouse warming" (p. 68).

Reliance on external aid and high levels of external debt relative to GNP would also increase vulnerability. The severe drought in southern Africa that ended in 1992 provides a graphic example: recent U.N. assessments of the import and food aid requirements of ten southern African countries indicate that four countries— Malawi, Mozambique, Tanzania, and Zambia—will require food aid to meet more than half of their import requirements. These four countries have low GNP per capita ($80–420 in 1990), a high percentage of GNP attributable to official development assistance (14–66%), and a high ratio of debt to GNP (86–385%). In contrast, Botswana, with a much higher GNP per capita and lower levels of external assistance and debt, is expected to cover all but 6% of its import requirements without direct food aid, even though the U.N. projects its 1992 food production at only one-third of normal, one of the lowest levels of the ten countries (GIEWS, 1992).

The presence of violent conflict is another sign of lack of resilience. The social and economic upheaval caused by intra- or interstate conflict disrupts normal mechanisms of adjustment and adaptation, including the provision of external humanitarian and development assistance. All five countries that suffered from famine in 1990 were torn by conflict (Millman *et al.*, 1991). Unfortunately, in both developed and developing countries conflict continues to interfere with humanitarian aid and with recovery and reconstruction efforts. Conflict may also contribute to further environmental degradation—sometimes intentionally, as evidenced recently by Iraq's efforts to create oil fires and spills during the Persian Gulf War.

Perspectives on Future Vulnerability

Much of the current concern about global environmental change has arisen in the developed countries of the world. As a result, most environmental and impacts research has focused on the problems and conditions of the quarter of the world's population living in the developed world. Few studies have carefully examined potential impacts and vulnerabilities in the developing world, especially as they may affect the poorest countries and the poorest people within poor countries.[8]

In contrast, discussions of how to control future emissions and other sources of global environmental change have quickly seized on the issue of future population growth, development, and industrialization in the developing world. Some argue that developing countries need not follow the development path that the industrialized countries—and many rapidly industrializing economies—have taken (see Huq, this volume). However, many in developing countries question the relative importance of global environmental change compared with pressing short-term human needs and longer-term development objectives. They perceive not only that developed countries have been the primary contributors to possible environmental change to date, but also that most of the concern about environmental impacts centers on adverse effects in the developed world. One exception to this is

---

[8] The recent IPCC reports did address the potential impacts of climate change for developing countries, but its review simply revealed how little is known about such impacts.

the threat of future sea-level rise, which has clearly motivated a number of island and low-lying states to take active roles in international environmental discussions. Nevertheless, these perceptions are fueled by fears that near-term efforts to deal with environmental change may compete or interfere with other priorities such as industrial and agricultural development and poverty reduction. In countries of limited natural and human resources, placing environmental change high on today's domestic agenda means displacing other critical needs.

Developing a balanced and integrated perspective of human vulnerability in both developed and developing worlds is thus a key task. It is important not only as a way of convincing developing countries that they should cooperate in global efforts to deal with environmental change, but also to ensure that such efforts complement and support other social and economic priorities as much as possible. It is *not* sufficient simply to develop technologies with lower environmental impacts as alternatives to fossil fuels, refrigerants, or wet rice production. A complementary need is to develop strategies for using and introducing these technologies in ways that *enhance* economic development, energy and food security, research and technical capabilities, and other aspects of societal resilience. It will be especially important to assess how such alternative technologies might increase or decrease present and future vulnerability with respect to different threats. For instance, how will modifications to rice cropping to reduce methane emissions affect fertilizer and water use? sensitivity to increased UV-B radiation or changes in temperature and solar radiation? vulnerability to climatic extremes or pest and disease outbreaks? Will such modifications affect future potential yield increases? Will they be equally accessible to small and large farmers and have comparable effects on production and profitability? Can they be coordinated with other changes in technology and behavior to improve nutritional quality, reduce food losses, lower input requirements, and increase access to extension services and markets? If the answers to most of these questions are pessimistic, it may be preferable to focus control efforts on methane sources other than rice production.

It is clear that both environmental and human vulnerabilities will change greatly in the future. Some of the largest changes in vulnerability may well result from trends in poverty and income distribution; but, unfortunately, these are difficult to measure and predict (Chen and Pitt, 1991). International initiatives to improve welfare in the poorest countries will also affect the vulnerability of marginal groups.

On the pessimistic side, it seems certain that rapid population growth—especially among the poorest and most vulnerable subgroups and in areas already threatened by environmental degradation—will contribute to greater stresses on the environment and to even larger numbers of people exposed to environmental hazards. Indeed the most disturbing possibility is the potential for feedbacks between environmental and human vulnerability mediated by conflict or population displacement. Population growth and environmental change could exacerbate already difficult situations of resource scarcity and competition, potentially heightening international tensions, spurring violent conflict, and reducing

101

cooperation in responses to disasters or in optimizing sustainable resource use. Plans by Turkey to construct a huge complex of dams and irrigation projects along the upper Euphrates River have led to increased tensions with downstream Syria, including a threat by Turkey to impound water if Syria continued to support Kurdish insurgents in Turkey (Homer-Dixon, 1991). The possible displacement of many millions of people by increasing sea levels or storm surge threatens not only the people and countries directly affected but also their neighbors and other areas which would be the recipients of immigrations.

More optimistically, Ausubel (1991) argues compellingly that technological and associated socioeconomic change will continue to reduce many forms of vulnerability to the environment. However, it is worth noting here that much of the world's population lags well behind the technological "leading edge." As Ausubel himself points out, the trend toward lower reliance on agricultural employment in developing economies is some 50–100 years behind that experienced in the developed world. Green Revolution techniques developed and applied in Asia since the 1960s have yet to succeed widely in Africa. In many poor countries, few but the urban elite have access to air conditioning, refrigeration, and modern health care. Developing countries cannot afford the levels of environmental protection taken for granted in most developed countries. On the other hand, it is quite possible that continuing rapid advances in global integration will accelerate the pace of change, even for the poorest and most remote peoples of the world.

How this wide range of global socioeconomic changes will affect the future evolution of environmental and human vulnerability is difficult to foresee. It is easy to be either optimistic or pessimistic. Optimists can point to rapid advances in technology and social organization, long-term improvements in rates of poverty and hunger, increased international cooperation and reduced tensions, and the diversity of options available for mitigation of and adaptation to environmental variation. Pessimists can emphasize the slowing of fertility decline in some developing countries, spreading environmental degradation at local and regional scales, indications that growth in crop yields may be leveling off, faltering economic progress in sub-Saharan Africa, and continuing regional conflicts and population displacement associated in some cases with resource competition and environmental degradation. Both can find both hope and dismay in the surprises of recent years: the end of the Cold War, the ozone hole, the rapid spread of AIDS. It is likely to take much careful research to resolve which of these trends and surprises are the most significant in the long run, and which of them are only temporarily or regionally important. In the end, it may be that Roger Revelle's "large-scale geophysical experiment" will be accompanied by an equally large-scale "social experiment" to test the true human dimension of vulnerability.

Acknowledgments

I am grateful to Dan Deudney and Olga Gritsai for their useful comments and suggestions.

References

ACC/SCN (Advisory Committee on Coordination–Subcommittee on Nutrition). 1992. Statement on nutrition, refugees and displaced persons. In *Summary Report of the Nineteenth Session of the Sub-Committee on Nutrition*, World Food Programme, Rome, Italy, 24–29 February 1992. ACC/SCN, Geneva, Switzerland, 1–3.

Ausubel, J. 1991. Does climate still matter? *Nature 350*, 649–652.

Bowden, M. J., R. W. Kates, P. A. Kay, W. E. Riebsame, R. A. Warrick, D. L. Johnson, H. A. Gould, and D. Weiner. 1981. The effect of climate fluctuations on human populations: Two hypotheses. In *Climate and History* (T. M. L. Wigley, M. J. Ingram, and G. Farmer, eds.), Cambridge University Press, Cambridge, U.K., 479–513.

Burton, I., R. W. Kates, and G. F. White. 1993. *The Environment As Hazard*. Second edition, Guilford Press, New York.

Chen, R. S. 1990a. Refugees and hunger. In *The Hunger Report: 1990*, HR-90-1, ASF World Hunger Program, Brown University, Providence, Rhode Island, 49–70.

Chen, R. S. 1990b. The state of hunger in 1990. In *The Hunger Report: 1990*, HR-90-1, ASF World Hunger Program, Brown University, Providence, Rhode Island, 1–26.

Chen, R. S. 1990c. Global agriculture, environment, and hunger: Past, present, and future links. *Environmental Impact Assessment Review 10*, 335–357.

Chen, R. S. 1992. Hunger among refugees and other people displaced across borders. In *Hunger 1993: Third Annual Report on the State of World Hunger—Uprooted People* (Marc J. Cohen, ed.), Bread for the World Institute on Hunger and Development, Washington, D.C., 15–36.

Chen, R. S., and M. B. Fiering. 1989. *Climate Change in the Context of Multiple Environmental Threats*, Research Report RR-89-1, ASF World Hunger Program, Brown University, Providence, Rhode Island, 23 pp.

Chen, R. S., and M. M. Pitt. 1991. *Estimating the Prevalence of World Hunger: A Review of Methods and Data*, Research Report RR-91-5, ASF World Hunger Program, Brown University, Providence, Rhode Island, 74 pp.

Chisholm, M. 1990. The increasing separation of production and consumption. In *The Earth As Transformed by Human Action: Global and Regional Changes in the Biosphere over the Past 300 Years* (B. L. Turner II, W. C. Clark, R. W. Kates, J. F. Richards, J. T. Mathews, and W. B. Meyer, eds.), Cambridge University Press, Cambridge, United Kingdom, 87–101.

Downing, T. E. 1991. *Assessing Socioeconomic Vulnerability to Famine: Frameworks, Concepts, and Applications*, Research Report RR-91-1, ASF World Hunger Program, Brown University, Providence, Rhode Island, 102 pp.

Duncan, M., and K. Webb. 1980. *Energy and American Agriculture*, Federal Reserve Bank of Kansas City, Kansas City, Missouri.

FAO (Food and Agriculture Organization of the United Nations). 1991. *FAO Production Yearbook 1990, Vol. 44*, FAO Statistics Series No. 99, FAO, Rome, Italy, 286 pp.

GIEWS (Global Information and Early Warning System on Food and Agriculture). 1992. *Food Supply Situation and Crop Prospects in Sub-Saharan Africa, Special Report 2 (June)*.

Gleick, P. H. 1989. The implications of global climatic changes for international security. *Climatic Change 15(1/2)*, 309–325.

Headrick, D. R. 1990. Technological change. In *The Earth As Transformed by Human Action: Global and Regional Changes in the Biosphere over the Past 300 Years* (B. L. Turner II, W. C. Clark, R. W. Kates, J. F. Richards, J. T. Mathews, and W. B. Meyer, eds.), Cambridge University Press, Cambridge, United Kingdom, 55–67.

Hinnawi, E. E. 1985. *Environmental Refugees*, United Nations Environment Programme, Nairobi, Kenya, 41 pp.

Homer-Dixon, T. F. 1991. Environmental changes as causes of acute conflict. *International Security 16(2)*, 76–116.

IPCC (Intergovernmental Panel on Climate Change). 1990. *Potential Impacts of Climate Change*, Report prepared for IPCC by Working Group II (June), World Meteorological Organization and United Nations Environment Programme, Geneva.

Kates, R. W. 1985. The interaction of climate and society. In *Climate Impact Assessment* (R. W. Kates, J. H. Ausubel, and M. Berberian, eds.), SCOPE 27, John Wiley and Sons, New York, 3–36.

Kates, R. W., and V. Haarmann. 1991. Poor people and threatened environments: Global overviews, country comparisons, and local studies. *Research Report RR-91-2*, ASF World Hunger Program, Brown University, Providence, Rhode Island, 70 pp.

Kates, R. W., and V. Haarmann. 1992. Where the poor live: Are the assumptions correct? *Environment 34(4)*, 4–12, 25–28.

Kates, R. W., R. S. Chen, T. E. Downing, J. X. Kasperson, E. Messer, and S. R. Millman. 1989. *The Hunger Report: Update 1989*. HR-89-1, ASF World Hunger Program, Brown University, Providence, Rhode Island, 70 pp.

Leonard, H. J. 1989. Environment and the poor: Development strategies for a common agenda. In *Environment and the Poor: Development Strategies for a Common Agenda* (H. J. Leonard, ed.), U.S.–Third World Policy Perspectives No. 11, Overseas Development Council, Washington, D.C., 3–45.

Liverman, D. M. 1989. Vulnerability to global environmental change. In *Understanding Global Environmental Change: The Contributions of Risk Analysis and Management* (R. E. Kasperson, K. Dow, D. Golding, and J. X. Kasperson, eds.), Clark University, Worcester, Massachusetts, 11–13 October 1989. Clark University, Worcester, Massachusetts, 27–44.

Matthews, R. W., D. Anderson, R. S. Chen, and T. Webb. 1990. Global climate and the origins of agriculture. In *Hunger in History: Food Shortage, Poverty, and Deprivation* (L. F. Newman, gen. ed.), Basil Blackwell, Oxford, United Kingdom, 27–55.

Millman, S. R., S. M. Aronson, L. M. Fruzetti, M. Hollos, R. Okello, and V. Whiting Jr. 1990. Organization, information, and entitlement in the emerging global food system. In *Hunger in History: Food Shortage, Poverty, and Deprivation* (L.F. Newman, gen. ed.), Basil Blackwell, Oxford, United Kingdom, 307–330.

Millman, S. R., R. S. Chen, J. Emlen, V. Haarmann, J. X. Kasperson, and E. Messer. 1991. *The Hunger Report: Update 1991*, HR-91-1, Alan Shawn Feinstein World Hunger Program, Brown University, Providence, Rhode Island, 26 pp.

NRC. 1983. *Environmental Change in the West African Sahel*, National Academy Press, Washington, D.C., 96 pp.

NRC. 1991. *Policy Implications of Greenhouse Warming*, National Academy Press, Washington, D.C., 127 pp.

NRC. 1992. *Policy Implications of Greenhouse Warming: Report of the Adaptation Panel*, National Academy Press, Washington, D.C.

Parry, M. 1990. *Climate Change and World Agriculture*, Earthscan Publications Ltd., London, United Kingdom, 157 pp.

Refugee Policy Group. 1992. *Conference Report, Migration and the Environment*, Refugee Policy Group, Washington, D.C.

Rickman, G. E. 1980. *The Corn Supply of Ancient Rome*, Oxford University Press, Oxford, United Kingdom.

Riebsame, W. E. 1990. The United States Great Plains. In *The Earth As Transformed by Human Action: Global and Regional Changes in the Biosphere over the Past 300 Years* (B. L. Turner II, W. C. Clark, R. W. Kates, J. F. Richards, J. T. Mathews, and W. B. Meyer, eds.), Cambridge University Press, Cambridge, United Kingdom, 561–575.

Sandler, R. H., P. R. Epstein, R. M. Cook-Deegan, and A. Shukri. 1991. Initial medical assessment of Kurdish refugees in the Turkey–Iraq Border Region. In *Journal of the American Medical Association 266(5)*, 638–640.

Schwarz, H. E., J. Emel, W. J. Dickens, P. Rogers, and J. Thompson. 1990. Water quality and flows. In *The Earth As Transformed by Human Action: Global and Regional Changes in the Biosphere over the Past 300 Years* (B.L. Turner II, W.C. Clark, R. W. Kates, J. F. Richards, J. T. Mathews, and W. B. Meyer, eds.), Cambridge University Press, Cambridge, United Kingdom, 253–270.

Seshu, D. V., T. Woodhead, D. P. Garrity, and L. R. Oldeman. 1989. Effect of weather and climate on production and vulnerability of rice. In *Climate and Food Security*, International Rice Research Institute, Manila, Philippines, 93–113.

Stern, P. C., O. R. Young, and D. Druckman (eds.). 1992. *Global Environmental Change: Understanding the Human Dimensions*, National Academy Press, Washington, D.C., 308 pp.

Tabibzadeh, I., A. Rossi-Espagnet, and R. Maxwell. 1989. *Spotlight on the Cities: Improving Urban Health in Developing Countries*, World Health Organization, Geneva, Switzerland, 174 pp.

Timmerman, P. 1981. *Vulnerability, Resilience, and the Collapse of Society*, Environmental Monograph No. 1., Institute for Environmental Studies, University of Toronto, Toronto, Canada.

Turner II, B. L. 1990. The rise and fall of population and agriculture in the central Maya lowlands: 300 BC to present. In *Hunger in History: Food Shortage, Poverty, and Deprivation* (L.F. Newman, gen. ed.), Basil Blackwell, Oxford, United Kingdom, 178–211.

UNFPA (United Nations Population Fund). 1991. *Population, Resources and the Environment: The Critical Challenges*, UNFPA, New York, 154 pp.

USCR (US Committee for Refugees). 1992. *World Refugee Survey 1992*, U.S. Committee for Refugees, Washington, D.C., 116 pp.

Venkateswarlu, B. 1989. Vulnerability of rice to climate. In *Climate and Food Security*, International Rice Research Institute, Manila, Philippines, 115–121.

White, G. F., and J. E. Haas. 1975. *Assessment of Research on Natural Hazards*, MIT Press, Cambridge, Massachusetts, 487 pp.

Whitmore, T. M., B. L. Turner II, D. L. Johnson, R. W. Kates, and T.R. Gottschang. 1990. Long-term population change. In *The Earth As Transformed by Human Action: Global and Regional Changes in the Biosphere over the Past 300 Years* (B. L. Turner II, W. C. Clark, R. W. Kates, J. F. Richards, J. T. Mathews, and W. B. Meyer, eds.), Cambridge University Press, Cambridge, United Kingdom, 25–39.

WHO (World Health Organization). 1990. *Potential Health Effects of Climatic Change: Report of a WHO Task Group*, Report WHO/PEP/90.10, WHO, Geneva, Switzerland, 58 pp.

World Bank. 1992. *World Development Report 1992: Development and the Environment*, Oxford University Press, New York, 308 pp.

7

Global Industrialization: A Developing Country Perspective

Saleemul Huq

Abstract

A wide disparity exists in the consumption of the world's product between the North and the South. Countries in the South cannot expect to follow the same development path as have those in the industrialized countries of the North. Alternative paths must be identified and followed.

Introduction

Global industrialization over the past 200 years since the Industrial Revolution has followed a relatively similar pattern in country after country. The so-called developing countries of the South are continuing to follow the path of the industrialized or developed countries of the North. Even the very terms "developing" and "developed" connote two stages of industrial development, with the transition from developing to developed occurring when a country has achieved a certain level of industrialization.

There is, however, a real question as to whether the earth and its resources can sustain the transition of all the world's developing countries into developed or industrialized countries, particularly considering energy and other nonrenewable resource use and also waste production. This chapter outlines the consequences of present levels of industrial waste and energy consumption, as projected into the future. It suggests that present levels and patterns of consumption and industrialization in the developed countries are inappropriate and indeed impossible for the developing countries to follow. Hence it will be necessary to look at new and alternative technologies, industries, and development paths in order for the poor countries of the South to offer a better quality of life to their present and future generations.

The world's population, at present just over 5 billion people, could double within the next 40 years, and could stabilize at roughly 14 billion people (base case) or much higher if fertility rates decline more slowly (Figure 1). The fastest population growth is in urban areas, particularly in Asia, Africa, and Latin America.

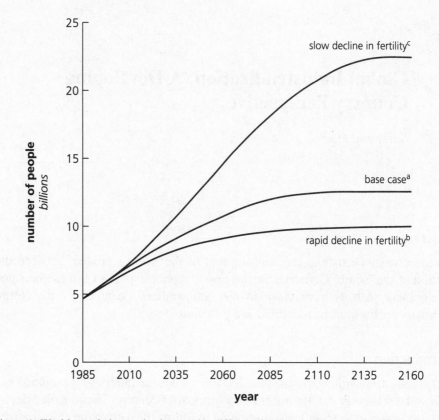

Figure 1. World population projections under different fertility trends, 1985–2160. (a) Countries with high nondeclining fertility levels begin the transition toward lower fertility by the year 2005 and undergo a substantial decline—by more than half in many cases—over the next 40 years. All countries reach replacement levels by 2060. (b) Countries not yet in transition toward lower fertility begin the transition immediately. For countries already in transition, total fertility declines at twice the rate for the base case. (c) Transition toward lower fertility (triggered when life expectancy reaches 53 years) begins after 2020 in most low-income countries. For countries in transition, declines are half the rate for the base case (from World Bank, 1992).

By most economic projections, per capita gross domestic product (GDP) in the high-income countries is expected to increase several-fold in the next 50 years, while showing much smaller increases in the low-income countries.

Quality of Life

A relationship between GDP and pollution is sketched qualitatively in Figure 2. It is postulated that pollution levels will rise with higher GDP, until cleaner and more efficient technologies are used, whereupon pollution per unit of output will gradually fall. This is further illustrated in Figure 3, where various environmental factors are plotted against per capita income levels. Figure 3 shows that safe water and sanitation become more available at early stages of development,

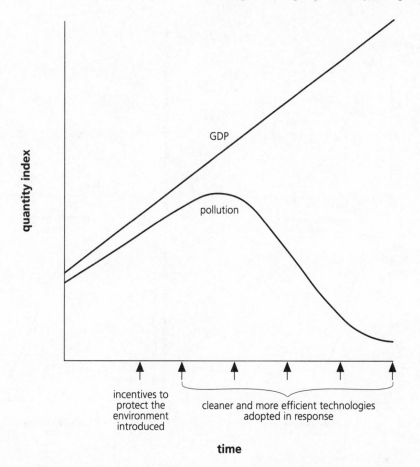

Figure 2. Breaking the link between growth in GDP and pollution (from World Bank, 1992).

and that concentrations of urban air pollutants eventually decrease with increasing income. But municipal wastes and carbon dioxide continue to increase for much longer.

It is important to look at the quality of life of the world's people, particularly the poor, in terms beyond simple GDP or GNP per capita. For instance, 1.5 billion people today have no access to health services or safe water, and over 2 billion have no access to sanitation; 1.2 billion are illiterate. Any worthy pattern of development must address these issues and offer the world's poor a better quality of life.

Yet we can only dimly perceive the components of a better quality of life. It is necessary to confront the problem of consumption, especially wasteful consumption, in the affluent parts of the globe. Figure 4 shows evidence that per capita direct protein intake (from plants) reaches a plateau once GDP per capita attains a certain level, but indirect protein intake (via cattle, poultry, and milk) keeps increasing. Apparently it is possible to satiate the human need for direct protein,

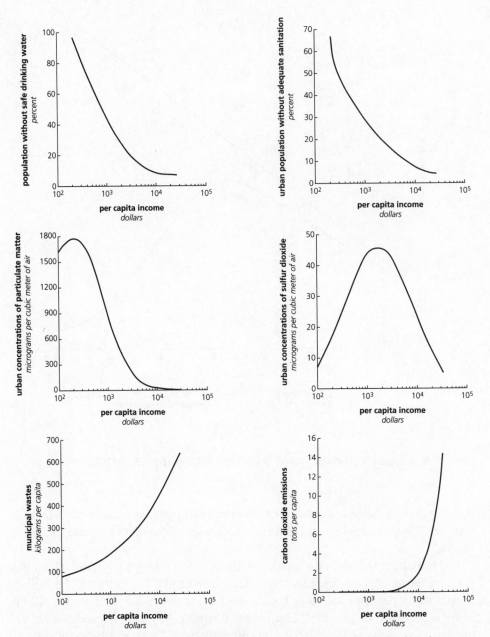

Figure 3. Environmental indicators at different country income levels. Estimates are based on cross-country regression analysis of data from the 1980s; carbon dioxide emissions are from fossil fuels (from World Bank, 1992).

but it is not possible to satiate the demand for indirect protein.

Many other consumption patterns are similar, as seen in Table 1. In all cases the developed countries are consuming much more per capita than the developing countries. Table 1 also presents ratios comparing per capita consumption in the

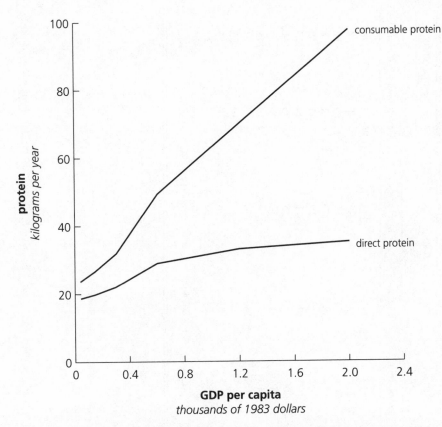

Figure 4. Direct and indirect protein intake and per capita GDP (reprinted by permission of the publisher from Linnemann *et al.*, 1977, © 1977 by Elsevier Science Publishing Co., Inc.).

United States and India; these ratios reach 320:1 for cars and 245:1 for copper.

Future Development

It is widely assumed that today's pattern of wasteful consumption in the developed countries must be replicated as an essential feature of the industrialization of the developing countries. But such a path to industrialization is not environmentally viable for the developing countries in particular because of the global climatic change effects, but also because of the higher expectations of publics in developing countries. The developing countries, therefore, must explore new and alternative development paths that improve the quality of life of their citizens. Improving water quality is vital. So is family planning and birth control, which are addressed especially effectively by promoting education, particularly of females. There is also a clear case for emphasizing renewable energy sources, which both lessen national dependence on fossil fuels and decrease pollution and greenhouse gas emissions.

Table 1: *Consumption patterns for selected commodities; distribution among developed and developing countries*

Category	Products	World Total (millions of tons)	Share		kg/cap (except vehicles)		Per capita ratio	
			Developed	Developing	Developed	Developing	Developed/ Developing	U.S./India
Food	Cereals[1]	1801	48%	52%	717	247	3	6
	Milk[2]	533	72%	28%	320	39	8	4
	Meat[3]	114	64%	36%	61	11	6	52
Forest	Round wood[4]	2410	46%	55%	388	339	1	6
	Sawn wood[5]	338	78%	22%	213	19	11	18
	Paper board[6]	224	81%	19%	148	11	14	115
Industry	Fertilizers[7]	141	60%	40%	70	15	5	6
	Cement	1036	52%	48%	451	130	3	7
Metals	Copper	10	86%	14%	7	0.4	19	245
	Iron & steel	699	80%	20%	469	36	13	22
	Aluminum	22	86%	14%	16	1	19	85
Chemicals	Inorganic	226	87%	13%	163	8	20	54
	Organic	391	85%	15%	274	16	17	28
Vehicles	Cars	370[8]	92%	8%	0.283[9]	0.012	24	320
	Commercial	105	85%	15%	0.075	0.0006	125	102

[1] Cereals data 1987.
[2] Milk data include cow, buffalo, and sheep milk (1987).
[3] Meat data include beef, veal, pork, mutton, and lamb (1987).
[4] Round wood includes fuel wood + charcoal and industrial round wood (1988).
[5] Sawn wood includes extracted from sawlogs and veneer logs (1988).
[6] Paperboard includes newsprint, printing and writing papers, and other paper + paperboard (1988).
[7] Fertilizer consumption data include nitrogen, phosphate, and potash fertilizers.
[8] Number of vehicles.
[9] Vehicles per capita.

Sources: FAO *Forest Products Year Book 1988*; U.N. FAO *Production Year Books 1988, 1989*; *Statistical Year Book 1987, 1988*; *Handbook of Industrial Statistics 1989*; *International Trade Statistics Year Book 1987*.

Economic incentives also must play a key role. With respect to domestic fossil fuel use, raising the price will encourage conservation. Environmental objectives can also be furthered by the proper design of the economic incentives to participate in international trade. Development objectives can be furthered by adopting market mechanisms to implement global environmental agreements that at the same time create income transfers from rich countries to poor: for example, a fixed number of emission rights for greenhouse gases would be allocated to all the countries of the world according to some formula, and those countries wishing to use more than their allocation would be required to purchase the additional rights from other (presumably poorer) countries that were not fully using theirs. Such an arrangement would be difficult to negotiate, but it would make available to many countries new ways of increasing the flow of modern, less wasteful, technologies to the developing world.

Exploring alternative pathways to development in some detail is an urgent matter. We must give special emphasis to finding ways to improve the quality of life of the poor in the developing countries without exhausting the earth's resources.

References

FAO (U.N. Food and Agriculture Organization). 1988. *Forest Products Year Book*. FAO, Rome, Italy.

FAO (U.N. Food and Agriculture Organization). *Production Year Book*. 1988. FAO, Rome, Italy.

FAO (U.N. Food and Agriculture Organization). *Production Year Book*. 1989. FAO, Rome, Italy.

Handbook of Industrial Statistics. 1989.

International Trade Statistics Year Book. 1987.

Linnemann, H., J. Dehoogh, M. A. Keyzer, and H. D. J. Vanheemst. 1977. Food for a growing world-population. *Technological Forecasting and Social Change 10(1)*, 27–51.

Statistical Year Book. 1987, 1988.

World Bank. 1992. *World Development Report, 1992: Development and the Environment*. World Bank, Washington, D.C.

PART 2

THE GRAND CYCLES: DISRUPTION AND REPAIR

8

Introduction

The Editors

In the grand cycles, atoms of carbon, nitrogen, sulfur, and phosphorus cycle back and forth from living plants and animals to the soil, atmosphere, and oceans. The winds and water currents of earth move these atoms global distances.

As the cycling of these four elements has developed over billions of years, biology has become deeply intertwined with geology. Biochemical processes, driven by sunlight and involving specialized molecules and bacteria, extract these elements from storage in inorganic reservoirs and make them available as nutrients for plants. Closely related biochemical and geophysical processes return these elements to storage and replenish the inorganic reservoirs. Energy in the form of high-temperature heat from the interior of the earth drives much slower cycling, which retrieves, in the form of volcanic emissions, the small fractions of nutrients that are not rapidly cycled.

All four grand cycles are now strongly influenced by human activity. Although the interactions are similar for sulfur and phosphorus, the chapters in this section consider only nitrogen and carbon.

The carbon cycle has been perturbed by fossil fuel use, deforestation, and changes in agriculture and animal husbandry. During the past two centuries these practices have led to more than a 25% increase in the concentration of carbon dioxide and more than a doubling of the concentration of methane in the atmosphere. The principal direct consequence, just at the edge of detectability today, is likely to be a modification of patterns of surface temperature and rainfall, as well as a rise in sea level.

The nitrogen cycle has been perturbed especially by worldwide fertilizer use, which has grown rapidly and steadily for more than 40 years. Secondary impacts on the nitrogen cycle have resulted from high-temperature combustion and from agricultural practices other than fertilizer use, such as land clearing and the deliberate planting of legumes. The current annual production of about 100 million metric tons of nitrogen in the form of ammonia compounds for nitrogen fertilizer approximately doubles the preindustrial rate of introduction of "new" nitrogen into the global nitrogen cycle. The consequences are far from fully understood, but they include an increase in the acidity of rain and snow, leading to the acidification of poorly buffered soil and water bodies, as well as the preferential stimulation of plant growth in previously balanced ecosystems. Both acidification and unbalanced plant growth have disruptive consequences for the services provided by ecosystems and for biodiversity.

The opening chapter by Ayres, Schlesinger, and Socolow, "Human Impacts on the Carbon and Nitrogen Cycles," compares the carbon and nitrogen cycles in preindustrial times and at present. Every aspect of the preindustrial picture is currently being modified by human activity. The authors suggest that nitrogen fertilizer, as an agent of global change, deserves greater attention.

The other five chapters in this section show some of the ways in which societies today are beginning to respond to the warnings of environmental scientists about disruptions of the carbon cycle. Moomaw and Tullis, in "Charting Development Paths: A Multicountry Comparison of Carbon Dioxide Emissions," develop a simple measure of the carbon intensity of national economies and provide an instructive survey of historical trends in carbon intensities across many industrialized, transitional, and developing countries. Their main conclusion is that carbon intensities correlate poorly with per capita gross domestic product, and that there is great variety in carbon trajectories during economic development. Some development paths produce rapid economic growth with relatively low carbon emissions.

Harriss, in "Reducing Urban Sources of Methane: An Experiment in Industrial Ecology," explains a novel device for identifying and monitoring urban methane emissions. Methane is a powerful greenhouse gas whose control will be an important part of any climate management regime. Harriss suggests that abatement of methane emissions may be achieved relatively cheaply with new technologies, and that the commercialization of these technologies could be accelerated by targeted policies.

Kononov, in "Reducing Carbon Dioxide Emissions in Russia," surveys the macroeconomic costs and effects on the Russian economy of meeting greenhouse gas reduction targets. He argues that substantial greenhouse gas abatement will accompany the structural changes now occurring in the Russian economy, in particular the secular shift towards real markets for energy. He cautions that investments aimed specifically at energy conservation must be looked at critically, since they may crowd out other socially useful capital investments.

Jiang, in "Energy Efficiency in China: Past Experience and Future Prospects," surveys policies to improve energy efficiency in China and argues that these have been only partially successful in the past. He identifies a number of mechanisms (investments in R&D, more regulation, rationalization of energy prices) that could be applied more effectively. Unlike Kononov, he does not perceive large opportunity costs in pursuing energy efficiency.

Williams, in "Roles for Biomass Energy in Sustainable Development," describes an imaginative global scenario in which biomass, grown renewably and converted into electricity and fluid fuels with advanced technologies, substitutes for fossil fuels and becomes a principal agent of global economic development. Williams argues that the benefits of this development path include not only reduced greenhouse gas emissions but also rural development, the stabilization of energy markets, and a mechanism for achieving large-scale land rehabilitation. Williams identifies several public policies that could accelerate the clarification of the risks and benefits of such a global biomass-to-energy strategy.

9

Human Impacts on the Carbon and Nitrogen Cycles

Robert U. Ayres, William H. Schlesinger, and Robert H. Socolow

Abstract

Human activities are substantially modifying the global carbon and nitrogen cycles. The global carbon cycle is being modified principally by the burning of fossil fuels, and also by deforestation; these activities are increasing the carbon dioxide concentration of the atmosphere and changing global climate. The nitrogen cycle is being modified principally by the production of nitrogen fertilizer, and also by the planting of legumes and the combustion of fossil fuels; these activities are more than doubling the rate of fixation of nitrogen and contributing to the unbalanced productivity and acidification of ecosystems. With the aim of quantifying these disruptions, the principal flows among reservoirs in preindustrial times and today are estimated in the framework of simplified models. The methane subcycle of the carbon cycle and the nitrous oxide subcycle of the nitrogen cycle are also discussed from this viewpoint.

The Grand Cycles

Carbon (C), nitrogen (N), sulfur (S), and phosphorus (P), the important biochemical building blocks of life, find their way to plants and animals, thanks to the interplay of biological and geochemical processes. Each of the four elements moves from one chemical state to another and from one physical location to another on the earth's surface in a closed loop, or "cycle." In view of their central role in life on this planet, the four cycles are here termed the "grand nutrient cycles."

The cycles are powered by solar energy, in conjunction with the earth's gravity and geothermal energy. The nutrients flow among "reservoirs." The reservoirs of interest are life forms (living and dead plants and animals), the soil, the oceans and other water bodies, the atmosphere, and rocks. The quantity of nutrient stored in a reservoir (the reservoir's "stock" of nutrient) changes whenever the total nutrient flows in and out of the reservoir are not equal.

In the grand nutrient cycles, one may identify three classes of reservoirs, shown in Figure 1:

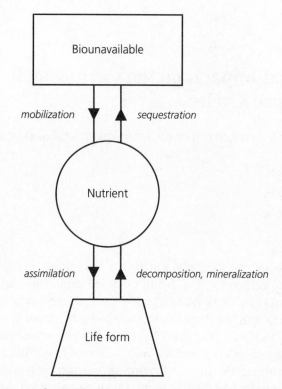

Figure 1. Basic elements of a grand nutrient cycle, presented as two subcycles. In the upper sub-cycle, the element is exchanged between a bio-unavailable reservoir and a nutrient reservoir. In the lower subcycle, the element is exchanged between a nutrient reservoir and a reservoir of life forms—living and dead plants and animals. Not shown, but possible, is the direct flow of an element from a life form to a bio-unavailable reservoir.

(1) *Bio-unavailable reservoirs*, which store nutrients in forms that cannot be incor-porated into plant or animal life without first being transformed chemically to some intermediate. (In the figures in this chapter, these reservoirs are shown as rectangles.)

(2) *Nutrient reservoirs*, which store nutrients in forms that are bioavailable, i.e., forms where the nutrients can be incorporated directly into plants without a chemical intermediate. (In the figures in this chapter, these reservoirs are shown as circles.)

(3) *Reservoirs of life forms*, plants and animals, either alive or dead. (In the fig-ures in this chapter, these reservoirs are trapezoids, reflective of the trophic levels of a food chain.)

Some important vocabulary is presented in Figure 1. An element passes from a bio-unavailable reservoir to a nutrient reservoir by *mobilization*; an example is the transformation of atmospheric nitrogen by nitrogen-fixing bacteria. An element passes in the reverse direction, from a nutrient reservoir to a bio-unavailable one by *sequestration*; an example is the return of nitrogen to the atmosphere from the nutrient reservoir by the action of denitrifying bacteria[1]. Transport of an element from a nutrient reservoir to plant or animal life is called *assimilation*; a special case, applying to carbon, is photosynthesis. Transport of the nutrients in plant and animal matter back to the nutrient reservoirs takes various forms, including (in the case of carbon) *decomposition* of plant matter and (in the case of nitrogen) *mineralization*.

To focus the discussion in this chapter, we will explore only the carbon and nitrogen cycles. (The sulfur and phosphorus cycles raise few new conceptual issues.) The chemical forms of carbon and nitrogen encountered in this chapter are grouped below.

Carbon

There are two bio-unavailable reservoirs: oxidized carbon in the form of carbonate (CO_3^{2-})in sedimentary rocks like limestone, and reduced organic carbon, known as "kerogen," also in sediments. Methane (CH_4) in the atmosphere is largely bio-unavailable: once in the atmosphere it is more likely to be oxidized to carbon dioxide than to be absorbed as methane by microbes. The carbon in the nutrient reservoir is present principally as bicarbonate ion (HCO_3^-) in water and carbon dioxide (CO_2) in the atmosphere. In the atmosphere, both CO_2 and CH_4 are important greenhouse gases. Newly photosynthesized plant matter may be represented by the hypothetical molecule CH_2O, which has approximately the same 1:2:1 ratio of carbon to hydrogen to oxygen atoms as average plant material.

Nitrogen

The principal bio-unavailable reservoir is atmospheric nitrogen, present almost exclusively in diatomic form (N_2); 78% of the molecules of the atmosphere are N_2 molecules. Mobilization is initiated by breaking the N≡N bond, producing available nitrogen in both oxidized and reduced forms: principally nitrate (NO_3^-) and ammonium ion (NH_4^+), respectively. Certain bio-available gases transfer nitrogen among reservoirs but remain in the atmosphere for only a short time before being rained out and returning to the nutrient pool: ammonia (NH_3) and several oxides of nitrogen (collectively referred to as NO_x), including nitrogen dioxide (NO_2) and nitric oxide (NO). Nitrous oxide (N_2O) is a bio-unavailable gas with a long residence time in the atmosphere; it is the most important nitrogen-containing greenhouse gas and is a useful tracer for biological processes. In plants nitrogen usually appears in an amine group ($-NH_2$).

Persistence and Stability of Cycles

If biological processes were to cease, the grand nutrient cycles would wind down, as the many chemical reactions proceeded toward chemical equilibrium—the

---

[1] In discussions related to the greenhouse effect, *sequestration* (usually referring only to carbon) is used differently: "carbon sequestration" is the accumulation of carbon in some reservoir that prevents carbon dioxide from reaching the atmosphere.

most stable (highest entropy) state of the atomic constituents. Gradually, for example, the chemically reduced carbon in life forms would combine with oxygen, until all the carbon became carbon dioxide (or, dissolved in water, carbonic acid). There is much more oxygen than reduced carbon available at or near the surface of the earth, so that if only reactions of oxygen with reduced carbon were proceeding toward chemical equilibrium, most of the atmospheric oxygen would remain.

However, a second process, the oxidation of atmospheric nitrogen, would also proceed, and it would indeed consume most of the atmospheric oxygen. The nitrogen would be gradually oxidized in the ocean, forming nitric acid, which is the most stable thermodynamic state of oxygen, nitrogen, and water in combination. The reaction is:

$$N_2 + (5/2)O_2 + H_2O \rightarrow 2\,HNO_3$$

If this reaction were to proceed to chemical equilibrium, nearly all of the oxygen in the atmosphere would be used up while consuming about ten percent of the atmosphere's nitrogen. The nitric acid would accumulate in the ocean and turn it into a strongly acidic solution with a pH of 1.5 (Lewis and Randall, 1923). Thus, sunlight-driven biological processes play a major role in driving the earth away from chemical equilibrium and sustaining the chemical environment on earth that is compatible with life as we know it.[2]

Each of the grand nutrient cycles is stabilized by negative feedback. We give two examples: a long-term and a short-term feedback loop operating within the carbon cycle. On a scale of millions of years the carbon dioxide concentration in the atmosphere is stabilized by a geological negative feedback loop powered by geothermal energy deep in the earth. The carbon dioxide in the atmosphere is slowly consumed by *weathering*, an aqueous process where silicate rocks are transformed to silica (SiO_2), while the carbon travels to the sea in bicarbonate ion (HCO_3^-). Later, in the oceans, bicarbonate is incorporated as calcium carbonate in the shells of zooplankton and phytoplankton. The shells then drift to the ocean floor, where they accumulate as sediments. The overall reaction is:

$$CaSiO_3 + CO_2 \rightarrow CaCO_3 + SiO_2$$

If there were no chemical reaction in the reverse direction, nearly all the carbon in the atmosphere and biosphere, and even the carbon dioxide in solution in the oceans, would slowly disappear.

The reverse process, producing silicates and carbon dioxide from carbonates and silica, does indeed occur under conditions of high temperature and pressure deep in the earth, where the sediments are carried by tectonic processes. The

---

[2] Moving toward chemical equilibrium in the absence of biological processes would involve other chemical reactions as well, such as the reduction of ferric oxides by buried organic carbon, releasing further carbon dioxide. Today, this is a biotic, anaerobic process..

carbon dioxide is then vented to the atmosphere through volcanic eruptions or hot springs. There is a negative feedback, because the rise in carbon dioxide concentration in the atmosphere (such as might result, for example, from a period of greater than average volcanic activity) accelerates the chemical weathering of silicate rocks, reducing the atmospheric carbon dioxide concentration toward its earlier level.

A second negative feedback loop, based on the biological process of *carbon fertilization*, is controversial, but it may have played a role in the past in maintaining a relatively constant distribution of carbon dioxide between the atmosphere and the biosphere over shorter periods of time. When there was a short-term increase in the atmospheric concentration of carbon dioxide (such as might have resulted from a higher than average biomass destruction by forest fires), the rate of photosynthesis may have increased, if adequate quantities of other necessary nutrients and water were available. If such an increase in the rate of photosynthesis produced a greater stock of carbon in plants and soil, the carbon dioxide concentration in the atmosphere would have gradually fallen toward its previous value. The controversy in this instance concerns whether the stock of carbon in biomass can substantially increase in the short term, or whether increased photosynthesis is more or less immediately accompanied by increased respiration and decomposition, with no net increase in reduced carbon.

This short-term feedback mechanism may also operate when fossil fuel is burned, but only partially. Carbon fertilization may remove some of the carbon dioxide initially emitted into the atmosphere, in favor of an increase in the stock of carbon in the biosphere, but the atmospheric carbon dioxide concentration will remain elevated above its earlier level.

Preindustrial Steady State—But Imbalance Today Due to Human Activity

We will assume in this chapter that in preindustrial times (from a few thousand to about one hundred years ago) each grand cycle maintained a dynamic steady state, which is very different from the static chemical equilibrium just described. In a dynamic steady state the flows among reservoirs are not zero; rather, the flows are constant and balance one another. An external energy source powers these flows.

None of the grand nutrient cycles is in steady state now. For example, as seen in the ice core data in Figure 2, nitrate deposition rates were approximately constant in Greenland for nearly 200 years, until about 1950, and have roughly doubled since then. The increases measured in the ice may reflect the growth of nitric oxide emissions in the United States.

In this chapter we will attempt to quantify the departure today from preindustrial conditions, for both carbon and nitrogen, by making quantitative estimates of stocks and flows, first in preindustrial times and then today. This is an exercise full of uncertainty, in spite of the considerable progress made in recent years, as local measurements have been extended to greater numbers of places and conditions,

The Grand Cycles

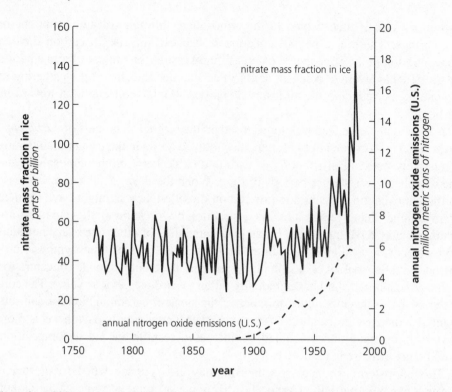

Figure 2. A 200-year record of nitrate in layers of an ice pack in South Greenland shows a relatively constant concentration until 1950 and more than a doubling in the next 30 years. Also shown (dashed line) are annual emissions of nitric oxides to the atmosphere from fossil fuel combustion in the United States (as estimated in U.S. Environmental Protection Agency, 1985, and graphed in Galloway, 1989). The trends in emissions and ice core concentrations are roughly parallel. (Modified from Mayewski et al., 1990. Reprinted with permission from *Nature*. © 1990 Macmillan Magazines Limited.)

and as modeling of processes has become more sophisticated. Our intent is to demonstrate some suggestive lines of argument.

It is instructive to develop indices that capture the degree of imbalance in the grand nutrient cycles brought about by human activities. Two kinds of indices are helpful:

(1) indices in the form of fractional annual change in the stock of a nutrient in some reservoir arising from human activity;
(2) indices in the form of ratios, with the nutrient transport resulting from current human activity in the numerator and the related nutrient transport resulting from *all* processes (natural + human) in the denominator.

As we will discuss further, a good and widely used index for the carbon cycle is the first kind: the rate of increase in the stock of carbon dioxide in the atmosphere. It has currently averaged almost one-half percent per year. One good index for the nitrogen cycle is the second kind: a ratio based on nitrogen fixation—the creation of bioavailable nitrogen from the reservoir of N_2 in the atmosphere. That ratio, today, is about one-half; that is, nitrogen fixation associated with current human activity accounts for about half of all nitrogen fixation.

The imbalance of the carbon cycle has received much attention in recent years, because increases in the carbon dioxide concentration in the atmosphere are closely coupled to potential global climate change. Difficult as it is to make comparisons across cycles, human activity has probably made the nitrogen cycle even more unbalanced than the carbon cycle at this time. The disruption of the global nitrogen cycle may have received less attention because mostly it leads to changes in nutrient balances (and thereby, to changes in ecosystems) rather than to changes in sea level and patterns of rainfall. People more easily empathize with climate change than with nutrient change, but the consequences for human well-being can be severe in either case.

For both cycles, if change occurs gradually, manifestations of disruption and strategies for mitigation are matched to a scale of decades or centuries. However, faster change from non-linear effects cannot be ruled out.

The Carbon Cycle

The Preindustrial Carbon Cycle

In preindustrial times the concentration of carbon dioxide in the atmosphere was roughly constant, but it has been rising for more than a century. In the middle of the last century about 280 of every million molecules in the atmosphere were carbon dioxide molecules (280 parts per million by volume, or 280 ppmv). Today the value has risen by 25%, to about 355 ppmv. The current rate of increase is about 0.4% per year.

In the carbon cycle, shown schematically in Figure 3, carbon dioxide is removed from the atmosphere and reduced carbon is incorporated directly into plants by the process of photosynthesis. A portion of the sunlight that drives photosynthesis is stored as chemical energy in the plant. Unlike the nitrogen cycle (see below), the carbon cycle involves no intermediary stages, which is why we call the carbon dioxide in the atmosphere a nutrient reservoir but the diatomic nitrogen in the atmosphere a bio-unavailable reservoir. The reduced carbon returns to carbon dioxide by respiration while the plant is alive and as part of the process of decomposition once the plant has died. The decomposition is accomplished by microfauna, bacteria, and fungi, acting upon plant litter and fine roots (Schlesinger, 1991).

The two key chemical reactions (each the reverse of the other) are:

$$CO_2 + H_2O + sunlight \rightarrow CH_2O + O_2 \qquad photosynthesis$$
$$(CH_2O = plant\ proxy)$$

$$CH_2O + O_2 \rightarrow CO_2 + H_2O + energy \qquad respiration,$$
$$decomposition$$

127

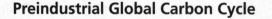

Preindustrial Global Carbon Cycle

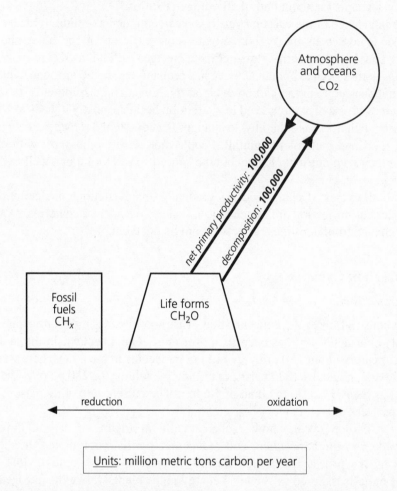

Figure 3. The preindustrial global carbon cycle. Flows are assumed in balance, with the reservoir of fossil fuels still untapped. The atmosphere is shown as a nutrient reservoir, rather than a bio-unavailable reservoir, because plants draw nutrient carbon directly from the reservoir of atmospheric carbon dioxide, without an intermediate step.

In quantifying the rate of carbon exchange between the biosphere and the atmosphere, one must distinguish *gross primary productivity* from *net primary productivity*. The process of photosynthesis is accompanied by a substantial amount of the reverse reaction—nearly simultaneous respiration that releases some of the incident energy to support plant metabolism. Gross primary productivity measures the rate of photosynthesis, and net primary productivity measures the actual rate of increase of the stock of organic carbon. Net primary productivity is generally the quantity of greater interest, because it measures the build-up of plant matter during a growing season.

As displayed in Figure 3, the global net primary productivity is about 100,000 million metric tons[3] of atmospheric carbon per year. The data shown in Figure 4 are evidence that for the eight-hundred-year period from 1000 to 1800, the carbon flow due to global net primary productivity was in balance with an equal flow of carbon from the land and oceans to the atmosphere due to plant decomposition; these ice core data show carbon dioxide levels that varied only within about 10 ppmv, or 4% (Siegenthaler and Sarmiento, 1993).

Disaggregating the annual net primary productivity for land and sea, 60,000 million metric tons of carbon is incorporated into land vegetation and 40,000 million metric tons of carbon is incorporated into phytoplankton in the oceans (Schlesinger, 1991). These flows from the atmosphere to land and sea are matched by approximately equal flows back to the atmosphere[4]—but seasonally displaced, so that there is a net increase of standing biomass throughout the northern hemisphere spring and summer and an equal net decrease during its fall and winter. This seasonal cycle is seen in measurements of the atmospheric carbon concentration. Figure 5 shows the famous data from Mauna Loa, Hawaii, where carbon dioxide concentrations in the atmosphere have been measured continuously since 1958; the concentration is about two percent greater at the spring peak than at the minimum the following autumn.

A net primary productivity on land of 60,000 million metric tons of carbon per year can be converted into an average productivity by dividing by the land area on earth (15 billion hectares[5]), obtaining 4 metric tons of carbon per year per hectare. Relative to this average rate, the productivity of marshes is about triple, the productivity of estuaries and tropical forests is about double, the productivity of grassland is about half, and the productivity of desert scrub is less than one tenth (Harte, 1988). Agriculture in most climates, managed with the help of fertilizers, other chemicals, and perhaps irrigation, can achieve a net primary productivity comparable to estuaries—i.e., 8 metric tons of carbon (about 20 metric tons of dry biomass, as CH_2O) per hectare per year.

Methane

Methane (CH_4) has its own subcycle. Methane flows from land and sea to the atmosphere as a consequence of *methanogenesis*, or methane production from organic matter. Two of the principal processes involving methanogenesis are (1) the decay of plant matter a few centimeters below the water surface in rice paddies, bogs, and other wetland environments, and (2) the conversion of cellulose to

---

[3] A metric ton is 1000 kilograms, or about 2200 pounds, about 10% larger than the familiar 2000-pound ton.

[4] Even in a dynamic steady state, the flow from atmosphere to the land will exceed the flow from the land to the atmosphere, because of the flow of carbon nutrient from land to sea (in *runoff*). However, less than one percent of the carbon photosynthesized on land flows into the sea before being oxidized. We will see below that for nitrogen the corresponding fraction (flow to the sea in chemically reduced form, divided by fixation rate on land) is much higher.

[5] A hectare is 10,000 square meters, or 2.47 acres.

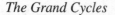

The Grand Cycles

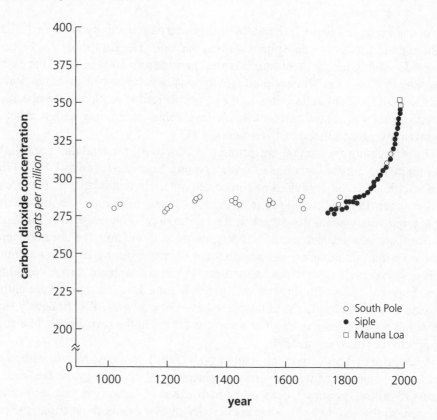

Figure 4. A 1000-year record of atmospheric carbon dioxide concentration. Data are from ice cores at the South Pole (open circles) and at the Antarctic base in Siple (closed circles), except for the rightmost two points, which are from Mauna Loa, Hawaii. (Source: Siegenthaler and Oeschger, 1987. Reprinted with permission from Munksgaard International Publishers Ltd., Copenhagen, Denmark, © 1987.)

carbon dioxide and methane by bacteria working in the stomachs and intestines of cattle, followed by belching.[6]

Very little of the methane in the atmosphere is metabolized by living organisms directly. The fate of most methane molecules is to be oxidized abiotically to carbon dioxide in the atmosphere, slowly and through many steps. The cycle is closed when photosynthesis again produces organic matter.

We know the size of the preindustrial reservoir of atmospheric methane quite accurately—about 1500 million metric tons, corresponding to a concentration of about 700 methane molecules for every billion molecules of atmosphere (0.7 ppmv). The rates of flow in and out of this reservoir must have been nearly equal, because over many centuries the methane concentration in the atmosphere was nearly constant. However, we do not know the preindustrial rates of flow of methane as accurately as the stock. Today, the *residence time* of methane in the

---

[6] Again representing plant matter by the artificial molecule, CH_2O, the reaction, schematically, is:
$2\,CH_2O \rightarrow CO_2 + CH_4$.

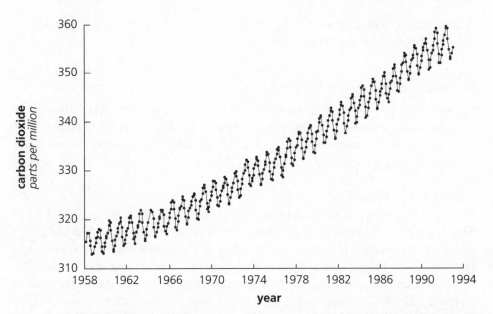

Figure 5. Monthly average concentration of atmospheric carbon dioxide at Mauna Loa Observatory, Hawaii. The data show both a seasonal variation (presumed to be due to the exchange of carbon between the atmosphere and vegetation) and a steady rise in annual average value (presumed to be due principally to human activity). The concentration is in parts per million by volume (ppmv) of dry air. A good model of the seasonal variation uses a shape that is exactly the same for each year, except for an increasing amplitude that doubles in about 100 years. The interannual variation is fitted by an unconstrained smooth curve. Data at other observing stations show nearly the same interannual variation, but each station shows a different seasonal pattern. (Data and model are in Keeling et al., 1989; data are updated in Keeling and Whorf, 1993.) The line shown here simply connects successive data points; it is not a model.

atmosphere is 10 years (that is, about one tenth of the methane in the atmosphere is oxidized to carbon dioxide in one year). If the residence time of methane in the atmosphere was also ten years in preindustrial times, the rate of flow of carbon in the form of methane in and out of the atmosphere would have been 150 million metric tons of carbon per year. However, in preindustrial times the oxidation mechanism that removes methane from the atmosphere may have been stronger than today, in which case the preindustrial residence time would have been shorter and the preindustrial flows of methane in and out of the atmosphere would have been larger.

For every methane molecule in the atmosphere in preindustrial times there were about 400 carbon dioxide molecules, and today the ratio is just over 200, but these ratios understate the relative importance of atmospheric methane. In the atmosphere today, an additional molecule of methane makes a much greater contribution to the greenhouse effect than an additional molecule of carbon dioxide, because of differences in the absorption of infrared radiation by the two gases, and because there is much less methane to begin with. Adding a methane molecule to the current atmosphere increases the greenhouse effect about 20 times as much as adding a molecule of carbon dioxide.

Human Impacts on the Carbon Cycle

Figure 6 presents rough estimates of the additions to the preindustrial carbon cycle brought about by human activity. Especially prominent are: (1) the combustion of fossil fuels, and (2) land-use changes that affect the global stock of biomass and the rates of biotic processes.

Combustion of Fossil Fuels

There is a huge reservoir of buried, dispersed, organic carbon, some of which has been aggregated by geological processes and transformed by heat and biological activity to form coal, petroleum, and natural gas. Humans are adept at finding and extracting these buried hydrocarbons wherever large amounts are in one place. We are commercializing such resources at increasing depths below the surface and at increasing depths of water when offshore. Extraction and combustion of these fuels have contributed significantly to altering the atmospheric CO_2 balance.

In 1991, approximately 5900 million metric tons of carbon were transferred physically from underground hydrocarbon reservoirs to the atmosphere, while being transformed chemically from reduced carbon to carbon dioxide. The 5900 is the sum of 2600 from oil, 2300 from coal, and 1000 from natural gas (Marland, 1993). Approximately three-fourths of the world's primary energy consumption derives from the combustion of these fuels. (See the chapter by Williams in this volume, especially its Figure 1, which gives global 1985 data. Then, the remaining one-fourth of primary energy was, roughly, 15% plant biomass, 5% hydropower, and 5% nuclear power.)

Coal combustion delivers less energy with each carbon atom than does oil, and oil less than natural gas. The reason is that coal has less than one hydrogen atom per carbon atom, oil has about two, and natural gas (which is largely methane) has almost four, and energy is released from the oxidation of both carbon and hydrogen. Hence, an intermediate-term strategy for reducing carbon transfer to the atmosphere from fossil fuels is to shift fossil fuel use away from coal and toward natural gas (provided leakage of unburned natural gas to the atmosphere can be kept very low).

A longer range strategy, explored in the chapter by Williams (this volume), is to substitute plant matter (biomass) for fossil fuel. Williams suggests that five to ten percent of global biomass oxidation could be arranged to occur within energy conversion facilities that produce electricity and gaseous and liquid fuels, permitting an equivalent amount of fossil fuel to be left below ground. Without human intervention, this biomass oxidation would occur in a dispersed manner (decay on the forest floor, for example), producing energy in forms too dilute to harness to human technology. The biofuels industry could be based on renewable plantations, where carbon absorption from the atmosphere in new plant growth stays in approximate

Human Additions to the Global Carbon Cycle

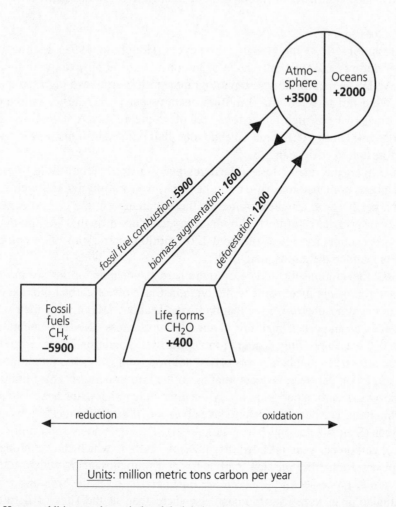

Figure 6. Human additions to the preindustrial global carbon cycle. Changes in reservoir stocks are shown by entries inside the boxes and sum to zero. Human impacts, dominated by the combustion of fossil fuels and by deforestation, lead to an increase in carbon dioxide in the atmosphere and bicarbonate in the oceans. The estimates shown here assume the "missing" carbon is in land plants (see text). The chemical structure of fossil fuels is stated as CH_x to emphasize that the ratio (x) of hydrogen atoms to carbon atoms in fossil fuels is variable, ranging from 4 for natural gas to less than 1 for coal. The rate of combustion of fossil fuels and the rate of increase in the atmospheric stock are known accurately; the other estimates are highly uncertain.

balance with carbon emission to the atmosphere from biofuel use. Then there would be no net transfer of carbon dioxide to or from the atmospheric reservoir.

Land-Use Changes

Deforestation affects the global carbon cycle (Houghton, 1993). Estimates of the rate of deforestation globally have been considerably improved with the help of the LANDSAT satellites, whose photographs of the earth have high spatial resolution. We estimate a rate of 10 million hectares per year. (Earlier estimates were sometimes twice as high.)[7] The total area of tropical forest remaining on the earth is about 1800 million hectares, so about one-half percent of remaining forest is lost to deforestation each year.

To convert the rate of loss in hectares per year to a carbon flow from land to atmosphere in metric tons of carbon per year, we multiply by 120 metric tons of carbon per hectare, a typical value for the net change in the carbon content of a hectare of primary rainforest, allowing for secondary growth[8]. We find that deforestation accounts for a flow of about 1200 million metric tons of carbon per year from the land to the atmosphere.

Fossil fuel combustion and deforestation are responsible for the two most important anthropogenic increments to the preindustrial flow of carbon dioxide from the land to the atmosphere. We see that their sum is about 7100 million metric tons per year. It is an important measure of the incompleteness of environmental science today that we cannot fully account for the fate of this additional flow. From Figure 5 we can infer that roughly half this excess anthropogenic flux, or 3500 million metric tons of carbon per year, is accumulating in the atmosphere, so about 3600 million metric tons of carbon per year must be accumulating either in the oceans or in terrestrial biomass. The oceans appear to take up about 2000 million metric tons of carbon per year (Siegenthaler and Sarmiento, 1993). The remainder, 1600 million metric tons of carbon per year, is called the "missing" carbon, which can be in one of only two places. Either the "missing" carbon has gone into the oceans, which would mean that carbon processes in the ocean are misunderstood, or the "missing" carbon is accumulating in terrestrial biomass elsewhere than at the sites of deforestation, where small changes in total terrestrial biomass are very difficult to detect.

The missing carbon may be going into the temperate forests of North America and Russia, increasing their stock of biomass in spite of increased timber and woodpulp harvesting. Arguments in support of the growth of temperate forests include: (a) *history*—the North American forests were cut down in recent centuries

---

[7] The largest single region of deforestation is the Amazon basin, where the estimated deforestation rate from the clearing of closed-canopy primary forest is about 2 million hectares per year (Skole and Tucker, 1993).

[8] We assume the carbon content of the biomass before and after deforestation to be, respectively, 160 and 40 metric tons per hectare. A more accurate calculation would take into account the delay between deforestation and plant decomposition: only about one-half of the biomass on deforested land will decompose in the first year (Houghton, 1991). It would also take into account year-to-year changes in regional deforestation rates and the multiple fates of land after deforestation.

and are growing back, still approaching their full size; (b) *CO₂ fertilization*—plants often grow more quickly when the concentration of carbon dioxide in their immediate environment is increased, a fact often exploited in greenhouses; and (c) *nitrogen fertilization*—plants often grow more quickly when additional nitrogen is supplied. In drawing Figure 6, for specificity, we have assumed the "missing" carbon is augmenting the global stock of biomass.

The anthropogenic component of atmospheric CO_2 influx may seem small: the 7100 million metric tons of carbon per year injected into the atmosphere from fossil fuel burning and deforestation is about 12% of the 60,000 million metric tons per year of terrestrial net primary productivity. But the anthropogenic term is not balanced, and so the carbon dioxide concentration in the atmosphere keeps growing. The build-up of carbon dioxide in the atmosphere accounts for roughly half of the total anthropogenic greenhouse effect, with the other half assigned to changes in the concentrations of several other trace gases in the atmosphere. Thus, a 12% perturbation within a grand nutrient cycle can have significant impacts.

Modifications of the Methane Subcycle

Human perturbations have also disrupted the methane subcycle. In the past two decades, the rate of increase in the concentration of methane in the atmosphere has averaged about 1% per year, more than twice the rate of increase in the concentration of carbon dioxide. In the last hundred years the methane concentration has more than doubled: the stock of carbon in the form of methane in the atmosphere today is about 3600 million metric tons, corresponding to a concentration of 1.7 ppmv.

The preindustrial methane flux into the atmosphere from anaerobic bacterial action has been augmented by an expansion of wet rice cultivation in the Orient and of cattle and sheep husbandry worldwide. Another important anthropogenic source, discussed by Harriss in this volume, is methane leakage from the natural gas system, at the wellheads and along the transmission and distribution systems.

Moreover, the methane destruction mechanism in the atmosphere is being diminished by an increase in emissions of carbon monoxide (CO). The hydroxyl radical (OH) is the principal scavenger of both molecules, so an increase of CO emissions into the atmosphere results in fewer OH molecules available to scavenge methane, thereby increasing methane's residence time and concentration in the atmosphere[9].

The Nitrogen Cycle

The Preindustrial Nitrogen Cycle

The four-reservoir model of the nitrogen cycle presented in Figure 7 captures the cycle's most essential components. There are two subcycles. In the subcycle of "new" nitrogen, the flows on land are between atmosphere and soil. In the subcycle of "recycled" nitrogen, the flows on land are between soil and plants.

---

[9] The electrically neutral hydroxyl radical (OH) that is chemically active in the atmosphere is not to be confused with the electrically negative hydroxide ion (OH⁻) found in aqueous solutions.

Preindustrial Global Nitrogen Cycle

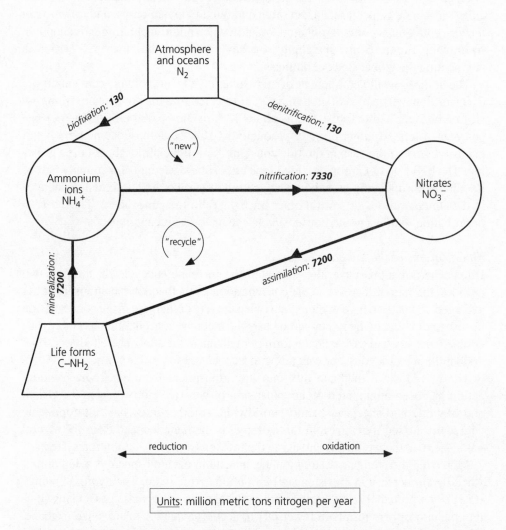

Figure 7. The preindustrial nitrogen cycle, modeled with four reservoirs and two subcycles. The subcycle of "new" nitrogen exchanges nitrogen between the atmosphere and the nutrient pool. This subcycle is shown unidirectional, reflecting the fact that biological organisms can neither convert atmospheric N_2 directly to nitrate nor convert reduced nitrogen to N_2. Nitrogen fixation directly to nitrate, as in lightning and fires, is small and not shown (see text). The subcycle of "recycled" nitrogen, involving much larger flows, is an oversimplification, in that most plants can and do receive a portion of their nutrients directly from the reduced nitrogen in the soil (a flow running in the opposite direction from the flow marked "mineralization"). The sum of land and ocean flows is shown; Table 1 disaggregates these estimates for land and sea separately.

In the first subcycle, the N≡N bond is broken and restored. The nitrogen flowing through this subcycle is called "new" nitrogen, because in this subcycle molecular nitrogen from the atmosphere replenishes the nutrient pool of ammonium and nitrate ions. In the second subcycle, nitrogen is "recycled," and the total stock of nutrient nitrogen remains unchanged. In a well established ecosystem the flows of "recycled" nitrogen are much larger than the flows of "new" nitrogen. We elaborate on these subcycles below.

The Subcycle of "New" Nitrogen
In the subcycle of "new" nitrogen, diatomic nitrogen (N_2) is transformed into available N by *nitrogen fixation*, principally *biofixation*. Because plant and animal life can utilize as nutrients both nitrates and ammonium ions, but not diatomic nitrogen, all life ultimately depends on nitrogen fixation. Most plants and animals get their nitrogen from other plants and animals. But the whole process cannot get started without specially adapted nitrogen-fixing microorganisms that catalytically disassemble and chemically reduce N_2 to ammonium ion (NH_4^+). Biofixation is accomplished by a few specialized bacteria and blue-green algae. The most important of the bacteria is *Rhizobium*, which fixes nitrogen symbiotically after attaching itself to the roots of legumes, such as alfalfa. An important nitrogen-fixing blue-green algae is *Anabaena*, a leftover from the early evolutionary history of the earth. We estimate that preindustrial biofixation occurred at a rate of about 130 million metric tons of nitrogen per year: 100 on land and 30 in the sea.

A second, much smaller, non-biological pathway to nitrogen fixation, not shown in Figure 7, involves the fixation of nitrogen by lightning. Lightning bolts oxidize atmospheric N_2 to nitric oxide (NO), which is rained out as nitrate (NO_3^-) within days. Lightning fixes about 3 million metric tons of nitrogen per year (Borucki and Chameides, 1984). There are about 100 flashes per second (or 3 billion per year), a rate well estimated by satellites; each flash fixes about 1 kilogram of nitrogen as nitric oxide, a mass known with an uncertainty of about a factor of three (Lawrence *et al.*, 1993).

Nitrogen in plant matter is largely found in reduced form, as amine ($-NH_2$) groups incorporated into organic molecules. *Nitrification*—the conversion of an amine or ammonium ion into a nitrate ion—plays a crucial role in increasing the movement of nitrogen in the environment. Nitrification is an aerobic process involving bacteria of the genera *Nitrosomonas* and *Nitrobacter*. The oxidation of the nitrogen is generally accompanied by the reduction of carbon dioxide, thereby assimilating carbon and linking the carbon and nitrogen cycles.

In *denitrification*, which closes the "new" nitrogen cycle, nutrients are removed from the nutrient pool as a result of the restoration of the N≡N bond and the transformation of nitrogen back into N_2, or, infrequently, nitrous oxide (N_2O)[10].

---

[10] It is unfortunate that *nitrification* and *denitrification* are the names of two processes that in no sense reverse one another. The two processes that reverse one another—in the sense of disassembling and reassembling the N≡N bond—are *nitrogen fixation* and *denitrification*.

Biological denitrification requires anaerobic environments, and is accomplished by a specialized set of bacteria, notably in the genus *Pseudomonas*. These bacteria are able to oxidize organic matter using oxygen extracted from the nitrate ion. Thus, denitrification creates a second link between the carbon and nitrogen cycles.[11]

Forest and brush fires are a pathway to non-biological denitrification known as *pyrodenitrification*. In pyrodenitrification, two nitrogen atoms that were in separate molecules in plants or soil bind to one another to make N_2, as the result of a sequence of high-temperature reactions[12] (Bowman, 1991; Kuhlbusch *et al.*, 1991). In the absence of a fire, most of the nitrogen in living organisms is recycled, and only a small fraction is denitrified; but in a fire much of the nitrogen is denitrified (Logan, 1983). A forest or brush fire, therefore, short-circuits the nutrient recycling system[13].

The preindustrial rates of global denitrification and global nitrogen fixation should have been approximately equal, assuming a steady state. From our earlier assumptions about fixation (130 million metric tons of nitrogen via biofixation and almost negligible net nitrogen fixation via lightning and fires), about 130 million metric tons of nitrogen would have been returned to the atmosphere each year (see Figure 7). Uncertainties abound, because measurements of denitrification rates are as difficult as measurements of biofixation rates[14].

Consider the nitrogen flows, separately, for the land and the sea. In preindustrial times there would have been a net nutrient flow from land to sea, yet, by our assumption of a steady state, no net nutrient accumulation on land or in the sea. Bioavailable nitrogen is transferred, as *runoff*, from land to sea, principally via rivers. In a balanced land–sea cycle one finds: (1) greater fixation than denitrification on land, leading to net nutrient availability; (2) transport of this net nutrient from land to sea (runoff); and (3) less fixation than denitrification in the sea—less by an amount equal to the runoff.

---

[11] Additional sources of oxygen for bacteria in anaerobic environments are sulfates ($SO_4{}^{2-}$), manganese dioxide (MnO_2) and ferric hydroxide ($Fe(OH)_3$). The corresponding processes link carbon cycling to sulfur, manganese, and iron cycling. Bacteria have greater difficulty extracting oxygen from the phosphate ion $PO_4{}^{3-}$ (which yields phosphine (H_3P), a gaseous compound), because the energy required to break the oxygen–phosphorus bond is very large; the strength of this bond accounts for the fact that phosphate molecules are the major energy carriers in living organisms.

[12] Atmospheric chemists and experts in combustion have focused principally on the production not of the principal gaseous nitrogen product of a fire, N_2, but of the gas that carries away most of the rest of the nitrogen from a fire, nitric oxide (NO). Their focus is explained by the fact that NO in the atmosphere participates in chemical reactions that affect the balance of other gases in the atmosphere, while N_2 is inert. However, from the perspective of the subcycle of "new" nitrogen, the passage of a nitrogen atom from a plant or animal to a molecule of N_2 in the atmosphere is more consequential than the passage of the same atom to a molecule of NO. In only a few days the NO is washed out of the atmosphere as nitrate in rain or snow, and if it falls on the land, it becomes a nutrient for another plant (Schlesinger, 1991).

[13] *Pyrofixation* is still another nitrogen process related to biomass fires. In pyrofixation nitric oxides are produced not from nitrogen originally in the biomass but (as with lightning) from N_2 in the atmosphere.

[14] Comparing the carbon cycle in Figure 3 with the subcycle of "new" nitrogen in Figure 7, we see that in nature reduced carbon is oxidized directly to the dominant atmospheric gas, CO_2, but reduced nitrogen is not oxidized directly to the corresponding gas, N_2. The biological subcycle of "new" nitrogen is a three-step cycle (fixation-nitrification-denitrification-fixation), while a direct analog of the carbon cycle would be a two-step cycle (fixation-denitrification-fixation) that does not occur.

We estimate that preindustrial runoff carried 20 million metric tons of nitrogen per year from land to sea. The rates of preindustrial denitrification on land and in the oceans would have been, in million metric tons of nitrogen per year, 80 (100 minus 20) and 50 (30 plus 20), respectively. Land and sea values are disaggregated in Table 1.

One year of such flows can affect only a tiny fraction of the 3,900,000,000 million metric tons of nitrogen contained in the reservoir of the global atmosphere. A cycle involving an annual flow in and out of the atmosphere of 130 million metric tons removes and returns just one part in 30 million of the stock in a year. Because the nitrogen in the atmosphere is nearly all bio-unavailable N_2, however, life depends on this trickle of fixed "new" nitrogen. Changing the size of the trickle will affect plant and animal life.

The Subcycle of "Recycled" Nitrogen
In the subcycle of "recycled" nitrogen, the flows of nitrogen from reservoir to reservoir do not break or restore the N≡N bond, and therefore do not change the amount of nitrogen in the global nutrient pool. (In local ecosystems, of course, the amount of nutrient nitrogen can either increase as a result of inputs such as acid precipitation, or decrease as a result of loss mechanisms such as leaching.) In Figure 7, the subcycle of recycled nitrogen, like the subcycle of new nitrogen, has three steps. As drawn, Figure 7 oversimplifies what is found in nature by showing plants obtaining all nitrogen from nitrates, not from nitrogen in chemically reduced form. In fact, the processes of nitrogen *assimilation* by plants can begin with nitrogen in either oxidized or reduced form. Nitrogen in reduced form—as the ammonium ion (NH_4^+) or amine group ($-NH_2$)—is acceptable to most plants. However, many plants absorb nitrogen preferentially in the form of nitrate ion (NO_3^-), and some plants cannot absorb the ammonium ion at all and must wait for nitrification to present the nitrogen as nitrate.

In the subcycle of recycled nitrogen, nutrients are retrieved from life forms by *mineralization*, a strange name for a process where decomposers, as a byproduct of obtaining energy from the oxidation of reduced carbon in plantlife, transform the reduced nitrogen in plant and animal matter into simpler molecules (especially NH_4^+) that still contain nitrogen in reduced form, but that no longer have carbon–nitrogen bonds. The process of *nitrification* completes the subcycle of recycled nitrogen, just as it completes the subcycle of new nutrients; it is the one process common to the two subcycles.

The total flow of recycled nitrogen is estimated in Figure 7 to be 7200 million metric tons of nitrogen per year. In Table 1 this flow is disaggregated: 1200 on land and 6000 in the sea. The rate on land is about twelve times the magnitude of flow of new fixed nitrogen, so in preindustrial times about 92% of the nitrogen on land was recycled. In the oceans, an even greater percentage was recycled; the recycling rate was 200 times the biofixation rate.

The rates of assimilation of carbon and nitrogen must be related, because plant growth involves a fairly predictable ratio of carbon to nitrogen uptake. Our numbers

Table 1: *Preindustrial and contemporary nitrogen cycle (units: million metric tons of nitrogen per year)*

Process	Preindustrial Flow			Human Additions			Flow Today		
	Land	Sea	Total[1]	Land	Sea	Total[2]	Land	Sea	Total
Subcycle of "new" nitrogen									
Fixation									
Life forms (reduced N)	100	30	130	40	—	40	140	30	170
Fertilizer (reduced N)	—	—	—	90	—	90	90	—	90
Combustion of fossil fuels (oxidized N)	—	—	—	20	—	20	20	—	20
Runoff (reduced N)	-20	20	0	-20	20	0	-40	40	0
Nitrification[3]	80	50	130[3]	100	20	120[3]	180	70	250
Denitrification	80	50	130	50	20	70	130	70	200
Subcycle of "recycled" nitrogen									
Assimilation	1200	6000	7200	160	—	160	1360	6000	7360
Mineralization	1200	6000	7200	130	—	130	1330	6000	7330
Nitrification[3]	1200	6000	7200[3]	130	—	130[3]	1330	6000	7330
Build-up in reservoirs									
Life forms	—	—	—	30	—	30	30	—	30
Amine nutrients	—	—	—	10	—	10	10	—	10
Nitrate nutrients	—	—	—	40	—	40	40	—	40
Atmosphere and oceans as N_2	—	—	—	-80	—	-80	-80	—	-80

[1] See Figure 7.
[2] See Figure 9.
[3] The two rows are combined in Figures 7 and 9.

for global annual uptake on land—60,000 million metric tons of carbon and 1200 million metric tons of nitrogen—are consistent with an "average" plant that incorporates carbon and nitrogen at a C/N mass ratio of 50 to 1 (about 60 to 1 on an atom for atom basis). There is a steady enrichment for carbon as plants grow and woody tissue develops, however, so these C/N ratios underestimate the ratio for carbon and nitrogen *stored* in plants. (In other words, the residence time of carbon in plants is longer than the residence time of nitrogen.) As a result, the older the plant tissue, the higher the ratio of carbon to nitrogen. The trunks of trees have C/N mass ratios of 150 or more, leaves less than 50, and algae in the oceans perhaps 10. Soil on average has a C/N mass ratio of 12 to 15 (Schlesinger, 1991).

Nitrous Oxide

There are 2.5 million N_2 molecules for every nitrous oxide (N_2O) molecule in the atmosphere. Still, nitrous oxide makes a powerful contribution to the greenhouse effect (molecule for molecule, additional N_2O is several hundred times as effective as additional carbon dioxide), while minute fractional changes in N_2 are unimportant. Accordingly, much effort has been devoted to understanding the global cycle of nitrous oxide (N_2O) and to documenting its build-up in the atmosphere in recent years.

The N_2O concentration in air up to 2000 years old trapped in layers of permanent ice has been analyzed (see Figure 8). The data reveal that until about two hundred years ago the concentration of nitrous oxide was relatively constant at about 285 ppbv, or 285 molecules of N_2O per billion molecules of atmosphere (Khalil and Rasmussen, 1992). This concentration corresponds to a reservoir of 1400 million metric tons of nitrogen as N_2O in the atmosphere.

Nitrous oxide enters the atmosphere as a minor byproduct of nitrification and denitrification. These flows must have been balanced by removal processes, or *sinks*. The only known sinks for N_2O are in the stratosphere, where destruction by solar ultraviolet photons or by activated atomic oxygen yields either N_2 or nitric oxide (NO). The destruction rate by these two mechanisms is such that one out of every 150 N_2O molecules in the atmosphere is destroyed every year. (An equivalent statement is that the residence time of N_2O in the atmosphere is 150 years.) Unlike the situation with methane, the strength of the destruction mechanism for N_2O in the atmosphere has hardly changed since preindustrial times. This permits an accurate estimate of the preindustrial rate of transfer of N_2O into and out of the atmosphere; it is equal to the size of the reservoir of atmospheric N_2O divided by the residence time, or about 10 million metric tons of nitrogen per year.

Human Impacts on the Nitrogen Cycle

A Doubling of Nitrogen Fixation

Figure 9 and Table 1 present rough estimates of the many additions to the preindustrial nitrogen cycle brought about by human activity. Every aspect of the cycle has been modified in the industrial age. Probably the most important modification

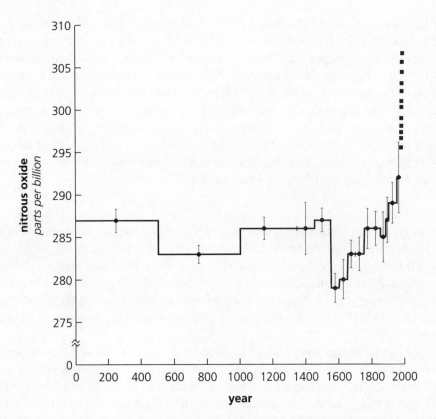

Figure 8. A 2000-year record of the atmospheric concentration of nitrous oxide (N_2O). The data since 1975 are direct measurements in the atmosphere. Earlier data are obtained from bubbles trapped in layers of polar ice. Data are averages from ice cores in the Arctic (at the Crete and Camp Century sites) and the Antarctic (Byrd Station). There is a 90% chance that a larger data set from similar cores for the same time period would give an average value within the vertical error bars. The data show little variation around an average concentration of about 285 N_2O molecules in every billion molecules of atmosphere (285 ppbv) in preindustrial times and a rise at a rate of about 1 ppbv per year in recent years. Modified from Table 2 and Figure 3 of Khalil and Rasmussen, 1992.

is the doubling of the rate of nitrogen fixation. We will trace the doubling of nitrogen fixation to three human activities. In decreasing order of importance they are:

(1) the production of nitrogen fertilizer,
(2) the increased planting of leguminous crops, and
(3) the combustion of fossil fuels.

What are the likely impacts of this doubling of preindustrial global fixation rates, and of almost certain further increases in fixation rates in the future? Worldwide, local ecosystems are being affected in two ways: by changes in nutrient balances and by changes in environmental acidity.

Nitrogen is a limiting ingredient in many ecosystems, so an increased input of fixed nitrogen generally increases net primary productivity. Unfortunately, in natural ecosystems, increased net primary productivity generally means ecosystem

Human Additions to the Global Nitrogen Cycle

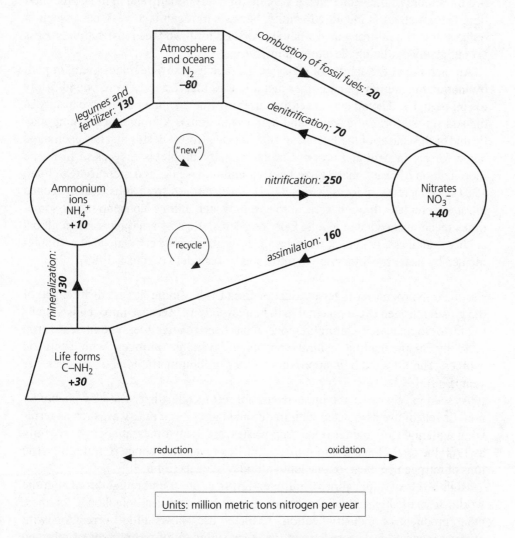

Figure 9. Human additions to the preindustrial global nitrogen cycle. Changes in reservoir stocks are shown by entries inside the boxes and sum to zero. Human impacts on the nitrogen cycle are dominated by the use of fertilizers, the planting of leguminous crops, and the combustion of fossil fuels. A build-up of nitrogen stocks in life forms and as nutrients is assumed, and detailed flows are modeled using additional arbitrary, but plausible assumptions (see text). Table 1 gives a further disaggregation that distinguishes land and sea.

disruption, because there will be differential growth of some species relative to others.

An increase in fixed nitrogen input does not necessarily increase local acidity. At locations of increased concentrations of chemically reduced nitrogen (ammonium ions and amines), the environment becomes less acidic, while at locations of increased concentrations of oxidized nitrogen (nitrates), the environment becomes more acidic. Both changes are disruptive. Globally and in most precipitation, the net effect is increased acidity, because nitrogen is a weak base when in reduced form but a strong acid when in oxidized form, and the nitrification process is continually oxidizing the ammonium ion.

An increase in the acidity of ecosystems, such as lakes, degrades them, in part, by mobilizing aluminum and other metallic ions that formerly were bound firmly to soil particles. The increased acidity "uses up" the alkaline buffering capacity in the soil (in the form of calcium and magnesium ions). Aluminum poisoning may be one of the causes of the European *Waldsterben* (forest death) that has decimated some forests in central Europe. Similar toxicity problems may arise from the mobilization of heavy metals like lead, cadmium, arsenic, and mercury (Stigliani, 1988). An increase in local acidity also leaches nitrates from soils; when leached nitrates percolate through soils into groundwater, nitrate concentrations sometimes reach unacceptable levels. One long-term effect of a net increase in global acidity is an increase in rock weathering, as weathering by carbonic acid is supplemented by faster processes due to nitric and sulfuric acid (Cronan, 1980).

Fertilizer. Nitrogen fertilizer production creates the single largest disturbance of the global nitrogen cycle assignable to human activity. Because nitrogen is a limiting factor in many agricultural regions, it has been relatively easy to increase crop yield by supplementing natural sources of available nitrogen with synthetic sources. The flow today is approximately 90 million metric tons of nitrogen per year (Smil, 1991).

As seen in Figure 10, the intensive use of nitrogen fertilizer is a new phenomenon. Levels of use associated with traditional agricultural practices, such as fertilizing with nightsoil and recycled crop wastes, are many times smaller. As recently as 1960 the total global use of nitrogen fertilizer was only about 10 million metric tons of nitrogen per year, or one tenth of today's levels (Smil, 1991).

High levels of application of nitrogen fertilizer are found today throughout the world. Remarkably, unlike the use of energy, automobiles, telephones, or most other products of industrialization, fertilizer use shows little correlation with average income levels across countries. One summary of recent rates of nitrogen fertilizer use across countries (in kilograms per hectare of arable land per year) finds a global average rate of 46—with the United States at 55, Japan at 145, and Egypt at 295 (Smil, 1991). It is widely believed that to feed a growing population the global rate of nitrogen fertilizer production must continue to increase for many decades.

Nitrogen fertilizer production usually begins with the reaction, at high pressure, of atmospheric nitrogen (N_2) with hydrogen from the methane (CH_4) in natural gas to produce ammonia (NH_3). The fertilizer is applied on the land both as ammonium (reduced nitrogen) and as nitrate (oxidized nitrogen). Each form has advantages. Ammonium fertilizer binds more tightly to clay and so is not as easily leached; nitrate fertilizer is more easily absorbed by plants, but is easily lost in

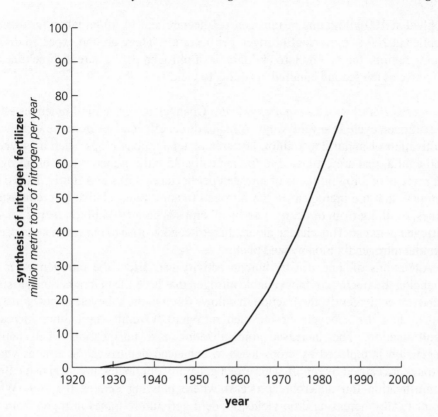

Figure 10. The global production of nitrogen fertilizer, 1920 to 1985. The cumulative production over this 65 year period is one billion metric tons; half of the production was between 1976 and 1985. (Source: Smil, 1991. Reprinted with permission of Cambridge University Press.)

runoff. Mostly, farmers opt for ammonium, delivered either as ammonia gas or, increasingly, as urea $(CO(NH_2)_2)$. Nitrogen is mineralized from urea slowly enough that, like a time-release capsule, urea can have its positive effect over a longer period of time. Moreover, urea is a convenient, easy way to transport reduced nitrogen: it is 47% nitrogen by weight. Farmers frequently opt for applying the nitrogen in both reduced and oxidized form at the same time, in the form of ammonium nitrate.

Nitrogen fertilizer is used especially intensively on corn. In the United States in 1987 corn was grown on 21% of all the cropland but received 41% of all nitrogen fertilizer. On average, in that year's corn production, 153 kilograms of nitrogen was applied per hectare (137 pounds per acre), more than double the rate in 1965 (National Research Council, 1993).

A simple model of a "typical" cornfield, developed by the Committee on Long-range Soil and Water Conservation of the Board on Agriculture of the National Research Council, is helpful in understanding these rates of use (National Research Council, 1993). The model assumes a grain yield of 4 metric tons of corn per hectare (t/ha) in the absence of fertilizer, growing to 7 t/ha when fertilizer is

145

applied at 100 kilograms of nitrogen per hectare and to 8 t/ha when fertilizer is applied at 200 kilograms of nitrogen per hectare[15]. There are the expected diminishing returns: the first hundred kilograms of nitrogen per hectare is three times as effective as the second hundred in adding to yield[16].

Increased Planting of Leguminous Crops. Changes in agricultural practices affect the nitrogen cycle in several ways. A major direct effect arises from the worldwide cultivation of about 250 million hectares of leguminous crops, such as soybean and alfalfa, that are estimated to fix, annually, 35 million metric tons of nitrogen, an average of 140 kilograms of nitrogen per hectare (Burns and Hardy, 1975). We estimate that the increment in the nitrogen fixation rate, relative to preindustrial times, resulting from direct fixation by all crops is about 40 million metric tons of nitrogen per year. This change alone, therefore, adds 40% to our estimate of preindustrial nitrogen fixation by land plants.

Alterations of land use by human activity can affect the nitrogen cycle by changing the fraction of bioavailable nitrogen that is denitrified rather than assimilated (or, equivalently, the fraction that flows through the subcycle of new nitrogen rather than the subcycle of recycled nitrogen). Overall, agriculture increases denitrification. The increased mineralization and nitrification when natural vegetation is replaced by crops leads to enhanced denitrification at sites where nitrates are carried by runoff. This effect generally dominates any local decrease in denitrification (an anaerobic process) where plowing aerates the soil. While dramatically increasing crop yields, modern agriculture mines nitrogen from the soil.

Combustion of Fossil Fuels. The combustion of fossil fuels mobilizes about 20 million metric tons of nitrogen per year, by injection into the atmosphere as nitrogen oxides (Logan, 1983; Müller, 1992). Two distinct classes of combustion processes are about equally involved: (1) processes where most of the nitrogen has originated in the fossil fuel as *fuel-bound nitrogen*, as is the case with coal combustion at electric power plants; and (2) processes where most of the nitrogen has originated as N_2 in the atmosphere, as is the case with combustion in gasoline and diesel engines. In both cases the global pool of bioavailable nitrogen is being augmented by mobilization from a bio-unavailable pool of nitrogen—in the first case from nitrogen in the pool of buried hydrocarbons, in the second case from the pool of atmospheric N_2. The rate of mobilization of nitrogen through fossil fuel combustion is about five times smaller today than the rate associated with nitrogen fertilizer production.

---

[15] There are 40 bushels of corn in a metric ton. Thus, yields of 4, 7, and 8 metric tons per hectare are, respectively, 64, 112, and 144 bushels per acre, and application rates of 100 and 200 kilograms per hectare are, respectively, 89 and 178 pounds per acre.

[16] The model can be taken further. Corn contains about 1.3% nitrogen. Thus the harvested grain retains 39 of the first 100 kilograms of nitrogen added as fertilizer, but only 13 of the second 100.

During the combustion of coal with one percent by weight of fuel-bound nitrogen (a typical concentration), about three-fourths of the fuel-bound nitrogen is emitted to the atmosphere as N_2 and one-fourth is mobilized as nitrogen oxides (NO_x) (Bowman, 1991, Figure 9)[17]. The production of NO_x from fuel-bound nitrogen is usually much larger than the production of NO_x from the N_2 in the air. Thus, we can estimate the rate at which fuel-bound nitrogen enters the global atmosphere today as NO_x, as a result of the combustion of coal. Coal combustion transfers 2300 million metric tons of carbon to the atmosphere (see above), and coal is about 70% carbon by weight (Harte, 1988), so about 3000 million metric tons of coal, containing about 30 million metric tons of nitrogen, are burned. Assuming that one fourth of the fuel-bound nitrogen becomes NO_x, about 8 million metric tons per year of nitrogen as NO_x are produced globally from coal combustion. This estimate is close to estimates obtained from detailed information about the actual range of coals and burning conditions (Müller, 1992).

Nitrogen oxides in the atmosphere, along with sulfur oxides, are the principal precursors of acid precipitation. The potential acidity carried to the land in precipitation today is the equivalent of about 17 million metric tons of hydrogen ion (H^+) per year—and the 17 is, roughly, 10 natural and 7 anthropogenic (Schlesinger and Hartley, 1992). Nitrogen oxides are also crucial precursors of photochemical smog. Accordingly, there is a large effort underway around the world to reduce NO_x formation in combustion, involving large investments in pollution control technology.

Reducing NO_x formation during coal combustion requires increasing the fraction of fuel-bound nitrogen that becomes N_2. This can be accomplished by staged combustion—burning a part of the fuel as a fuel-rich mixture, then adding air downstream and burning the rest of the fuel as a fuel-lean mixture (Bowman, 1992). The alternative of recovering the NO_x from the combustion waste stream, for input into fertilizer production, would be a demonstration of linking two industries in the spirit of industrial ecology; its current economics are dubious.

The second class of combustion processes that fix nitrogen involve combustion conditions where the temperature is so high that the N_2 molecule can be oxidized. The highest temperatures achievable for combustion of fuels in air occur at a particular ratio of fuel to air, known as the stoichiometric ratio, where, approximately, all the carbon and hydrogen in the fuel is oxidized to carbon dioxide and water, with neither excess air nor excess fuel. At the temperatures associated with stoichiometric fuel–air mixtures, the N≡N bond of atmospheric nitrogen is often broken, but at not much lower temperatures the N≡N bond is

---

[17] The lower the concentration of N in the fuel, the higher the NO fraction, because the key N_2 formation reaction (NO + N → N_2 + O) is quadratic in the concentration of single-N species and so becomes relatively more important when there is more N around. When fuel is 0.1% N by weight (a typical value for petroleum products), combustion converts roughly one-fourth of fuel nitrogen to N_2 and three-fourths to NO (Bowman, 1991, Figure 9).

147

likely to remain intact. Stoichiometric burning characterizes diesel and gasoline engines[18].

Human beings are probably also increasing the rate of combustion of biomass, relative to preindustrial times. Human beings deliberately burn biomass to cook and to heat houses, to clear land in slash-and-burn shifting agriculture, to prepare land for replanting by burning crop residues, and for other reasons. These activities probably increase pyrodenitrification more than they increase pyrofixation, and thus they decrease the global stock of bioavailable nitrogen.

We summarize our estimates of nitrogen fixation (in units of million metric tons of nitrogen per year): human beings add 150 today (90 from fertilizer, 40 from leguminous crops, and 20 from combustion) to 130 in preindustrial times, thereby approximately doubling the preindustrial global rate of nitrogen fixation.

Other Modifications of the Nitrogen Cycle

An increase in the global nitrogen fixation rate is contributing indirectly to the increase of three other important flows of nitrogen, each of which has other causes as well. Relative to preindustrial times one finds:

(1) a doubling of runoff of nitrogen from soils, reflecting erosion and other failures in land management;
(2) at least a doubling of the rate of ammonia volatilization, reflecting, especially, large increases in the world's population of cattle; and
(3) a 50% increase in nitrous oxide emissions, from 10 to 15 million metric tons of nitrogen per year, reflecting largely unidentified changes in land use.

A Doubling of Runoff. No symbol has been used more frequently to illustrate the misuse of land by human beings than the loss of topsoil through erosion. Nitrogen and other nutrients move with the topsoil and also percolate through the soil; the

---

[18] The fixation of atmospheric N_2 by U.S. automobile driving may be estimated with reference to the pollution control standards for nitrogen oxide emissions. These emission standards are written in *grams of nitrogen oxides per mile*, to which we must add: (1) the "grams" in question are grams of nitrogen dioxide (NO_2), so a "gram" of nitrogen oxide is 0.30 grams of nitrogen; and (2) the "mile" in question is not a mile of actual driving, but a mile of driving on a standard driving cycle intended to represent average driving. The current U.S. federal nitrogen oxide standard is 1.0 grams per mile (0.62 grams per kilometer); driving with no pollution controls produces about 4.0 grams per mile (2.4 grams per kilometer) (Bowman, 1992); and improved pollution control technologies, involving catalytic reduction of nitric oxides to N_2 in the exhaust gases, can probably achieve 0.2 grams per mile (0.12 grams per kilometer) (Bowman, 1992). In the United States today, cars are driven about two trillion miles per year (more than 100 million cars driven more than 10,000 miles each per year). If, on average, U.S. driving exactly meets the current federal nitrogen oxide emission standard, approximately 0.6 million metric tons of nitrogen fixation per year would occur, and the investments in pollution control would have avoided another 1.8 million metric tons of nitrogen fixation.

An estimate of a decade ago assigning about 8 million metric tons of global nitrogen fixation to all "mobile sources" (Logan, 1983) may still be correct today, because (1) the total number of vehicle-kilometers traveled globally has increased, but (2) stronger pollution controls have been imposed. An accurate estimate requires knowledge of actual levels of emissions per vehicle-kilometer (which may be higher than the emission levels achieved by new cars on specific driving cycles), as well as actual vehicle-kilometers of use for all vehicle types (cars, trucks, and planes) around the world.

result is elevated concentrations of nitrogen ions in rivers, lakes, and ground-water. These increased nitrogen ion concentrations are due not only to poor land practices that result in the loss of nitrogen previously in the soil, but also to increased inputs of nitrogen. Globally today, the nitrogen flow from land to the oceans in runoff is about 40 million metric tons of nitrogen per year (Schlesinger, 1991), approximately double the nitrogen flow in runoff in preindustrial times. It follows that the annual input to land of bioavailable nitrogen is greater today than in preindustrial times by 130 million metric tons—the difference between 150 million metric tons of increased fixation (making the approximation that all anthropogenic additions are inputs to land, not sea) and 20 million metric tons of increased runoff.

A Doubling of Ammonia Volatilization. Ammonia is volatilized from soils at a much higher rate today than in preindustrial times. Ammonia does not accumulate in the atmosphere, because, although not easily oxidized, it is extremely soluble and reactive (Warneck, 1988). The ammonium ion returns to earth when ammonia combines in water droplets with nitric acid to form ammonium nitrate and with sulfuric acid to form ammonium sulfate and bisulfate—water-soluble salts which are quickly washed out of the atmosphere by rain. Ammonia volatilization redistributes bioavailable nitrogen on the land, and it depletes the total nutrient pool available to land plants when the ammonium ions volatilized from the land are delivered to the sea.

Increasing the flow of ammonia in a side loop of the nitrogen cycle should increase the concentration of ammonium ion relative to nitrate ion, other things being equal, because the residence time of nitrogen in reduced form will be increased by cycles of volatilization, rainout, and return to the soil, that are unaccompanied by oxidation. Increased ammonia volatilization provides some mitigation of the acidification of the environment that accompanies increased nitrogen fixation, because the increased stock of ammonia, amines, and ammonium ions chemically neutralizes part of the increased stock of nitrate. Each local disruption of the nitrogen cycle, however, be it an increase in acidity or alkalinity, affects the local natural ecosystem and its biodiversity. When change is rapid, some species will adapt faster than others, stressing the ecosystem in countless ways. As discussed in the chapter by Schlesinger in this volume, a rapid increase in bioavailable nitrogen in any form is likely to lead to a loss of biodiversity.

The principal source of increased volatilization of ammonia comes by a curious route involving cattle. There are about 1.3 billion head of cattle on earth today, and, on average, each is a processor that delivers to the atmosphere, annually, about 16 kilograms of nitrogen as ammonia (Schlesinger and Hartley, 1992). The nitrogen is originally ingested in food, then excreted largely as urea; the urea slowly decomposes and gives off ammonia, just as happens by design when urea is used as a nitrogen fertilizer. The annual flow of nitrogen from land to the atmosphere as ammonia, associated with the chemical decomposition of the urea from cattle, is estimated by simple multiplication to be an amazing 20 million metric tons of nitrogen.

The estimated total global ammonia volatilization (in units of million metric tons of nitrogen per year) is 75; other principal contributors are: another 12 for urea from animals other than cattle, 10 from biological processes in unmanaged ecosystems, 9 from biological processes in fertilized soils, and 13 from processes at the surface of the ocean (Schlesinger and Hartley, 1992). The multiplier that best describes the volatilization of ammonia today relative to preindustrial times is obviously somewhat arbitrary (among other things, one needs to estimate the preindustrial animal population and its diet), but it seems safe to conclude that human activity has at least doubled the rate of ammonia volatilization.

A 50% Increase in Nitrous Oxide Emissions. Assuming the rate of removal of nitrous oxide (N_2O) from the atmosphere has not changed from its preindustrial value (10 million metric tons of nitrogen per year), its 150-year residence time implies that any increment in emissions relative to preindustrial times should appear as an almost equal increment in nitrous oxide in the atmosphere. In fact, the atmospheric concentration has increased from 285 to above 310 ppbv, and is climbing at a rate of about 1 ppbv per year, equivalent to a rate of increase in nitrogen in the atmosphere in the form of nitrous oxide of about 5 million metric tons of nitrogen per year. The additional nitrous oxide production due to human activity, therefore, should also be about 5 million metric tons of nitrogen per year, and the total emissions of N_2O from all sources today should be about 15 million metric tons of nitrogen per year. This is a 50% increase in emission rate, relative to pre-industrial times.

A Model of an Unbalanced Nitrogen Cycle

Is human activity now resulting in a net build-up of bioavailable nitrogen? To answer this question we need to compare how human activity has affected (1) the rate of mobilization of N_2 from the atmosphere (nitrogen fixation), and (2) the rate of return of N_2 to the atmosphere (denitrification). We earlier estimated that human activity (increased fertilizer production, planting of nitrogen-fixing crops, and combustion) has increased the global nitrogen fixation rate by 150 million metric tons of nitrogen per year. Can we also estimate the increase in the global denitrification rate?

One way to estimate the global denitrification starts from our estimate, immediately above, that human activity has increased the rate of N_2O emission to the atmosphere by 5 million metric tons of nitrogen per year. We assume that the increase in N_2O emission is due exclusively to denitrification and that in denitrification there is some definite ratio between the rates of emission of N_2 and N_2O. (In denitrification, N_2 is the principal nitrogen product and N_2O is a minor nitrogen product.) We see that there will be a net build-up of nitrogen in plant life unless the N_2/N_2O ratio in denitrification is higher than 30[19].

---

[19] The ratio here is grams of nitrogen as N_2 divided by grams of nitrogen as N_2O. It is, equivalently, a ratio of *molecules* of N_2 to molecules of N_2O. It is about 1.6 times the corresponding ratio of grams of N_2 to grams of N_2O, since the molecular weight of N_2O is about 1.6 times the molecular weight of N_2.

The key ratio, the relative production of N_2 and N_2O in denitrification, is not well known, but it is probably less than 30. By one estimate, for denitrification on land, this ratio is 16 (Council for Agricultural Science and Technology, 1976). We conclude that one likely result of human activity is an increase in the amount of nitrogen available to the earth's biota.[20]

Figure 9 and Table 1 present a simple model of the increases in flows and stocks of nitrogen since preindustrial times. The specific numerical estimates in this model require several arbitrary, but plausible assumptions about the subcycles of "new" and "recycled" nitrogen. For the subcycle of "new" nitrogen, we assume: (1) the global increase in the annual denitrification rate is 70 million metric tons of nitrogen; (2) the resulting annual increase in nitrogen in biota, 80 million metric tons of nitrogen, occurs only on land; and (3) this annual increase, in millions of metric tons of nitrogen per year, is apportioned 30 to plants, 40 to nitrates, and 10 to amines.

As for the subcycle of "recycled" nitrogen, we assume that recycling of the anthropogenic increment of nitrogen has been established only to the extent that on land the flows of nitrogen into the pool of bioavailable reduced nitrogen from the "new" subcycle and from the "recycle" subcycle are equal; thus, incremental mineralization is assumed to be 130 million metric tons per year. Recall that for the preindustrial nitrogen cycle on land, mineralization was estimated to exceed biological fixation of reduced nitrogen by a ratio of 12 to 1; it seems reasonable that the time required to establish such efficient recycling would substantially exceed the time since large human disturbances began.

Is the Nitrogen Build-Up Related to the "Missing" Carbon?
An increase in the global stock of bioavailable nitrogen could well contribute to an increase in the stock of reduced carbon in terrestrial biomass, by the nitrogen fertilization effect discussed above. For a rough estimate of the rate at which nitrogen would be incorporated in terrestrial biomass to account for the 1600 million metric tons of "missing" carbon per year discussed earlier, we assume that all the missing carbon is in plants (rather than in soil carbon) and assume a C/N mass ratio for assimilation of 50 to 1 (see above). An annual build-up in plants of 1600 million metric tons of carbon would be accompanied by a build-up of 30 million metric tons of nitrogen. Such a rate is roughly consistent with the data above. A more careful comparison would take into account carbon and nitrogen flows into soils, and carbon and nitrogen flows associated with deforestation.

Over the next few decades, stock-and-flow models of the global nitrogen cycle similar to the one presented in Figure 9 and Table 1 will become better established. Such models will build on a more precise understanding of historical data and on

---

[20] The argument for a net build-up of nitrogen in plants today is even stronger when one takes into account that not all of the increase in N_2O in the atmosphere is due to increased N_2O emissions in denitrification. Other possible sources are (1) increased N_2O emissions during nitrification, and (2) direct N_2O emissions from industrial processes used to make nylon and explosives. If, for example, anthropogenic effects have increased the rate of emission of nitrogen as N_2O in denitrification by only 3 million metric tons of nitrogen per year, there will be a net build-up of nitrogen in plant life unless the N_2/N_2O molecular ratio in denitrification is higher than 50.

hundreds of careful field studies of C-to-N ratios, N_2-to-N_2O ratios, assimilation-to-sequestration ratios, and other factors affecting the flows of carbon and nitrogen in many kinds of ecosystems.

Policies to Address Disruptions in the Nitrogen Cycle—Preliminary Thoughts
One source of disruption of the nitrogen cycle already receives concerted attention in public policy—the formation of nitric oxides in the combustion of fossil fuels. The attention arises from concern for local and regional air pollution: nitrogen fixation in combustion contributes to photochemical smog and to acid precipitation. Soil erosion, a second source of disruption through its connection to nitrogen in runoff, has been a historic target of agricultural policy. However, policy addressing fertilizer production and use—the most important source of disruption of the nitrogen cycle—is much less developed.

Indeed, in the framework of global change, fertilizer use may well deserve as much attention as fossil fuel use, the principal agent of disruption of the global carbon cycle. The chapter by Gadgil (this volume) describes many of the strategies for energy efficiency (effectively, carbon efficiency) being developed and tested around the world. Each of the many current strategies to modify fossil fuel use has its analog for fertilizer.

Industrial ecology urges attention to be paid to policies that encourage the closing of loops in materials flows. In this instance, one would search for ways to recover the nitrogen in crop residues, in the wastes from feedlots, and in waste streams at sites of centralized food processing and food consumption.

More traditional policies would seek to promote technological innovations that would lead to a future food production, distribution, and consumption system with a much higher nitrogen utilization efficiency. Some examples on the production side include: (1) improved soil management (including new crops or varieties and new crop combinations) to increase yield at a given level of fertilizer use; (2) improved mechanisms to get the nitrogen in fertilizer incorporated into plants; and (3) bioengineered crops, like nitrogen-fixing corn. Policies addressing food distribution would seek to achieve the reduction of pre-consumer spoilage. Policies addressing consumption—end-use nitrogen efficiency—would consider changes in diet. Analogous policies would address crops grown for purposes other than food, including crops for energy. Policies could be market-based or regulatory and could operate at the local, regional, national, or international level. One could, for example, imagine a regime of tradable permits to fix nitrogen.

Conclusions

The simplified modeling presented throughout this chapter carries several messages. First, the principal processes that govern the global cycling of nutrients have been identified. Second, the roles of biological and geochemical processes are remarkably intertwined. Third, although the quantitative details of reservoir models are highly uncertain, numerical estimates of stocks and flows are constrained by mass balances and by element ratios like C/N and molecular ratios like N_2O/N_2 in particular processes. Fourth, research in environmental science is steadily deepening our understanding of every aspect of the grand nutrient cycles. Fifth, the human impacts on the grand nutrient cycles are significant disturbances.

It is not yet possible to assess accurately the significance for human beings of the disturbance of the global nitrogen cycle—either in absolute terms or relative to the disturbance of the global carbon cycle. The disturbance of the global carbon cycle is probably leading to global warming, and the disturbance of the global nitrogen cycle is probably leading to global fertilization. At an earlier time of anthropocentricity and ecological innocence, one might have supposed that warmer is better and more fertile is better, too. Ecology teaches otherwise: that the earth's ecosystems have adapted to today's climate and today's nutrient sources and are likely, on balance, to be damaged by any rapid change in environmental conditions. As environmental science becomes more deeply understood, the nitrogen cycle may well take center stage in studies of global change.

Comparing the global carbon and nitrogen cycles has particular messages for technology and policy. The global nitrogen cycle has strayed far from its preindustrial steady state, yet the nitrogen cycle is scarcely part of public discourse. The global carbon cycle, by contrast, is well known to the public via the greenhouse effect and deforestation. The principal human activity disturbing the carbon cycle, fossil fuel use, has long been the subject of innovations in technology and policy designed to reduce environmental impact. The principal human activity disturbing the nitrogen cycle, fertilizer use, has had less attention of this kind. Reasoning by analogy, it may be productive to think about fertilizer use in terms of the quantification of environmental externalities, the delivery of end-use efficiency, and the creation of competitive markets—ways of thinking about fossil fuel use that have led to some partial successes in the grand task of conducting the human enterprise within environmental constraints.

Acknowledgments

We have been helped immeasurably by general guidance and expert advice from Allison Armour, Tom Bowman, Francis Bretherton, Bill Chameides, Tom Graedel, John Harte, Wendy Hughes, Sivan Kartha, Aslam Khalil, Ann Kinzig, Pamela Matson, Steve Pacala, Vern Ruttan, David Skole, and Valerie Thomas.

References

Borucki, W. J., and W. L. Chameides. 1984. Lightning: estimates of the rates of energy dissipation and nitrogen fixation. *Reviews of Geophysics and Space Physics 22(4)*, 363–372.

Bowman, C. T. 1991. Chemistry of gaseous pollutant formation and destruction. In *Fossil Fuel Combustion: A Source Book* (W. Bartok and A. F. Sarofim, eds.), Wiley.

Bowman, C. T. 1992. Control of combustion-generated nitrogen oxide emissions: technology driven by regulations. In *Twenty-Fourth Symposium (International) on Combustion/ The Combustion Institute, 1992.*, 859–878.

Burns, R. C., and R. W. F. Hardy. 1975. *Nitrogen Fixation in Bacteria and Higher Plants*. Springer-Verlag, New York.

Council for Agricultural Science and Technology. 1976. *Effect of Increased Nitrogen Fixation on Stratospheric Ozone*. CAST Report (53), Council for Agricultural Science and Technology, Ames, Iowa.

Cronan, C. S. 1980. Solution chemistry of a New Hampshire subalpine ecosystem. A biogeochemical analysis. *Oikos 34*, 272–281.

Galloway, J. N. 1989. Atmospheric acidification: Projections for the future. *Ambio 18*, 161–166.

Harte, J. 1988. *Consider a Spherical Cow: A Course in Environmental Problem Solving*. University Science Books, Mill Valley, California.

Houghton, R. A. 1993. The role of the world's forests in global warming. In *World Forests for the Future: Their Use and Conservation* (K. Ramakrishna and G. M. Woodwell, eds.), Yale University Press, New Haven, Connecticut.

Keeling, C. D., R. B. Bacastow, A. F. Carter, S. C. Piper, T. P. Whorf, M. Heimann, W. G. Mook, and H. Roeloffzen. 1989. A three-dimensional model of atmospheric CO_2 transport based on observed winds: Observational data and preliminary analysis. In *Aspects of Climate Variability in the Pacific and the Western Americas* (D. H. Peterson, ed.), Geophysical Monograph, Vol. 55, American Geophysical Union, Washington, D.C.

Keeling, C. D., and T. P. Whorf. 1993. Atmospheric CO_2 records from sites in the SIO air sampling network. In *Trends '93: A Compendium of Data on Global Change* (T. A. Boden, D. P. Kaiser, R. J. Sepanski, and F. W. Stoss, eds.), ORNL/CDIAC-65, Carbon Dioxide Information Analysis Center, Oak Ridge National Laboratory, Oak Ridge, Tennessee, 18–28.

Khalil, M. A. K., and R. A. Rasmussen. 1992. The global sources of nitrous oxide. *Journal of Geophysical Research 97(D13)*, 14,651–14,660.

Kuhlbusch, T. A., J. M. Lobert, P. J. Crutzen, and P. Warneck. 1991. Molecular nitrogen emissions from denitrification during biomass burning. *Nature 351*, 135–137.

Lawrence, M. G., W. L. Chameides, P. S. Kasibhatla, H. Levy II, and W. Moxim. 1993. Lightning and atmospheric chemistry: The rate of atmospheric NO production. *Handbook of Atmospheric Electrodynamics*.

Lewis, G. N., and M. Randall. 1923. *Thermodynamics*. McGraw-Hill, New York.

Logan, J. A. 1983. Nitrogen oxides in the troposphere: global and regional budgets. *Journal of Geophysical Research 88*, 10,785–10,807.

Marland, G. 1993. Presentation at the Office for Interdisciplinary Earth Studies Global Change Institute on the Carbon Cycle, Snowmass, Colorado, July 19–30, 1993.

Mayewski, P. A., W. B. Lyons, M. J. Spencer, M. S. Twickler, C. F. Buck, and S. Whitlow, 1990. An ice-core record of atmospheric response to anthropogenic sulphate and nitrate. *Nature 346*, 554–556.

Müller, J.-F. 1992. Geographical distribution and seasonal variation of surface emissions and deposition velocities of atmospheric trace gases. *Journal Geophysical Research 97(D4)*, 3787–3894.

National Research Council, 1992. *Soil and Water Quality: An Agenda for Agriculture*. National Academy Press, Washington, D.C.

Schlesinger, W. H. 1991. *Biogeochemistry: An Analysis of Global Change*. Academic Press, New York.

Schlesinger, W. H., and A. E. Hartley. 1992. A global budget for atmospheric NH_3. *Biogeochemistry 15*, 191–211.

Siegenthaler, U., and H. Oeschger. 1987. Biospheric CO_2 emissions during the past 2000 years reconstructed by deconvolution of ice core data. *Tellus 39B*, 140–154.

Siegenthaler, U., and J. L. Sarmiento. 1993. Atmospheric carbon dioxide and the ocean. *Nature 365*, 119–125.

Skole, D., and C. Tucker. 1993. Tropical deforestation and habitat fragmentation in the Amazon: Satellite data from 1978 to 1988. *Science 260*, 1905–1910.

Smil, V. 1991. Nitrogen and phosphorus. In *The Earth as Transformed by Human Action* (B. L. Turner, II, W. C. Clark, R. W. Kates, J. F. Richards, J. T. Mathews, and W. B. Meyer, eds.), Cambridge University Press, Cambridge, United Kingdom, 423–436.

Stigliani, W. D. 1988. Changes in valued "capacities" of soils and sediments as indicators of non-linear and time-delayed environmental effects. *Environmental Monitoring and Assessment 10*, 245–307.

Warneck, P. 1988. *Chemistry of the Natural Atmosphere*. Academic Press, London.

10

Charting Development Paths: A Multicountry Comparison of Carbon Dioxide Emissions

William Moomaw and Mark Tullis

Abstract

Research over the past 20 years has shown that the relationship between energy use and economic growth is not linear, as previously thought. These assessments are extended by analyzing the historical relationship between economic output and emissions of carbon dioxide for different countries. Carbon intensities are shown to differ widely, even among countries with similar levels of industrial activity. In industrial economies carbon intensities have, in general, continued to fall, whereas trends in some poorer, less stable economies of the south have been more chaotic.

The Need for New Measures of Development

Throughout history, human societies have organized themselves in diverse ways to meet the needs and wants of their members. Since World War II, a paradigm of development has evolved which has focused entirely on the economic component of human activities while largely ignoring many environmental and social consequences. As an exception, Western development assistance during the 1970s concentrated on assisting "the poorest of the poor" in developing countries through rural development projects which sought to address the basic human needs of the rural poor (Morss and Morss, 1986). More recently, the World Commission on Environment and Development (1987) has focused attention on "sustainable development" that emphasizes the interrelationship between the environment and economic development. It has been difficult to provide an operational measure that effectively illustrates this useful concept in practice (Lélé, 1991). In this chapter we suggest an approach that links traditional economic measures with their environmental consequences, and we illustrate the methodology by examining the specific example of the relationship between economic development and fossil fuel carbon dioxide emissions.

Although only about one-fifth of the world's present population is firmly imbedded in an industrial economy, development is to a large extent measured by the degree of industrialization a society has achieved. According to the conventional dogma, the world is divided into two classes of nations, the developed and the

157

developing, by which is primarily meant industrialized and industrializing. Nonindustrialized or subsistence agricultural economies were formerly referred to as underdeveloped, but in recent years have come to be known as "least developed countries" or LDCs. It has also become customary to rank order nations in terms of their per capita gross national products (GNP), as "First World" to refer to the "developed" market economies, "Second World" to mean the formerly centrally planned economies of Central and Eastern Europe, "Third World" to represent the newly industrializing countries such as India, China, and Brazil, and "Fourth World" to refer to the "least developed" nonindustrial nations. Development economists also recognize explicitly the structural difference between GNP growth in an economy like that of South Korea and that of Kuwait.

This unfortunate categorization of nations dominates the thinking and practices of national governments in their own economic policies, as well as the international aid and development agencies. The choice of language implies that there is an objective definition of a "developed" economy, and that it represents the ultimate stage of social progress towards which all other economies must inevitably move. It is of course not the case that achieving industrial status represents some final, static stage of development. The structures of the "developed" economies of the United States, Western Europe, and Japan, for example, have changed significantly during the past 20 years, and some analysts even speak of postindustrial societies. In other words, economic development is a dynamic process, and in a real sense, all nations have developing economies. It is our goal to develop a methodology that will illuminate the relationship between a particular region's economic development and its environmental impact, regardless of its stage of economic development, and to provide a framework by which one can follow the development process through time.

Economic Measures of Development

Measuring levels of national economic performance has come under considerable scrutiny during recent years. Until recently, GNP has been the favored measure of determining the aggregate output of a nation's goods and services. Indeed, once one peels back the cover of development discussions, economic development is seen as exceedingly one-dimensional, with progress measured principally by growth in per capita GNP. More recently, the measure gross domestic product (GDP), which excludes profits and interest from overseas investments, and worker's remittances and other payments from abroad, but is limited to measuring economic activity within national boundaries has been seen as a more useful measure for comparing national economic outputs.

For our purposes, GDP has two particular advantages over GNP. First, GDP comes closer than GNP to accounting for all of the economic activity, including the production of exports, that creates environmental impacts directly ascribable to a particular country. Second, countries with extensive foreign investments and operations show a disproportionately higher level of economic wealth when measured by GNP compared to those nations which lack such investments. For comparative purposes, GDP therefore appears to provide a better measure of economic status among very different kinds of countries.

A number of critics have pointed out that all gross productivity measures suffer from a number of significant shortcomings, especially if one is interested in measuring the long-term sustainability of an economy (Repetto, 1992). Since traditional economics treats environmental problems as externalities to the economic system, GDP does not account for losses even when major disruptions to vital natural processes occur. In fact, the very costs of addressing environmental problems add to GDP rather than being charged against it. GDP also fails to measure the depletion of natural capital stocks, and therefore substantially overstates real economic growth for countries that depend heavily on the production of energy, mineral, or forest resources.

An additional problem invariably arises when using one currency for international comparisons (such as the U.S. dollar). For example, the strong appreciation of the U.S. dollar through 1985 masks real growth in GDP and per capita GDP in some countries. Differential inflation rates also make cross-national comparisons difficult. To address this problem, some development economists have utilized surrogates such as per capita energy consumption in place of GDP to measure the degree of economic development and industrialization. It is now widely recognized that raw energy consumption is a poor measure since it ignores the efficiency of its application to various end uses, and as a measure of throughput, can hardly be an effective indicator of economic sustainability.

A much more effective approach that skirts problems associated with exchange rates, inflation, and different economic structural differences is the set of national GDP estimates produced by Robert Summers and Alan Heston. These data are particularly well suited for intercountry comparisons as they consider, among other things, purchasing-power parity for a wide range of common goods and services instead of simply using exchange rates. As a result, these estimates tend to raise GDP figures for developing countries and lower them slightly for most industrialized countries (Summers and Heston, 1991).

This chapter suggests that sustainable economic development represents far more than growth in per capita GDP. It is also necessary to include some simultaneous measure of environmental impact as well. This leads to the notion of a multidimensional development path that might incorporate several measures in addition to the conventional economic ones. Because the authors have been working on global climate change issues, the data presented here will be limited to carbon dioxide emissions associated with fossil fuel use, but we are also working on other environmental, economic, and social measures as well.

We begin with an attempt to better assess and compare both individual countries' and groups of countries' environmental performance by examining the historical relationship between CO_2 emissions and growth in GDP. Second, by applying various statistical tools to this model, we develop both efficiency measures for individual countries as they develop and measures by which countries can be rated at different stages of their development. Third, this model can be used to project future environmental efficiency both nationally and globally. Such projections could be used as a basis for alternative development strategies.

Environmental Measures of Development

Just as it is difficult to find a single comprehensive measure of economic activity, it is even more difficult to create a single environmental index that covers all aspects of economic activity. After examining a number of possibilities, we have concluded that rather than trying to quantitatively measure very different environmental impacts with a single measure, it is better to correlate economic activity with more focused environmental measures.

A useful approach for assessing the significance of different contributions to environmental impact was developed by Commoner (Commoner *et al.*, 1971). They attributed environmental impact to the following simple algebraic identity:

$$\text{Environmental impact} = \frac{\text{Impact}}{\text{Activity}} \times \frac{\text{Activity}}{\text{Capita}} \times \text{Population}$$

Environmental impact is itself often a product of an environmental release times a damage function. For climate change, the damage function is extremely complex (it may have both adverse and beneficial terms) and its quantitative details are often uncertain. On the other hand, national carbon dioxide emissions from fossil fuels are fairly well determined. We therefore use fossil fuel carbon dioxide emissions instead of impact, although it would be fairly simple to convert these emissions into direct radiative forcing, which is itself not a full damage function.

The first term on the right-hand side is referred to as the technology factor, because it is determined largely by the choice of technology to accomplish a given human activity. In our case, this becomes carbon dioxide emissions per sectoral activity.

The second term is referred to as the per capita affluence factor, while the third is simply the population multiplier. Dividing through by population allows one to evaluate the environmental impact per capita, which is particularly useful for comparing nations of vastly different size. The affluence factor is a measure of the average amount of a given activity or good that is utilized by each individual in a society. It may be thought of as the product of the amount of activity per unit of GDP times the per capita income:

$$\frac{\text{Activity}}{\text{Capita}} = \frac{\text{Activity}}{\text{GDP}} \times \frac{\text{GDP}}{\text{Capita}}$$

The initial term is a measure of economic efficiency for a particular input such as energy consumption (or perhaps more explicitly manufacturing or transportation energy), while the second represents a more conventional definition of affluence, namely, the one we have utilized in our study.

We have examined carbon dioxide emissions from fossil fuel use as one broad-based surrogate measure for environmental impact. Although climate change brought about by an enhanced greenhouse effect is not at the top of the environmental agenda for most countries, such as the newly industrializing nations of the South or the reindustrializing nations of Eastern Europe, carbon dioxide is closely proportional to many of the other less well characterized air pollutants such as acidifying sulfur dioxide and lung-damaging particulate matter. We are working to refine our methods to generate similar data sets for these and other additional air pollutants and solid and hazardous waste.

Estimates for CO_2 production from fossil fuel consumption between 1950 and 1990 were derived from Oak Ridge National Laboratory's comprehensive time series data for global and national CO_2 emissions which were calculated from energy statistics compiled by the U.N. Statistical Office and U.S. Bureau of Mines. This data includes emission estimates from gas, liquid, and solid fuels, and from cement production for more than 200 countries (Boden *et al*, 1991).

The Development Plane

In searching for a format in which to compare national economic and environmental data simultaneously, we have used a variation of resource–wealth correlation diagrams to produce a development plane (Moomaw, 1991). One axis measures affluence utilizing GDP per capita as a conventional index of economic well-being, and on the other is a measure of environmental impact or emissions per capita. In other words, this is the product of the technology and affluence factors cited above.

If the data are for a given year, then one obtains a snapshot of the state of a nation's development for that year. The existence of a plane, however, suggests a much more important concept. Instead of thinking of economic development as a one-dimensional process related solely to GDP, one can now consider a development path that simultaneously includes both economic and environmental factors. Figure 1 illustrates the relative positions on the plane for sixteen selected countries using 1990 data. While our diagrams are limited to one economic and one environmental variable, conceptually, it is possible to consider a larger number of factors in a multidimensional space. One might examine how a nation's economic development correlates simultaneously with atmospheric pollutants and toxic chemicals, as it follows its unique development trajectory. Or a third axis might be used to follow some other social factor such as health or literacy status as a nation.

By organizing information in national terms, the development plane (or, more generally, development space for multidimensional analysis) also implies a political dimension as well. Although we are increasingly aware of the fragility and mobility of political boundaries, nation-states remain the principal structures by which environmental and economic information is organized, and nation-states remain primary actors in the implementation of policies.

161

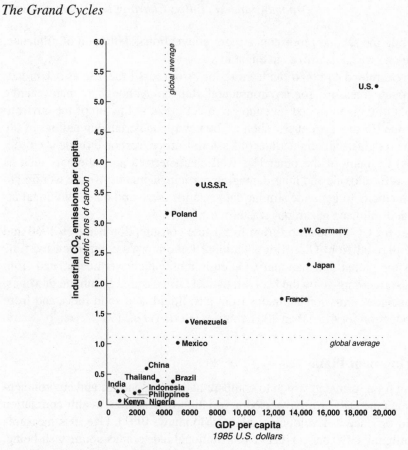

Figure 1. Static development plane for 16 countries.

Charting National Development Paths

There has been much discussion about the heavy dependence of economic well-being on the use of fossil fuels. The reluctance of the United States to sign the global warming treaty at Rio was based upon the belief that any reduction in the use of fossil fuels would be harmful to the economy. Similarly, development strategies for many newly industrializing nations assume that they too must substantially increase their CO_2 emissions if they are to "develop" economically. It seemed to us that it would be useful to examine these assumptions empirically by examining both the current situation and the historical development path of a variety of countries or specific clusters of countries (Tullis, 1992). In this study we have chosen to examine the relationship between per capita GDP and carbon dioxide emissions of 16 countries.

The nations are divided into four groups, and their relative position in the development plane for the year 1990 is shown in Figure 1:

- Planned economies: former USSR, Poland, China, India
- Members of the Organization for Economic Cooperation and Development (OECD): USA, Germany, Japan, France

162

- Low- and middle-income oil-producing countries: Venezuela, Mexico, Indonesia, Nigeria
- Industrializing countries: Brazil, Kenya, Thailand, Philippines.

The United States, as the largest national producer of CO_2 and with a high GDP, stands out in the upper right-hand corner of the plane with per capita emissions of about 5.4 metric tons. France, Japan, and West Germany have slightly lower, nearly equal per capita GDPs, but CO_2 emission levels that range from one-third to just over one-half of the U.S. levels.

Among the planned economies, Poland and the former USSR have per capita income that is comparable to Brazil, Mexico, and Venezuela, but at an environmental cost of several times their carbon dioxide emissions. The remaining nations are clustered in the lower left-hand corner of the plane, relatively poor, but still producing CO_2 at a fairly low per capita rate of less than 0.6 tonnes.

The annual global per capita CO_2 emission rate from fossil fuels is approximately 1.15 tonnes per person (Boden *et al.*, 1991). If atmospheric levels of carbon dioxide were to be stabilized at their present level of 354 parts per million (0.035%), annual CO_2 emissions would have to be reduced by more than 60% (IPCC, 1990). For the present population of just over 5 billion people, this implies a reduction in the annual global average CO_2 release rate to 0.5 tonnes per person. If the world population doubles to 10 billion, per capita emissions would have to drop to 0.25 tonnes per person.

In order to take full advantage of the insights provided by the development plane concept, let us examine the development paths of the 16 selected countries. The planned economies of both the Second and Third Worlds show a similar pattern, as shown in Figures 2a and 2b. The former USSR, Poland, and India roughly quadrupled their per capita fossil fuel carbon emissions between 1950 and 1990. While economic growth tripled in the Soviet Union in a very steady fashion during this period, it only doubled in Poland and increased a mere 50% in India. The political upheavals during the 1980s in Poland appear as economic stagnation and a slight drop in carbon emissions, reflecting the collapse of the state-supported heavy industrial base. New data from the former Soviet Union reveal that the steady 4.5% annual CO_2 increase between 1950 and 1989 has fallen in every year since 1989 (Flavin, 1992). While the former USSR remained the second-largest producer of CO_2 with 17% of the global total, the recent decline, combined with a decrease of as much as 15% in GDP, will add a sharp downward hook to the former Soviet Union's development path. Although economic and energy use data from the nations of Eastern and Central Europe may be questionable, all show a similar pattern to that of Poland and the former USSR.

China is interesting as the "Great Leap Forward" and Cultural Revolution of the 1960s translates into the "Great Loop Backward" in its development path during this period. Despite this setback, in 40 years, China has shown an astounding five-fold increase in per capita GDP while increasing its per capita and total fossil fuel carbon emissions 15 and 30 times, respectively. It currently ranks third in total

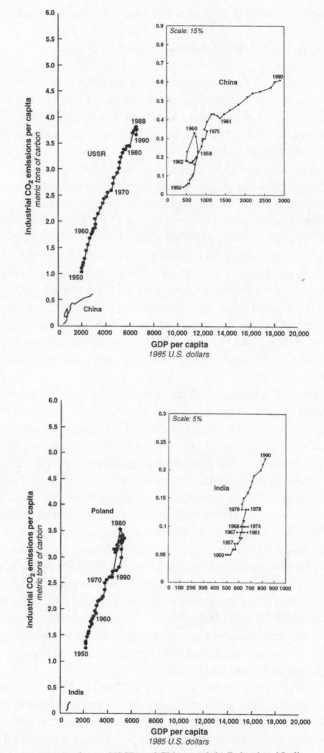

Figure 2. Planned economies for (a) USSR and China, and (b) Poland and India.

164

emissions and, as the world's largest consumer of coal, could well overtake the former Soviet Union during this decade.

Fuel choices also play a critical role in carbon emissions. Currently, oil and coal contribute equally to global CO_2 with natural gas accounting for one-fifth of the total. Within the former Soviet Union coal and oil use have largely remained constant while gas use has increased; whereas, in the United States the opposite has occurred. It is important to note that China and India use large quantities of biomass fuels (estimated to be about one-quarter of the total in China), which for the most part is not managed in a sustainable fashion. It is also interesting to note the lack of certain features in the development paths of all of these nations such as the effects of the oil price shocks of the 1970s that altered the energy consumption patterns of most of the rest of the world.

The market economies of the industrial countries show a remarkable and complex development pattern, which, while similar in overall pattern, differs in important details (see Figures 3a and 3b). All of the industrial market economies responded dramatically to the oil price shocks of the 1970s. For these countries, growth in carbon emissions came to an abrupt halt in 1973, and in most cases retreated further with the additional price hikes of 1979. Real oil prices had remained virtually constant for the 50 years preceding the fourfold increase, and fossil carbon emissions increased regularly. Japan's development path like that of post-war Germany shows a much steeper slope than that of the historically earlier industrialization period. The much higher levels of economic growth per unit of CO_2 occurs for these countries because they were able to industrialize or reindustrialize with far more efficient technology than was available earlier. Unfortunately, this pattern has been difficult to repeat by India and most other nations currently considered to be "developing."

Several differences among the responses of the four upper-income industrialized countries represented here should be noted. While all four instituted policies to improve energy efficiency, and markets responded to the price rise, the United States relied most heavily on the market and has suffered the widest swings in both its economy and post-oil shock carbon emissions. Specific policies and technology choices have produced little or no further growth in the case of Japan and Germany, and a decrease for France. In all three countries, energy efficiency gains are impressive, but much of the decline in France can be attributed to the aggressive introduction of nuclear power, which grew from 8% of electricity production in 1973 to 75% by 1989. It is interesting to note that the economic growth rate in France has been slower since 1979 than those of Germany and Japan.

It is also possible to extrapolate the pre-1973 fossil fuel intensive trends to the years 1990 and beyond to 2000 (see Figures 4a and 4b). All of the industrialized countries have substantially lower actual per capita carbon dioxide emissions than had the pre-1973 trends continued. What is especially interesting is that the actual per capita GDP in 1990 is somewhat lower for France than it would have been if the pre-1973 growth rate had held; for Germany it is slightly lower, for Japan it is slightly better, and for the United States, per capita GDP is significantly higher. It

165

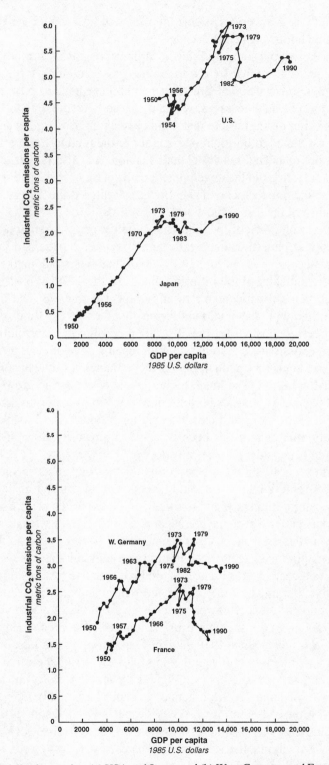

Figure 3. Industrialized countries: (a) USA and Japan, and (b) West Germany and France.

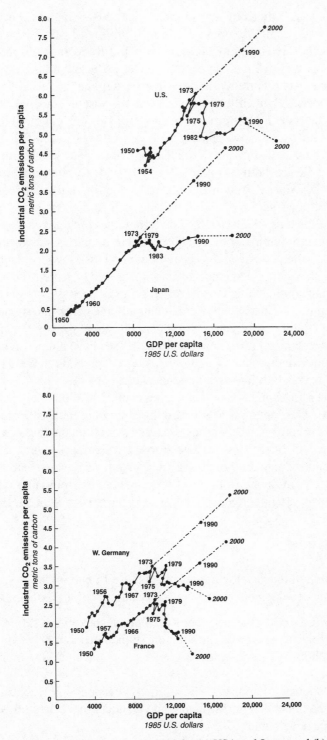

Figure 4. Industrial countries: extrapolation to 2000 for (a) USA and Japan, and (b) West Germany and France.

appears that energy efficiency gains in the United States increased economic productivity as well despite the fuel switching from low-carbon gas to high-carbon coal and the expensive move to nuclear power. In France, high cost nuclear power has proven to be an effective, but expensive CO_2 reduction strategy. A more complete factor analysis is currently under way to determine the relative contributions of changes undertaken by different countries in the post oil embargo period.

Newly industrializing countries that are also oil and gas producers provide another interesting set of development paths (see Figures 5a and 5b). The Mexican economy anticipates that of the Soviet Union with its steady growth ending in a downward hook as both economic growth and carbon emissions decline. Indonesia shows steady growth in both its economy and fossil fuel emissions, while Venezuela has suffered a wild, chaotic path as markets have been buffeted by internal political problems and the powerful forces of world energy prices.

Of the four low-fossil-resource industrializing countries shown in Figures 6a and 6b, only Thailand, which relies heavily on its skilled low-cost labor for manufacturing for the global economy, has shown a steady pattern characteristic of early industrialization. While Brazil started off on a steep growth curve with very low carbon emissions just as Japan did, it has suffered economic stagnation during the 1970s by failing to effectively address the altered energy market. Higher energy prices and the political upheaval of the 1980s have blunted economic gains in the Philippines and produced an erratic pattern of carbon releases. The development path of Kenya falls into a pattern that can only be described as chaotic, a pattern that is seen for a number of developing countries. While per capita GDP has doubled in 40 years, attempts at establishing a growing industrial base have been highly erratic, producing wild swings in carbon and pollutant emission levels. This chaotic pattern is likely due to hard currency shortages which constrain the amount of oil that can be purchased in a given year. However, much of the increase in CO_2 emissions from all of these nations is from automobiles, which (except in Brazil) are imported, and hence add little to economic development.

Conclusions

An examination of the development paths of these 16 nations at different stages in their economic development process reveals a number of important insights.

- Nations with comparable per capita wealth can have very different levels of CO_2 emissions depending on the structure of their economy.
- Market economies can continue to grow even while curtailing their CO_2 emissions.
- Market economies demonstrate a short-term response to price rises in fossil fuels, but the imbedded infrastructure and practices leave them in very different places in the development plane.
- There are many different development paths, and some are capable of producing rapid economic growth with relatively low carbon emissions.

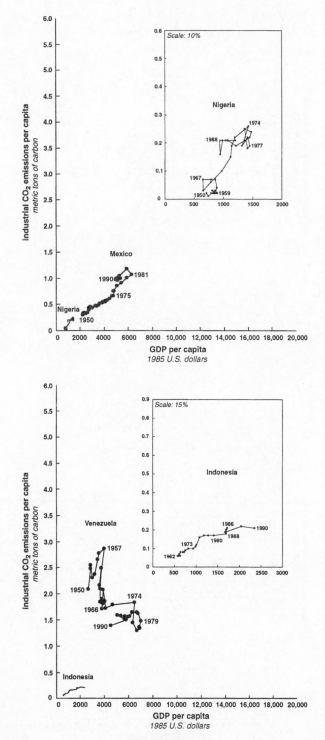

Figure 5. Low and middle income in oil-producing countries: (a) Mexico and Nigeria, and (b) Venezuela and Indonesia.

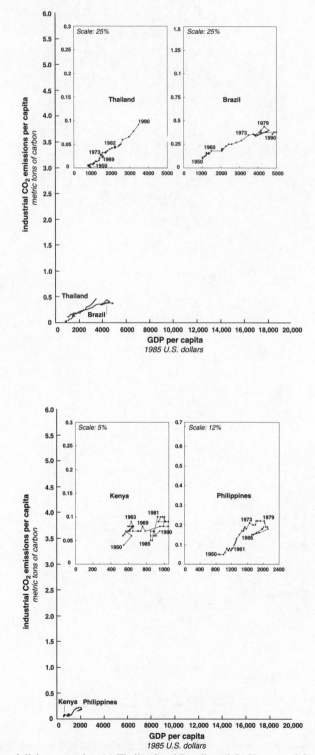

Figure 6. Industrializing countries: (a) Thailand and Brazil, and (b) Kenya and the Philippines.

170

- Efficiency gains are achievable in a relatively short period of time, and significantly reduce the growth rate of carbon emissions.
- Fuel and technology choices can make significant reductions in carbon emissions.
- Political upheaval, economic depressions, and war decrease carbon emissions at a very high cost to economic development.
- Predicting the future of either economic growth or carbon emissions by projecting current trends is a risky business.

What this analysis reveals most clearly is that it is necessary to recognize that all nations follow a unique development path that depends upon their own mix of the classical inputs of labor, capital and resources, plus the availability of technical and managerial capability. The political and economic structure of a nation is also crucial to creating the conditions necessary for moving along a development path towards a sustainable future.

The future development paths of individual nations need to converge in some sort of "circle of sustainability" that one hopes lies in the lower right-hand corner of the development plane. Because nations differ so strongly in their endowments of these inputs, there is little reason to assume that all will follow the paths of the North Americans, Western Europeans, or Japanese. If the world is really serious about implementing the Climate Treaty signed at the U.N. Conference on the Environment and Development in Rio de Janeiro to address carbon dioxide and other greenhouse gases, then governments, private companies, development agencies, and individual citizens must take seriously the lessons of industrial ecology and develop optimal development paths for all economies.

Acknowledgments

The authors wish to thank Thomas A. Boden and Gregg Marland of the Oak Ridge National Laboratory for assistance with the carbon dioxide data. We also wish to acknowledge the support of the National Science Foundation for initiating this work through grant number DIR-8911919, the U.S. Environmental Protection Agency under assistance agreement number GR 820301 to the Tufts University Center for Environmental Management, and the Northeast Regional Center for Global Environmental Change. Although this is supported research, the report reflects the views and findings of the authors and not necessarily those of any of the funding agencies.

References

Boden, T. A., R. J. Sepanski, and F. W. Stoss (eds.). 1991. *Trends '91: A Compendium of Data on Global Change*. Carbon Dioxide Information Analysis Center, Environmental Sciences Division, Oak Ridge National Laboratory, Oak Ridge, Tennessee, 665 pp.

Commoner, B., M. Corr, and P. J. Stamler. 1971. *Data on the United States Economy of Relevance to Environmental Problems*. Prepared for Committee on Environmental Alterations, American Association for the Advancement of Science, Washington, D.C.

Flavin, C. 1992. *Vital Signs 1992: The Trends That Are Shaping Our Future*. W. W. Norton & Company, New York, 60–61.

IPCC (Intergovernmental Panel on Climate Change). 1990. *Climate Change: The IPCC Scientific Assessment*. (J. T. Houghton, G. J. Jenkins, and J. J. Ephraums, eds.), Cambridge University Press, Cambridge, U.K.

Lélé, S. M. 1991. Sustainable development: A critical review. *World Development 19(6)*, 607–621.

Moomaw, W. R. 1991. Participating in the energy revolution. In *Technological Cooperation for Sustainable Development Seminar Papers*. Sixth Talloires Seminar on International Environmental Issues, sponsored by Tufts University, The Regional Environmental Center for Central and Eastern Europe, University of São Paulo in cooperation with The United Nations Conference on Environment and Development, Talloires, France, May 6–9, 1991, Tufts University, Medford, Massachusetts, 123–130.

Morss, E. R., and V. Morss. 1986. *The Future of Western Development Assistance*. Westview Press, Boulder, Colorado.

Repetto, R. 1992. Accounting for environmental assets. *Scientific American*, 94–100.

Summers, R., and A. Heston. 1991. The Penn world table (Mark V): An expanded set of international comparisons, 1950–1988. *Quarterly Journal of Economics 106*, 327–368.

Tullis, D. M. 1992. The Development Plane Model: Comparing The Soviet Union and China. MALD Thesis, The Fletcher School of Law and Diplomacy, Tufts University.

World Commission on Environment and Development. 1987. *Our Common Future*. Oxford University Press, Oxford, 43–66.

11

Reducing Urban Sources of Methane: An Experiment in Industrial Ecology

Robert Harriss

Abstract

Methane is a powerful greenhouse-forcing gas, and its main anthropogenic sources arise from leaks in natural gas distribution networks, landfills, and sewage treatment facilities. A new portable device for detecting and monitoring urban methane fluxes is described. Once detected, leaks can often be cheaply prevented, thus reducing costs and reducing urban carbon emissions. The redesign of cities to optimize industrial and domestic carbon metabolisms is suggested.

Reducing greenhouse gas emissions will be a primary task for the field of industrial ecology. Many nations have agreed to the principle of reducing greenhouse gas emissions, especially carbon dioxide emissions related to fossil fuel combustion. The Convention on Climate Change, endorsed by 153 countries at the U.N. Conference on Environment and Development (UNCED) in June 1992, requires industrial countries to develop national emission limits and emission inventories for greenhouse gases. Although the document signed at UNCED lacks emissions reduction targets and timetables, it embodies the goal of returning greenhouse gas emissions to "earlier levels" by the turn of the century (Parson *et al.*, 1992). Actions pledged by some industrial countries, principally in Western Europe, have called for either a freeze or a 20% reduction in carbon dioxide emissions by the year 2000 or soon thereafter. In February 1991, the U.S. government proposed that a comprehensive framework for greenhouse gases would be preferable to focusing on carbon dioxide. At present, the entire process of reaching an international consensus is bogged down in a mire of disagreements over goals, targets, timetables, and equity issues (e.g., Collins, 1991).

This chapter is based on the premise that realistic, cost-effective options are currently available to reduce methane emissions to a level which would stabilize the amount of methane in the global atmosphere within a few decades (Hogan *et al.*, 1991). In the longer term (50–100 years), it is likely that methane emissions could be further reduced to a level where the atmospheric concentration, and consequently the potential for climate change, could be significantly reduced. The best opportunities for near-term emissions reductions are in urban/industrial systems. By developing an industrial ecology approach to reducing methane emissions in cities of industrial countries, it may be possible to learn how to avoid future emissions in cities in developing countries where most of the world's future population growth will take place. Following a brief review of the role of methane as a greenhouse gas and as a driver of global environmental changes, the remainder of this chapter will focus on the design of an experiment using concepts of industrial ecology to identify realistic mitigation options for reducing methane emissions from cities.

Atmospheric Methane and Its Sources

Increasing concentrations of atmospheric methane are an important contributor to the potential for global warming, second only to carbon dioxide. In 1990, methane contributed approximately 18% of the annual increase in the thermal radiation trapping capacity (e.g., greenhouse effect) of the atmosphere. Methane is also about 35 times more effective at trapping heat per molecule than carbon dioxide for a time period of 20 years (Houghton *et al.*, 1992). Although the rate of increase in the global atmospheric concentration very recently seems to be slowing, it appears that concentrations in the atmosphere have doubled over the past two centuries and continue to rise at about 0.6% per year (Steele *et al.*, 1992).

There are several reasons for placing a high priority on searching for mitigation options to reduce methane emissions related to human activities. First, methane requires a reduction in emissions of only 10–15% to stabilize the global atmospheric burden. In contrast, carbon dioxide emissions would have to be reduced by 60 to 80% to stabilize the global atmospheric burden (Houghton *et al.*, 1992). Since fossil fuel energy, the primary source of increasing atmospheric carbon dioxide, is critical to the structure and functioning of industrial society, it is unlikely that major reductions of this magnitude will be possible in the near term. The strategy for methane emissions reduction discussed later in this chapter could achieve a 10–15% goal within a decade. Second, methane resides in the atmosphere for around 20 years, whereas carbon dioxide lasts about 200 years. Due to methane's higher global warming potential and shorter lifetime in the atmosphere, stabilization or reduction of methane emissions will have a rapid impact on reducing the risks of potential climate change. Finally, methane is a potential energy resource. If methane pollution related to human activities can be concentrated and captured in a cost-effective manner it can be used as a fuel, further increasing the benefits of the mitigation effort. Most of the methane derived from domestic and agricultural waste streams originates from biomass; if this methane replaces a fossil fuel and if the biomass is sustainably harvested, the net benefit includes reductions in both methane and carbon dioxide emissions to the atmosphere.

The most comprehensive assessment of global sources of atmospheric methane and their future dynamics has been conducted by the Intergovernmental Panel on Climate Change (IPCC). The IPCC data suggest that human activities are now responsible for approximately 70% of the global methane source (Houghton *et al.*, 1992). The documented natural sources include wetlands, termites, lakes, and coastal waters. Sources of methane related to human activities are illustrated in

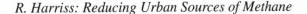

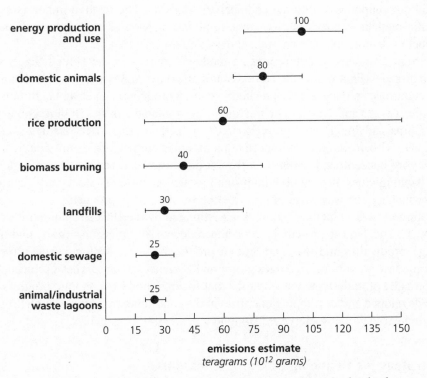

Figure 1. Global sources of atmospheric methane (best estimate and range) related to human activities (data from Houghton *et al.*, 1992).

Figure 1. Most of the methane associated with energy production and use, landfills, domestic sewage, and animal wastes can be linked directly to urban systems and their associated infrastructure. If approximately one-half of the methane leaking from natural gas production and distribution is lost in cities, and about half of the animal waste lagoons are associated with stockyards and meat-packing plants in close proximity to urban areas, the methane source associated with urban systems would be approximately 90 Tg/yr (1 Tg = 10^{12} grams), which would be 25% of the global anthropogenic source. In a preliminary attempt to estimate urban methane emissions, concentrations of methane in ambient air of a few cities have been compared to concentrations of industrial chlorofluorocarbons with relatively well known emission rates. This approach produced an estimate of approximately 30–60 Tg/yr from urban locations (Blake *et al.*, 1984). Whichever figure is taken, it is clear that methane stabilization could be achieved by cutting urban methane emissions alone. Stabilization could be achieved by cutting anthropogenic methane by about 30 Tg/yr, assuming other sources and the removal processes remain constant.

Estimates of methane emissions at global scales are not very useful for planning or implementing specific mitigation programs. Pollution control actions generally require a cost–benefit analysis based on quantitative data. The significant uncertainties and highly aggregated nature of the global source estimates in Figure 1 would not support a traditional cost–benefit analysis. The relative importance of specific methane sources at city, state, regional, or national scales depends on a number of demographic, economic, and ecosystem variables.

Current estimates of anthropogenic methane sources in the United States are summarized in Figure 2. In the United States and other highly urbanized and industrial countries, methane sources related to energy production and use, landfilling of organic wastes, and wastewater treatment are dominant. In developing countries like China and India, agricultural sources of methane (e.g., rice paddies, cattle) dominate. Urbanization is an important factor enhancing methane emissions due to the highly concentrated nature of urban solid wastes and sewage. Food systems provisioning cities also involve intensive production systems (stockyards, dairies, etc.) which require waste lagoons for treating sewage and byproducts.

Future growth in methane sources, as estimated by the IPCC, is summarized in Table 1. The largest growth is expected to occur in emissions from landfills, energy production and use, and animal production/waste systems. The sources with modest growth are rice agriculture and biomass burning. These projections reflect a set of underlying assumptions that include modest economic progress in both developed and developing countries and a continuing trend towards an urbanized/industrial global society (see Houghton *et al.*, 1992, Chapter A3).

A Strategy for Reducing Methane Emissions

General concepts and technical justification for reducing greenhouse gas emissions are becoming the focus of both international and national global change research planning. Top-down global and national analyses have indicated that considerable technical potential exists for reducing emissions of carbon dioxide and methane (e.g., Hogan *et al.*, 1991; Rubin *et al.*, 1992). However, the implementation of specific programs requires a strategy that can be subject to cost–benefit analysis at city-to-regional scales. In the case of methane, the urban sources, and perhaps coal mining, are the most likely targets for realistic mitigation actions. Agricultural sources are a much more complex area for mitigation. The scientific, engineering, and sociopolitical issues associated with changes in global food production systems are likely to require long-term research and discussion. In contrast, any realistic, cost-effective program which improves energy use or waste management in an urban environment, and at the same time reduces methane emissions, will encounter far fewer implementation problems.

Understanding Urban Methane Metabolism

Recent advances in measurement technologies and methods for quantifying urban methane emissions provide new opportunities for implementing an immediate program in the industrial ecology of this important greenhouse gas. The combination of fast-response laser detectors with high sensitivity to small variations in atmospheric methane concentrations (i.e., less than 0.3% of the amount of

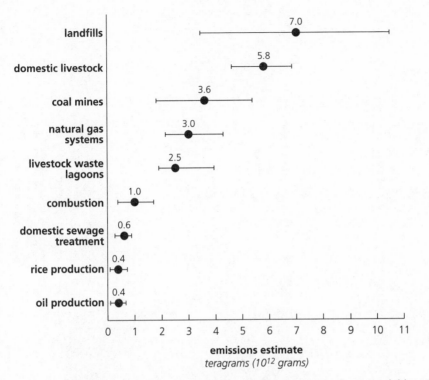

Figure 2. Preliminary U.S. methane emission estimates for sources related to human activities.

Table 1: *Present and projected global methane emissions related to human activities*

Source	1990		2050		2100		% Change
	Tg CH$_4$	%	Tg CH$_4$	%	Tg CH$_4$	%	1990–2100
Energy production and use	91	26	140	22	222	29	+144
Domestic animals	84	24	173	27	198	26	+136
Rice production	60	17	87	14	84	11	+40
Landfills	38	11	93	15	109	14	+187
Biomass burning	28	8	34	5	33	4	+15
Animal wastes	26	7	54	9	62	8	+138
Domestic sewage	25	7	47	8	53	8	+112
Total	351	100	630	100	762	100	+117

Data from Houghton *et al.*, 1992.

methane in ambient air) and global positioning system (GPS) technology, now makes rapid, accurate surveys of entire urban areas possible (McManus *et al.*, 1991). Figure 3 shows an example of a typical methane map for a medium-sized U.S. city with a population of 36,747 (1990). This map was derived from a single survey of the city during evening hours (10:00 p.m. to 2:00 a.m.), when traffic

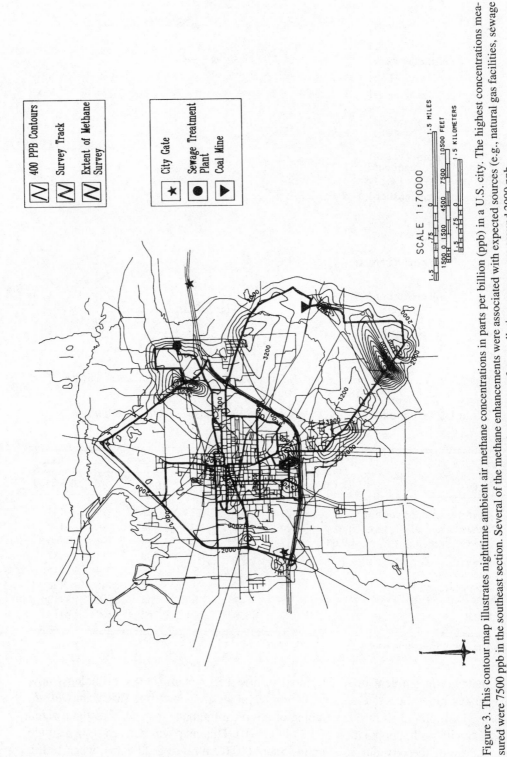

Figure 3. This contour map illustrates nighttime ambient air methane concentrations in parts per billion (ppb) in a U.S. city. The highest concentrations measured were 7500 ppb in the southeast section. Several of the methane enhancements were associated with expected sources (e.g., natural gas facilities, sewage treatment plant, and a coal mine). The concentrations at the edge of town, away from pollution sources, are around 2000 ppb.

was at a minimum and atmospheric conditions were relatively quiescent. A truck carrying atmospheric gas detectors for methane and carbon dioxide, and a GPS, conducted the survey to first locate all methane sources. Experience in a number of cities has demonstrated that city environmental and engineering officials are seldom aware of methane pollution sources. Even natural gas distribution companies have limited capabilities for locating and measuring small losses of methane (e.g., less than 1% of throughput) from their distribution systems. Emissions from obvious sources like landfills, anaerobic sewage sludge digestors, and gas compressor stations with pneumatic devices that are designed to leak are seldom known quantitatively. In the case of the city shown in Figure 3, ten potentially significant methane sources were identified for further study: eight components of the natural gas system, a landfill, and a sewage treatment plant. A tracer technique (Lamb *et al.*, 1992) was used to measure methane flux rates from each of the sources (Table 2). For this U.S. city it is clear that methane emissions from the landfill and sewage treatment plant were the most important urban methane sources. Leaks in the natural gas distribution system were equivalent to approximately 0.2% of the gas being put through the system. The relative importance of specific sources will vary in different communities, depending on factors like the age of the gas distribution system, waste management procedures and technologies, and attention to operation and maintenance of infrastructure. It is also important to keep in mind that repairing numerous small leaks in gas system pipeline connections may be a more cost-effective approach to reducing methane emissions than modifying a wastewater treatment plant. So, it is critical to have a mapping capability that locates and quantifies small and large emission sources.

The type of quantitative, source-specific data discussed in the above paragraph can support cost–benefit studies on options for controlling methane emissions. Such studies have not been conducted to date. However, these data do indicate the type of information and understanding that will be required for future studies on the industrial ecology of methane. Based on current data, a first step in any near-term strategy will be to conduct low-cost leak mitigation options related to natural gas production, transmission, and use. These programs are particularly important if natural gas use is going to increase substantially, as many experts expect. The U.S. Environmental Protection Agency is currently starting a voluntary program for gas companies to sign on for enhanced leak reduction programs. More ambitious leak reduction programs in the natural gas industry will involve infrastructure changes and modifications.

Table 2: *Methane emissions to the atmosphere from urban sources illustrated in Figure 3*

Source	Flux (liters/min)
Landfill	4600
Sewage treatment plant	2000
Natural gas system components	295

A second step in the mitigation of methane emissions is a considerably enhanced program of landfill gas recovery. Landfill gas-to-energy projects are currently active at approximately 157 sites around the U.S. (Governmental Advisory Associates, 1991). Best use of methane gas from landfills would require additional emphasis on optimizing methane production in large landfills to improve the economics of recovery (Barlaz *et al.*, 1990).

A strategy for reducing emissions from wastewater treatment would involve developing tight enclosures for facilities where anaerobic digestion processes produce methane concentrations of commercial value. The disposition of solid sludge from wastewater treatment is also a significant consideration, because of its potential as a substrate for methanogenesis. Studies should be conducted to consider how to best manage sewage sludge, domestic solid wastes, and certain industrial wastes as carbon resources on local to regional scales. A near-term industrial ecology approach to reducing methane emissions from this source may be to combine sewage sludge, domestic solid wastes, and certain industrial wastes in appropriate proportions to optimize methane production for energy. Methane mitigation options like these will have to be evaluated by multidisciplinary teams including engineers, economists, and others.

A longer-term strategy for reducing methane emissions from urban environments will focus on the redesign of cities to achieve optimal efficiency in carbon metabolism. This task will also require multidisciplinary studies at the local to regional scale, including engineering, socioeconomic, political, and ecosystem considerations. Several design issues should be given special attention. First, it seems obvious from the perspective of carbon metabolism that a compact, high-density urban design concentrates waste flows and improves the economics of waste-to-energy conversion. Second, a compact city provides a wide range of opportunities for reducing energy demands for transportation and space heating. The immediate negative response to a future world dominated by high-density megacities would include potential socioeconomic and pollution considerations. However, today's cities cannot be described as optimal either socially or environmentally, and the search for better solutions continues.

Summary

In this chapter, methane pollution has been treated as a "resource out of place." New technologies for accurately mapping and measuring methane leaks in urban environments provide an opportunity for intensive studies of the industrial ecology of methane. In the next decade, it is likely that some sources of methane pollution can be reduced with net economic benefit; reducing losses from natural gas systems and recovery of landfill gas for energy are two examples. In the medium to long term (50–100 years), it is suggested that the best strategy may be to enhance biogenic methane production as a primary pathway for urban and industrial carbon metabolism. This approach would ultimately lead to a redesign of cities and industrial production processes in order to most efficiently concentrate and integrate carbon-based waste flows. The reduction of methane emissions to the atmosphere, and increased substitution of biogenic methane-based fuels for oil and coal, will lead to significant reductions in the risk of potential climate changes through reductions in the global atmospheric greenhouse gas burden.

Acknowledgments

This chapter benefited greatly from the comments of two reviewers, Tom Graedel and Robin Cantor. The work mentioned on mapping urban sources of methane has been conducted in collaboration with colleagues at Aerodyne Research, Washington State University, and the University of New Hampshire. Diana Wright and Denise Blaha made important contributions to the typing and preparation of figures for the manuscript.

References

Barlaz, M. A., R. K. Ham, and D. M. Schaefer. 1990. Methane production from municipal refuse: A review of enhancement techniques and microbial dynamics. *Critical Review Environmental Control 19*, 557–584.

Blake, D. R., V. H. Woo, S. C. Tyler, and F.S. Rowland. 1984. Methane concentrations and source strengths in urban locations. *Geophysical Research Letters 11*, 1211–1214.

Collins, C. P. 1991. Climate change negotiations polarize. *Ambio 20*, 340–344.

Governmental Advisory Associates. 1991. *Methane Recovery from Landfill Yearbook.* Governmental Advisory Associates, Inc., New York, 408 pp.

Hogan, K. B., J. S. Hoffman, and A. M. Thompson. 1991. Methane on the greenhouse agenda. *Nature 354*, 181–182.

Houghton, J. T., B. A. Callander, and S.K. Varney (eds.). 1992. *Climate Change 1992: The Supplementary Report to the IPCC Scientific Assessment*, Cambridge University Press, New York, 198 pp.

Lamb, B., G. Allwine, L. Bamesberger, H. Westberg, B. McManus, J. Shorter, C. Kolb, B. Mosher, R. Harriss, and T. Howard. 1992. Measurement of methane emission rates from natural gas systems using a tracer flux approach. In *Proceedings of the International Gas Research Conference.* Orlando, Florida, 14 pp.

McManus, J. B., C. E. Kolb, P. M. Crill, R. C. Harriss, B. Mosher, B. Lamb, K. Maas, and J. Ryddock. 1991. Measuring urban fluxes of methane. *World Resource Review 3*, 162–183.

Parson, E. A., P. M. Haas, and M.A. Levy. 1992. A summary of the major documents signed at the earth summit and the global forum. *Environment 34*, 12–15.

Rubin, E. S., R. N. Cooper, R. A. Frosch, T. H. Lee, G. Marland, A. H. Rosenfeld, and D. D. Stine. 1992. Realistic mitigation options for global warming. *Science 257*, 148–266.

Steele, L. P., E. J. Dlugokencky, P. M. Lang, P. P. Tans, R. C. Martin, and K. A. Masarie. 1992. Slowing down of the global accumulation of atmospheric methane during the 1980s. *Nature 358*, 313–316.

12

Reducing Carbon Dioxide Emissions in Russia

Yuri Kononov

Abstract

The high energy intensity of the Russian economy has made it one of the world's largest emitters of atmospheric carbon. Macroeconomic models are used to assess the costs of reducing carbon dioxide emissions in Russia. The costs would be high, and would require major structural changes in the Russian economy.

Introduction

Growth in global fossil fuel consumption has led to the degradation of the atmosphere. Two particularly important strategies to counter these negative environmental trends may be identified: (1) a reduction of the energy intensity of the economy (the ratio of total energy consumption to gross national product, or GNP) and (2) a decrease of the share of energy consumption associated with fossil fuels, especially coal.

In the United States, Japan, and several other industrialized countries during the past two decades, energy consumption per unit of GNP has decreased on average by 1% to 2% annually. In the USSR and other countries where industrialization is in its initial phases, however, energy intensity has declined at a much slower rate. By 1990, consumption of primary energy per unit of GNP in the USSR was approximately 2.2 times higher than in the United States and 3 times higher than in Western Europe and Japan.[1]

At least half of the reduction in energy intensity achieved in the United States and Japan can be attributed to structural changes in the economy that did not occur in the USSR (Kononov *et al.*, 1992). The processes of structural change that reduce the energy intensity of a developing economy always proceed more slowly than the local and global degradation of the environment. Therefore, it is appropriate and urgent for countries to use economic tools to encourage environmentally responsible and energy-efficient technologies, and more generally to exert a positive influence on energy consumption and production.

---

[1] All ruble estimates in this article use early-1990 rubles, a time before the dramatic inflation of the ruble began. In early 1990 the real exchange rate between dollars and rubles was about 0.8–1.1.

Accordingly, to investigate policy interventions on behalf of environmental quality, we have been investigating policies that change the structure of the Russian economy, paying particular attention to feedback mechanisms between energy and the environment (Kononov, 1990a). Results for Russia, which accounted for 57% of Soviet GNP and 64% of its resource consumption in 1990, largely generalize to the Commonwealth of Independent States (CIS) as a whole, because the CIS and Russia, as well as the USSR of the past, all have almost identical economic and energy structures.

Effects of the Structure of the CIS Economy on Atmospheric Emissions

As seen in Table 1, a much larger share of total GNP is associated with industry, agriculture, and construction in the USSR than in the United States: 60% for the USSR (in 1989) but only 28% for the United States (in 1987). In fact, the structure of the USSR economy during the late 1980s resembled the U.S. economy of the late 1960s to early 1970s. The high inertia and low efficiency of the USSR economy is well captured by the slow pace of structural change.

Some of the consequences for carbon dioxide emissions of the structure of the economy captured in Table 1 can be inferred from Table 2, which displays, by economic sector, estimates for the USSR economy of the carbon dioxide emissions per unit of economic activity, arbitrarily normalized. Assuming that the economies of the United States and the USSR have similar carbon dioxide emissions intensities across sectors, and further assuming that the emissions for the service sector will be well below those of the reference sector (machine building), the service-intensive U.S. economy will produce much less carbon dioxide per unit of GNP. In fact, emissions of CO_2 are about 2.5 times higher per unit of GNP in the USSR than in the United States.

The emissions of the traditional air pollutants associated with energy use have been similarly high in the USSR: on a total mass basis, again about 2.5 times higher per unit of GNP in the USSR than in the United States. In 1990 the electric

Table 1: *Share of GNP by economic sector (%)*

Sector	USSR (1989)	USA (1987)
Industry	32	21
Agriculture	18	2
Construction	10	5
Transport and communication	6	9
Trade	12	16
Services	22	47
Total	100	100

Table 2: *Carbon dioxide emissions intensity by Soviet economic sector*

Sector	CO_2 Emissions per Unit of Output (% of reference sector)
Industry	
Machine building[1]	100
Chemistry	525
Metallurgy	835
Agriculture	295
Construction	145
Transport and communication	855
Trade	50
Services	n.a.

[1]Reference sector.
n.a. = not available.

power plants of the USSR released 56 million metric tons of pollutants: 12 million as solid particles, 16 million in the form of sulfur dioxide, 5 million as nitrogen oxides, 13 million as carbon monoxide, and 10 million in the form of hydrocarbons (*Narodnoye Khoziaistvo*, 1991). Atmospheric emissions from USSR automobiles, accounting for about 35% of total atmospheric emissions, were more than half of those in the United States in spite of arising from only about one-tenth as many automobiles. Moreover, pollution reduction in the CIS has occurred more slowly than in most industrialized countries.

Progress toward a market economy, including an increase of economic efficiency and an acceleration of economic restructuring, can change this picture. But history provides no clear precedents for the transformations that must occur in the CIS. Even assuming that this transition will be free from catastrophic ruptures, difficulties abound in establishing a "base case" scenario: a most likely path for the development of energy and economy in the CIS and its associated atmospheric emissions. Rather than choosing a single base-case scenario against which to test deliberate policy interventions, we have explored three variants, each covering the period from 1990 to 2010. Some characteristics of these base-case scenarios are summarized in Table 3.

The first (pessimistic) scenario is characterized by an enduring economic crisis and minimal expansion of oil and gas production and nuclear power. The national economy expands slowly, with GNP growth averaging 1.1% per year between 1990 and 2010. In the second (optimistic) scenario economic reforms are more successful and GNP growth is 2.7% per year. The third (maximum) scenario envisages active participation of foreign capital, intensive introduction of new technologies, more successful conversion of military industries, and more active and efficient foreign trade. All these factors combine to achieve a GNP growth of 3.3% per year.

Table 3: *Base-case scenarios*

Index	Scenario 1 (pessimistic)	Scenario 2 (optimistic)	Scenario 3 (maximum)
GNP	125	170	190
Energy/GNP	93	74	70
Primary energy[1]	116	126	133
Production			
Electricity	136	154	160
Oil	84	88	100
Natural gas	142	150	155
Coal	119	127	129
Nuclear	129	185	185
Emissions			
CO_2	102	108	115
NO_x	70	71	73
SO_2	60	64	65

[1]Derived as a product of the two previous entries.

Index in 2010 as percent of index in 1990.

The differences among scenarios for fuel and energy consumption are much smaller than for GNP, because the energy intensity of the economy falls more quickly as the rate of economic growth increases: an economy growing rapidly can allocate more resources to energy efficiency. Higher economic growth rates promote investments in technologies with low energy intensity and also stimulate market mechanisms for energy conservation.

All three of our scenarios show reduced emissions of nitrogen oxides, sulfur dioxide, and other traditional air pollutants, because all three allow for the introduction of pollution abatement technologies, and all three involve an increasing role for natural gas. But emissions of carbon dioxide by 2010 exceed those in 1990 by 2.5% in the pessimistic scenario, by 8% in the optimistic scenario, and by 15% in the maximum scenario.

Macroeconomic Effects of Meeting Emission-Reduction Targets for Carbon Dioxide

National environmental policies can greatly influence the prices of energy and other commodities, the relative rates of development across the sectors of the economy, and the evolution of capital and labor markets. Policies deliberately designed to reduce carbon dioxide emissions, for example, may elicit capital investment in energy conservation and in new energy sources greater than those envisioned in the base scenarios. Such additional investment, in turn, may lead to additional production of specialized energy equipment, construction materials, and other goods, as well as further investments in infrastructure, stimulating further demand for labor, capital, energy, and other goods. To the extent that the equipment and materials required for these investments are imported, the compensating exports will require still further capital investment and energy.

The effects of energy development on the economy and on energy consumption are nonlinear, involving multiple feedback loops. Ratios of rates of change (elasticities) are especially sensitive to the overall condition of the economy (its rate of growth and its flexibility) and to the balance of investments.

Each path of energy development, over time, brings about substantial changes in prices, in the pattern of consumption of goods and services, and in the state of the environment. While it would be very useful to create an aggregate index, a "quality of life," whose dynamics could be used to assess social costs and to compare scenarios, we have not yet developed an index of this kind in which we have confidence. Instead, for now, we are using final consumption as a surrogate index.

Strategies to Reduce Carbon Dioxide Emissions

Our modeling begins with a search for a set of efficient measures to reduce carbon dioxide. We use a system of simulation models, MAKROEN, developed at the Siberian Energy Institute (Kononov, 1990b), which maximize private and governmental consumption within constraints on the labor supply, the rate of development of each sector of the economy, and the rate of change of the structure of final consumption. Our models quantify the impact of specific measures on the production structure of the national economy, the energy supply system, and air quality.

Carbon dioxide emissions can be reduced by energy conservation, by substitution of natural gas for coal at power plants and boilers, and by substitution of nuclear or renewable energy for fossil fuels. We discuss each method in turn.[2] Incremental or more deep-seated technical improvements that would lead to large efficiency gains are not considered in any of the strategies modeled in this chapter. If such technical improvements also take place, much larger carbon dioxide reductions could be achievable.

Energy conservation is the most efficient method of reducing carbon dioxide emissions. Inasmuch as in all three base scenarios the lowest-cost energy conservation measures are already implemented by the year 2010, strategies to accelerate the reduction of carbon dioxide emissions involve more costly energy conservation measures, among which are thermal insulation of buildings and reductions of losses in urban heat supply systems.

To be sure, each particular conservation measure has its own optimal level of deployment, above which investments in additional energy supply are preferred. Marginal costs rise with the level of energy conservation accomplished, and the marginal value of the fuel saved decreases, so that economically justified

---

[2] A fourth method, sequestering the carbon dioxide produced in fossil fuel combustion instead of allowing it to reach the atmosphere, is judged to be far too expensive.

reductions of carbon dioxide emissions via energy efficiency investments are always bounded. We estimate that the removal from the year 2010 economy of the first 100 million metric tons of atmospheric carbon emissions (as carbon dioxide), relative to the base-case scenarios, can be achieved at a marginal cost of from 100 to 200 rubles per ton of carbon, the next 100 million metric tons at 400 to 800 rubles per ton, and a third 100 million metric tons at 1500 to 2200 rubles per ton— a highly nonlinear relationship between incremental capital investments in energy efficiency and incremental reductions in carbon dioxide emissions.[3]

Because the three base scenarios already envisage very high rates of development of the gas industry, our optimization models explore only relatively modest additional investments in the gas industry beyond those envisioned in the base-case scenarios. We estimate that, relative to the base-case scenarios, atmospheric carbon dioxide emissions can be reduced in 2010 by 40 to 80 million metric tons of carbon, corresponding to an increase in gas production of 100 to 200 billion cubic meters (4 to 8 EJ) per year, at marginal costs comparable to those incurred to achieve the second 100 million metric tons of carbon reductions from conservation (just discussed). However, such an increment in natural gas production might not be as advantageous for the climate as our simple analysis predicts if, at the same time, there are increases in the emissions of another greenhouse gas: methane. Unless the leaky Russian gas supply system (including wells and distribution networks) is improved, methane leakage may range from 3% to 7% of total gas production (Rabchuk and Ilkevich, 1991) (see Table 4).

Our models suggest that, at comparable marginal costs, even under the most favorable conditions, the installed capacity of hydropower and solar photovoltaic power plants can be increased in 2010 (relative to levels envisioned in the base-case scenarios, by no more than about 10 million kW) corresponding to a reduction of annual carbon dioxide emissions of no more than about 10 million metric tons of carbon. Rapid development of nuclear power might yield greater reductions of carbon dioxide emissions at less cost, but even if public opinion were to become more positively disposed toward nuclear energy, comparable constraints on marginal costs would limit additions to capacity to no more that another 35 million kW, reducing annual emissions by only another 50 to 60 million metric tons of carbon.

Macroeconomic Costs of Meeting Carbon Dioxide Reduction Targets

The negative consequences for the economy and society that result from introducing a constraint on carbon dioxide emissions grow nonlinearly as these constraints become more strict. Looking across scenarios, however, we find that these negative consequences are less severe when the economy is growing more rapidly,

---

[3] Since the Soviet economy emits roughly 1000 million metric tons of carbon to the atmosphere each year as carbon dioxide, each 100 million metric tons of carbon reduction is approximately a 10% reduction.

Table 4: *Methane leakage rates from the Soviet gas supply system (1989)*

Gas Supply Subsystem	Percent of Total Gas Production
Production	
Wells	0.8–1.8
Cleaning, drying, compressing	0.2–0.5
Collection networks	0.1–0.2
Subtotal	1.1–2.5
Transportation	
Underground storage	0.1–0.2
Compressor stations	0.2–0.5
Linear parts of main pipelines	0.9–1.7
Subtotal	1.2–2.4
Distribution and consumption	
Distribution networks	0.1–0.5
Industrial consumers	0.6–1.2
Residential and commercial consumers	0.1–0.1
Subtotal	0.8–1.8
Total routine losses	3.1–6.7
Accidents	0.2–0.3
Total losses including accidents	3.3–7.0

Data from Rabchuk and Ilkevich, 1991.

because then the economy is more flexible and can more easily adapt to a capital investment deficit. But energy demand and carbon dioxide emissions grow hand in hand, making it costly in terms of quality of life, no matter what the rate of economic growth, to meet a strict emission reduction "target." The target we have explored particularly closely is one where Russian emissions are reduced by 20% of 1990 emissions in 2010.

To achieve a 20% reduction of carbon dioxide emissions by the year 2010, relative to 1990 levels, will be painful. As seen in Table 5, such a 20% reduction would require an additional 160 billion rubles to be invested in the national economy if the economy is growing slowly (Scenario 1), or an additional 290 to 830 billion rubles if it is growing more rapidly (Scenarios 2 and 3). These investments exceed (by 30–60%) the total capital investments in both energy supply and energy conservation envisaged in the base scenarios. To achieve the target of 20% reduction by 2010 will also require immediate capital investments to expand the production of energy-saving equipment and associated materials, as well as major changes in the structure of production, with much greater levels of savings and much reduced private and government consumption. By contrast, to achieve a 10% reduction by 2010 would require capital investments three to eight times smaller: 50, 70, or 100 billion rubles in Scenarios 1, 2, and 3, respectively.

Table 5: *Incremental investments to reduce carbon dioxide emissions by 20%*[1]

Type of Investment	Scenario 1	Scenario 2	Scenario 3
Direct investments			
Energy conservation	+240	+370	+860
Energy supply	−80	−120	−180
NO_x and SO_2 reduction	−10	−20	−20
Indirect investments			
Energy conservation	+40	+80	+210
Energy supply	−20	−30	−50
Total investments	+160	+290	+830
Consumption in 2010 vs. base case	−6%	−8%	−20%

[1]In billions of early-1990 rubles.

Index in 2010 as percent of index in 1990.

The last row in Table 5 shows the sensitivity of economic growth rates to the severity of emission reduction targets adopted in Russia. The reductions in final (private plus state) consumption that accompany specific reductions of carbon dioxide emissions in 2010, relative to 1990 levels, are displayed for the three GDP growth scenarios. These show that a 20% reduction in CO_2 would lead to a 6% reduction in final consumption in a low-growth scenario, an 8% reduction in CO_2 in a moderate-growth scenario, and an almost 20% reduction under a high-growth scenario.

Conclusions

Investing in energy conservation is the most effective way to achieve large decreases in emissions of both carbon dioxide and traditional air pollutants. However, because even energy conservation investments are subject to diminishing returns, Russia must be alert to investments in energy efficiency that crowd out other socially useful capital investments, with adverse consequences for the economic welfare of the population.

High rates of economic growth and successful solutions of environmental problems can be achieved only with deep structural shifts in the Russian economy. The more ambitious the goals for environment quality, the faster must economic reforms be introduced that permit full development of market mechanisms, and the speedier must be Russia's transition to a postindustrial society.

Acknowledgments

Thanks are due to Elizabeth Economy for editing (beyond the call of duty) and to Battelle Pacific Northwest Laboratories for financial support.

References

Kononov, Yu. D. 1990a. Impact of economic restructuring on energy efficiency in the USSR. *Annual Review of Energy 15*, 502–512.

Kononov, Yu. D. 1990b. Trends and peculiarities in applying models to the planned-economy countries. *Energy 15(7/8)*, 715–727.

Kononov, Yu. D., H. G. Huntington, E. A. Medevedeva, *et al.* 1992. The effects of changes in the economic structure of energy demand in the USSR and the United States. In *International Energy Economics* (T. Sterner, ed.), Chapman and Hall, London, 47–64.

Narodnoye Khoziaistvo SSSR v 1990 (Statistical Annual of the Soviet Economy). 1991. Moscow, USSR, 273 pp.

Rabchuk, V. I., and N. I. Ilkevich. 1991. A study of methane leakage in the Soviet natural gas supply system. Working paper prepared for the Battelle Pacific Northwest Laboratory, Contract No. 139368-AK1, Washington.

13

Energy Efficiency in China: Past Experience and Future Prospects

Jiang Zhenping

Abstract

The Chinese economy is currently undergoing rapid growth. Continued wasteful use of energy (chiefly coal) is neither economically nor environmentally feasible. Thus there is an urgent need for increased energy efficiency in China. This chapter reviews recent efficiency trends and proposes new initiatives including further energy price reform and better information for consumers.

Introduction

During the 1980s the Chinese national economy grew rapidly, while at the same time there was a serious attempt to increase energy efficiency and improve the environment. Averaged over the decade, the annual growth rate of gross national product (GNP) was 8.9%, while annual growth rate of energy consumption was 5.1%. Capital was invested in environmental protection in the years 1986 to 1990 at an average rate of 0.7% of GNP.

Between 1949 and 1990 commercial energy production (consisting of coal, oil, natural gas, and hydropower) grew more than 40 times. In the same period the dominance of coal decreased somewhat, to 73% of commercial consumption. Total commercial energy consumption in China was 29.0 EJ (990 million tons of coal equivalent, or Mtce, in the common unit of energy policy discourse in China[1]), with the industrial sector dominating the transportation and buildings sectors to a much greater extent than in more highly industrialized countries. China is a self-contained energy economy: in fact, in 1990 energy production exceeded energy consumption by 5%.

Pollution from coal burning is still increasing, because pollution control measures are not keeping up with expanded use. The control of smoke and dust at the stack or chimney is incomplete, as is the control of emissions of sulfur dioxide—leading some regions to suffer from acid rain.

---

[1] One "ton of coal equivalent," or tce, is 7×10^9 calories, or 29.3×10^9 Joules. One EJ = 10^{18} Joules.

Nonetheless, environmental protection efforts are both qualitatively and quantitatively improving. During the 1980s, the emphasis in pollution control changed from treatment of point sources to comprehensive regional pollution control, and from treatment of pollution by dispersal to reduction in total quantity of pollution emitted. Whereas before pollution was treated as a purely administrative matter, now one finds the integration of legislation, economic incentives, technology, and science. In addition, pollution control is more centralized and uniform across the country.

In parallel with these changes in environmental protection strategy, there has been a strong, complementary focus on energy efficiency.

Energy Efficiency Achievements in the 1980s

China first articulated energy conservation targets only in 1977, about 5 to 7 years after most other nations. The consequences of policy inattention and structural differences can be seen in data on energy intensity (the ratio of energy use to GNP) in the period from 1972 to 1977: during this five-year interval energy intensity fell at an annual rate of 1.4% in developed countries, while in China it rose 2.8%. In 1980 Chinese authorities adopted an energy policy that gave equal emphasis to the development of energy supplies and to conservation. Shortly after 1978, the rate of change of energy intensity in China crossed through zero, and energy intensity has continued to decrease ever since.

Energy conservation policy in China in the 1980s took several forms. Energy efficiency was listed explicitly as an objective in the Sixth and Seventh Five-Year Plans (1981–85 and 1986–90). A network for the management of energy conservation was established throughout China. All enterprises were required to implement technical innovations that improved performance and quality of products while consuming less energy and raw materials during production. Rewards and penalties were introduced, related to energy management. An energy conservation fund was established in the national budget to subsidize energy conservation projects with coincident social benefits, where the enterprises could not afford these investments themselves.

There have been systematic efforts to improve energy efficiency in industry. Priority has been given to improving energy conversion equipment, such as industrial boilers, furnaces, and kilns, but also district heating systems, cogeneration systems (producing both heat and electricity), and stoves. Active financial support has been given to environmentally beneficial projects like town gas and district heating. More than a thousand new technologies of energy conservation have been introduced from abroad. The renovation of specific industries has been targeted, including fertilizer, cement, and steel, with uneven success: between 1980 and 1990 the energy used to produce a ton of steel fell from 60 GJ (2.04 tce) to 47 GJ (1.62 tce), but the energy used to produce a ton of cement fell only from 6.0 GJ (0.205 tce) to 5.9 GJ (0.200 tce).

In rural areas energy conservation strategies and renewable energy strategies were intertwined. By 1990 the installed capacity of mini-hydropower stations reached

12.4 GW and the number of households using biogas had reached 5 million. Wood-saving stoves were in use in 120 million households, half of all rural households.

Strategies to Improve Future Energy Efficiency

In the next ten years China will face the dual tasks of developing its economy and protecting its environment. Crucial to its success will be further strides in improving energy efficiency. Some improvements will be achieved with relative ease, as the structure of the Chinese economy becomes more heavily weighted to service industries and, generally, begins to resemble more closely the structure of highly industrialized countries.

But other improvements in energy efficiency will have to be achieved deliberately. There are still large gaps between the efficiency of energy use in many industries in China and abroad—gaps that can often be traced to specific pieces of equipment like fans, compressors, and boilers, or to systems control by sensors and computers. Within specific industries China's lagging industrialization also shows up in differences in product mix: in steel, more iron and less steel, more ordinary steel and less alloy steel, more plate and less pipe; in chemicals, a low proportion of production of synthetic rubber and tires. Highly decentralized, small-scale production facilities account for an anomalously high share of production in industries like steel, copper, aluminum, and calcium carbide; such facilities are frequently more polluting and less energy efficient, per ton of output, than larger plants. The share of such facilities in total production will have to decrease.

Industrial boilers, industrial heat supply systems, and industrial ovens all must receive targeted attention, for they account for 40% of total energy use today. Again, one strategy will be to replace smaller units with larger ones. Another strategy will be to extend the role of cogeneration facilities. A third strategy will be to improve waste heat recovery. Industrial integration will also have to be improved, because in China today the design, manufacturing, and use of industrial ovens and furnaces are not coordinated, and there is no coherent implementation of research and development.

There are also important opportunities for improvements in the efficiency of production of electricity, which accounts today for 20% of China's primary energy consumption. Currently, the average coal consumption per unit of electricity produced is 20% higher in China than in developed countries, and even for the most advanced units in China the coal consumption is 10% higher. Here, too, China must deal with a very large number of small plants: power plants with capacity exceeding 200 MW constitute only 38% of total installed capacity, whereas in Japan the corresponding figure is over 80%.

Conclusions

The successful implementation of the next steps in national energy efficiency will depend especially on five developments:

- Investments in energy conservation research and development should be increased. Conservation projects must be demonstrated and popularized, and technologies of energy efficiency must be promoted.
- An energy conservation fund should be established, whose purposes should include receiving and managing foreign investments, offering low-interest or interest-free loans for enhanced energy conservation projects, and conducting analyses and evaluations of investments in energy efficiency.
- Existing energy conservation rules and regulations should be enforced, and new laws should be developed. Of particular importance are laws that encourage a new service industry providing energy conservation services.
- Stronger efforts are needed to popularize the opportunities available to achieve energy efficiency and to connect energy efficiency with environmental quality in the public consciousness.
- The rationalization of energy prices is essential, inasmuch as prices are the most sensitive means for adjusting supply and demand, controlling waste, and promoting energy efficiency. Today in China, in spite of ten years of efforts like multitrack pricing that have raised average energy prices part of the way toward world market prices, the energy price system is still severely distorted. Prices for coal, oil, and electricity are still set within an economic plan that keeps primary energy costs low relative to the costs of downstream products. The result is to weaken the incentives for energy conservation in industry. In the years ahead, as China becomes more and more open to the outside world, its economy will become more and more market-based, and energy prices will increase step by step. Some price changes will be large, and these will bring about significant and necessary adjustments in the Chinese economy as a whole.

Suggested Readings

Bai, X. The position of China in signing the Treaty of Global Climate Change and Counter-measures to Decrease the Emission of Carbon-Dioxide. In *Proceedings of Coal and Environment International Conference*, Beijing, December 1991. UN, State Science and Technology Commission of China, in press.

Fang, L., Y. Wang, X. Bai, and R. Hu. 1992. *National Report of the People's Republic of China on Environment and Development*, China Environmental Science Press, Beijing, 81 pp.

Jiang, Z., X. Zhuang, K. Jiang, J. Li, and Z. Xie, *et al. Energy Data Base of China* (in preparation).

Liu, X., Z. Jiang, and X. Hu. 1988. *Energy Intensity in China and the World*, China Bright Daily Press, Beijing, 298 pp.

Lu, Q., J. Zhao, and L. Cui. 1991. *Energy Statistics Yearbook*, China Statistics Press, Beijing, 413 pp.

Qu, G. Control over the coal-burning environmental pollution. In *Proceedings of the Efficiency of Coal Using and Environmental Imports in China*, World Bank and Ministry of Energy, Beijing, April 1992, in press.

Sinton, J., Z. Jiang, M. Levine, K. Jiang, F. Liu, and X. Zhuang. *China Energy Source Book*, Lawrence Berkeley Laboratory, Berkeley, California, in press (in English).

Xia, G., W. Song, S. Qu, and Z. Jiang. 1990. *Energy in China (Special Issue, Part 2). Journal of the Energy Exploration and Exploitation 8(4)*, Multi-science Publishing Co. Ltd., England, 263–308.

Xing, D., B. Wang, Z. Jiang, and D. Zhou. 1987. *Energy Conservation of China in the Period of the Sixth Five-Year Plan (1981-1985)*, China Energy Press, Beijing, 211 pp.

Yi, Q., L. Zhu, Z. Jiang, and J. Zhao. *Handbook for Resources Comprehensive Utilization*, China Science and Technology Press, 1129 pp.

Zhou, D., and Z. Liu. *Energy Conservation and Its Policies in China*, in press.

Zhou, F., D. Qiu, and X. Liu. 1991. *Regional Study of Environmental Considerations in Energy Development Project (PRC)*, Final Report (Draft). Asian Development Bank T.A. No. 5357-Regional.

14

Roles for Biomass Energy in Sustainable Development

Robert Williams

Abstract

Advanced technologies such as gasifier/gas turbine systems for electric power generation and fuel cells for transportation make it possible for biomass to provide a substantial share of world energy in the decades ahead, at competitive costs. While biomass energy industries are being launched today using biomass residues of agricultural and forest product industries, the largest potential supplies of biomass will come from plantations dedicated to biomass energy crops. In industrialized countries these plantations will be established primarily on surplus agricultural lands, providing a new source of livelihood for farmers and making it possible eventually to phase out agricultural subsidies. The most promising sites for biomass plantations in developing countries are degraded lands that can be revegetated. For developing countries, biomass energy offers an opportunity to promote rural development.

Biomass energy grown sustainably and used to displace fossil fuels can lead to major reductions in carbon dioxide emissions at zero incremental cost, as well as greatly reduced local air pollution through the use of advanced energy conversion and end-use technologies. The growing of biomass energy crops can be either detrimental or beneficial to the environment, depending on how it is done. Biomass energy systems offer much more flexibility to design plantations that are compatible with environmental goals than is possible with the growing of biomass for food and industrial fiber markets. There is time to develop and put into place environmental guidelines to ensure that the growing of biomass is carried out in environmentally desirable ways, before a biomass energy industry becomes well established.

Introduction

Biomass (plant matter) has been used as fuel for millennia. In the 18th and 19th centuries it was widely used in households, industry, and transportation. In the United States, as late as 1854, charcoal still accounted for nearly half of pig iron production, and throughout the antebellum period wood was the dominant fuel for

199

both steamboats and railroads (Williams, 1989). Biomass dominated global energy consumption through the middle of the 19th century (Davis, 1990). Since then biomass has accounted for a diminishing share of world energy, as coal and later oil and natural gas accounted for most of the growth in global energy demand. Today biomass is not much used by industry, though it is still widely used for domestic applications in developing countries—especially in rural areas (Hall *et al.*, 1993). Still, biomass accounts for about 15% of global energy use, only slightly less than the share of global energy accounted for by natural gas (see Figure 1).

Although the trend has been away from biomass as an energy source, there are strong reasons for revisiting biomass energy:

- Dependence on gasoline and diesel for transport fuels has led to urban air pollution problems in many areas that cannot be solved simply by mandating further marginal reductions in tailpipe emissions. California has adopted a policy mandating the phased introduction of low- and zero-emission transport vehicle/fuel systems. Other jurisdictions are likely to pursue similar policies (Wald, 1992), and in fact 12 eastern U.S. states in 1994 collectively asked the U.S. Environmental Protection Agency to impose the California regulations on them (Wald, 1994). Some biomass-based transport energy options could effectively address this challenge (Johansson *et al.*, 1993; Williams, 1994).
- The prospect of declining future production of conventional oil in most regions outside the Middle East (Masters *et al.*, 1990) once more raises concerns about the security of oil supplies. Fluid fuels derived from biomass substituted for imported oil can help reduce energy security risks (Johansson *et al.*, 1993).
- Responding to concerns about global warming may require sharp reductions in the use of fossil fuels (IPCC, 1990). Biomass grown sustainably and used as a fossil fuel substitute will lead to no net buildup in atmospheric carbon dioxide, because the CO_2 released in combustion is compensated for by the CO_2 extracted from the atmosphere during photosynthesis.
- A major challenge facing developing countries is to find ways to promote rural industrialization and rural employment generation, to help curb unsustainable urban migration (Goldemberg *et al.*, 1988, 1987). Low-cost energy derived from biomass sources could support such activities (Johansson *et al.*, 1993).
- There are large amounts of deforested and otherwise degraded lands in tropical and subtropical regions in need of restoration (Grainger, 1988). Some of these lands could be restored by establishing biomass energy plantations on them. Part of the revenues from the sale of biomass produced on such lands could be used to help pay for these land restoration efforts (Hall *et al.*, 1993; Johansson *et al.*, 1993).
- In industrialized countries, efforts to provide food price and farmer income stability in the face of growing foodcrop productivities has led to a system of large-scale agricultural subsidies. Despite mounting economic pressures to reduce or eliminate such subsidies, so doing is difficult politically (OECD, 1992). However, converting excess agricultural lands to biomass production for energy

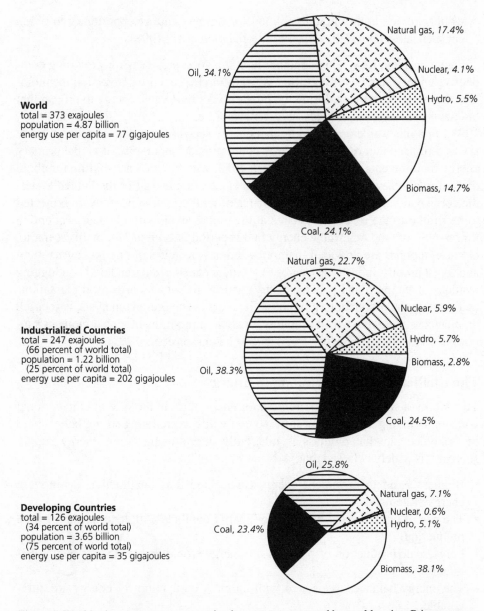

World
total = 373 exajoules
population = 4.87 billion
energy use per capita = 77 gigajoules

Industrialized Countries
total = 247 exajoules
  (66 percent of world total)
population = 1.22 billion
  (25 percent of world total)
energy use per capita = 202 gigajoules

Developing Countries
total = 126 exajoules
  (34 percent of world total)
population = 3.65 billion
  (75 percent of world total)
energy use per capita = 35 gigajoules

Figure 1. World primary energy consumption by energy source and by world region. Primary energy consumption is shown for the world (top), industrialized countries (middle), and developing countries (bottom) in 1985. Data from all energy sources except biomass are from Johansson *et al.* (1993). Biomass energy data are estimates based on surveys, from Hall *et al.* (1993).

The primary energy associated with electricity produced from nuclear and hydroelectric sources is assumed to be the equivalent amount of fuel required to produce that electricity, assuming the average heat rate (in MJ per kWh) for all fuel-fired power-generating units.

would provide both a new livelihood for farmers and an opportunity to phase out such subsidies (Hall *et al.*, 1993; Johansson *et al.*, 1993).

Such considerations, taken together with the good prospects for providing competitive energy supplies from biomass using modern energy conversion technologies, led a recent study exploring the prospects for renewable energy to project that biomass can have major roles as a renewable energy source (Johansson *et al.*, 1993). In a renewables-intensive global energy scenario constructed for that study it was estimated that renewable energy could provide about 45% of global primary energy requirements in 2025 and 57% in 2050, with biomass accounting for about 65% of total renewable energy in both years (see Figure 2). For the United States, the corresponding renewable energy shares of total primary energy were projected to be similar to the renewable shares at the global level, with biomass accounting for 55–60% of total renewable energy in this period (see Figure 3). In this scenario, biomass supplies are provided mainly by biomass residues of ongoing agricultural and forest product industry activities (e.g., sugar cane residues and mill and logging residues of the pulp and paper industry) and by feedstocks grown on plantations dedicated to the production of biomass for energy. The present analysis, is focused on plantation biomass, which accounts for about three-fifths of global biomass supplies in this scenario in the period 2025–50 (Johansson *et al.*, 1993).

The Challenges Posed by Biomass Energy

The notion of shifting back to biomass for energy flies in the face of conventional wisdom. Bringing about such a shift would require overcoming strong beliefs held by many people that biomass is inherently unpromising as an energy supply source. It is widely believed that:

- Biomass is an inconvenient energy carrier and thus unattractive for modern energy systems.
- The use of land to grow biomass for energy conflicts with land needs for food production.
- Large-scale production of biomass for energy would create environmental disasters.
- The energy balances associated with biomass production for energy are unfavorable.
- Biomass energy is inherently more costly than fossil fuel energy.
- Resource constraints will limit biomass to a minor role in a modern global energy system.

In what follows each of these concerns is dealt with in turn.

Attracting Consumer Interest by Modernizing Biomass Energy

Biomass is often called "the poor man's oil" (Goldemberg *et al.*, 1988, 1987). This characterization arises in part from the low bulk density of biomass fuels. Freshly

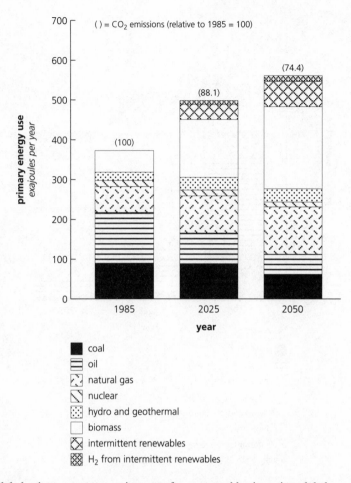

Figure 2. Global primary energy requirements for a renewables-intensive global energy scenario. This figure shows global primary energy requirements for the renewables-intensive global energy scenario developed in Johansson *et al.* (1993) in an exercise carried out to indicate the future prospects for renewable energy for each of 11 world regions. In developing this scenario, the high economic growth/high energy efficiency demand projections for solid, liquid, and gaseous fuels and electricity developed by the Response Strategies Working Group of the Intergovernmental Panel on Climate Change (Response Strategies Working Group of the Intergovernmental Panel on Climate Change, 1990) were adopted in Johansson *et al.* (1993) for each world region. For each region a mix of renewable and conventional energy supplies was constructed in Johansson *et al.* (1993) to match these demand levels, taking into account relative energy prices, regional endowments of conventional and renewable energy sources, and environmental constraints.

The primary energy associated with electricity produced from nuclear, hydroelectric, geothermal, photovoltaic, wind, and solar thermal-electric sources is assumed to be the equivalent amount of fuel required to produce that electricity, assuming the average heat rate (in MJ per kWh) for all fuel-fired power-generating units in a given year. This global average heat rate is 8.05 MJ per kWh in 2025 and 6.65 MJ per kWh in 2050.

For biomass-derived liquid and gaseous fuels the primary energy is the energy content of the biomass feedstocks delivered to the biomass energy conversion facilities.

Primary energy consumption in 1985 includes 50 EJ of noncommercial biomass energy (Hall *et al.*, 1993). It is assumed that there is no noncommercial energy use in 2025 and 2050.

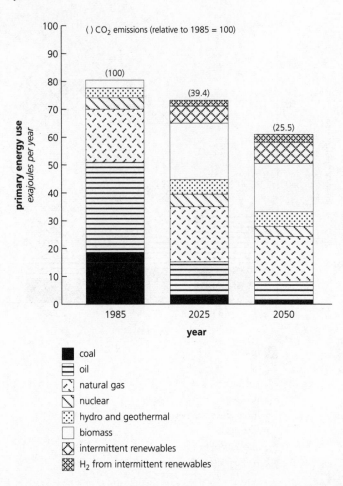

Figure 3. Primary energy requirements for the United States in a renewables-intensive global energy scenario. This figure shows primary energy requirements for the United States in the renewables-intensive global energy scenario developed in Johansson *et al.* (1993) in an exercise carried out to indicate the future prospects for renewable energy for each of 11 world regions, one of which is the United States. In developing this scenario, the high economic growth/high energy efficiency demand projections for solid, liquid, and gaseous fuels and electricity developed by the Response Strategies Working Group of the Intergovernmental Panel on Climate Change (1990) were adopted in Johansson *et al.* (1993). For the United States and other industrialized countries, this demand scenario involves a slow decline in primary energy demand as the economy expands, as a result of the emphasis given to improved energy efficiency. The mix of renewable and conventional energy supplies shown was constructed in Johansson *et al.* (1993) to match these demand levels, taking into account relative energy prices, endowments of conventional and renewable energy sources, and environmental constraints.

The primary energy associated with electricity produced from nuclear, hydroelectric, geothermal, photovoltaic, wind, and solar thermal-electric sources is assumed to be equivalent to the amount of fuel required to produce that electricity, assuming the average heat rate (in MJ per kWh) for all fuel-fired power-generating units in a given year. The U.S. average heat rate is 8.07 MJ per kWh in 2025 and 6.42 MJ per kWh in 2050.

For biomass-derived liquid and gaseous fuels the primary energy is the energy content of the biomass feedstocks delivered to the biomass energy conversion facilities.

cut wood typically has an energy density of about 10 GJ per tonne—compared with 25–30 GJ per tonne for various coals and more than 40 GJ per tonne for oil; it is thus both difficult and costly to transport biomass fuels long distances; in rural areas of developing countries women and children spend considerable time gathering fuelwood for cooking. Wood cookstoves also pollute—generating in rural kitchens of developing countries total suspended particulates, benzo-a-pyrenes, and other pollutants—often at levels far in excess of ambient air quality standards (Smith and Thorneloe, 1992).

As incomes rise, consumer preferences shift toward energy carriers of higher quality. The higher the quality of the fuel, the more convenient is its use and the less pollution is generated. This phenomenon is well-known in cooking fuels: charcoal is preferred to wood, kerosene is preferred to charcoal, and clean gaseous fuels such as liquid petroleum gas (LPG) are preferred to kerosene. This "energy ladder" is often invoked to show that consumers will shift away from biomass fuels as their incomes rise. For instance, data show that biomass accounts for 38% of energy use in developing countries (used mostly by poor people in rural areas), but just 3% of energy use in industrialized countries (see Figure 1).

However, with modern technologies, biomass can be converted into liquid or gaseous fuels or into electricity, in cost-effective ways. It is in these forms that biomass becomes an acceptable energy source at high-income levels.

Addressing the Food Vs. Fuel Controversy

The renewables-intensive global energy scenario developed in Johansson *et al.* (1993) calls for establishing worldwide some 400 million hectares of biomass plantations for energy by the second quarter of the 21st century—a land area that is not small compared with the nearly 1500 million hectares now in cropland (WRI, 1992). Because the world population is expected to nearly double by that time, the potential for conflict between biomass production for food and biomass production for energy warrants careful scrutiny. Because land is needed to grow food, but energy can be provided in many ways, food production should have priority. The key questions are: How much land is needed for food production? And how does this need compare with the arable land resource? In addressing these questions it is useful to consider the industrialized and developing country situations separately.

Industrialized Countries

Because their population growth is slow and food yields have been increasing, the amount of land needed for food production is declining in industrialized countries.

In the United States, more than one-fifth of total cropland, some 33 million hectares, was idled in 1990, either to keep food prices high or to control erosion. The U.S. Department of Agriculture forecasts that an area of over 50 million hectares may be idle by 2030 as a result of rising crop yields, despite an expected doubling of exports of corn, wheat, and soybeans (Soil Conservation Service, 1989). The urgency of addressing the challenge of excess agricultural lands was the major theme of the 1987 report of The New Farm and Forest Products Task Force to the Secretary of Agriculture:

> The productive capacity of U.S. agriculture is greatly underutilized. The country today has carryover stocks of between six months and one years production of major commodities, with productivity continuing to increase at a faster rate than demand. Estimates of land in excess of production needs to meet both domestic and export market demand range as high as 150 million acres [61 million hectares]—with about one-third of that already available from the Conservation Reserve Program. This represents an enormously wasted national asset which, if transformed into a more productive one through new products, would have a profoundly positive impact on the Nation's economy.

In the European Union more than 15 million hectares of land will have to be taken out of farming by the year 2000, if surpluses and subsidies associated with the Common Agricultural Policy are to be brought under control (Hummel, 1988). By 2015, according to a Dutch study, the land needed for food production in the community could be 50 to 100 million hectares less than at present (Netherlands Scientific Council for Government Policy, 1992). In the United States and the European Union together, therefore, 100 million hectares of farmland or more could be idle by the second decade of the next century, which is more than one-third of the total land dedicated to agricultural production today.

While the conversion of excess cropland in the industrialized countries to energy plantations presents an opportunity to make productive use of these lands, such a conversion cannot be easily accomplished under the present policies. In many countries farmers are deterred by a subsidy system that specifies what crops the farmer can produce in order to qualify for a subsidy; and energy crops are not allowed.

In 1991 this subsidy system transferred about $320 billion to farmers (170 billion in Western Europe, 80 billion in the United States, 60 billion in Japan, and 10 billion in Canada, in current U.S. dollars [OECD, 1992]). These subsidies, amounting to almost $400 per capita for the 800 million people living in the countries of the Organization of Economic Cooperation and Development, actually rose between 1987 and 1991, in spite of serious political efforts to reverse the tide (see Table 1). The calculation of these subsidies combines costs to consumers in the form of higher food prices (about $200 billion) and direct payments to farmers from taxpayer revenue (about $140 billion), and subtracts revenues from tariffs (about $20 billion). The $80 billion subsidy in the United States is about one-fifth of the total retail expenditure on energy sources in the United States (Energy Information Administration, 1991).

Gradual conversion of surplus cropland to profitable biomass energy production would make it possible to phase out many of these subsidies. As long as a system of subsidies continues, however, the bias against energy crops should be removed.

206

Table 1: *Subsidies to agricultural producers in OECD countries (current dollars)*

	Total Transfers ($ billions)		Per Capita Transfers ($)	
	1987	1991	1987	1991
Western Europe	139	166	390	440
EU (12 countries)[1]	119	142	370	410
Non-EU (5 countries)[2]	20	24	630	740
U.S.	81[3]	81	330	320
Japan	66	63	540	510
Canada	9	10	340	350
Australia and New Zealand	1	1	40	60
Total OECD	295	321	360	380

[1] European Union countries: Belgium, Denmark, France, Germany, Ireland, Italy, Luxembourg, Netherlands, United Kingdom, Spain, Greece, and Portugal.
[2] Non-European Union countries: Austria, Finland, Norway, Sweden, and Switzerland.
[3] For comparison, U.S. retail expenditures on energy were $394 billion in 1987 (Energy Information Administration, 1991).

From OECD, 1992.

Developing Countries

For developing countries the situation is quite different. Because of expected population growth and rising incomes, it is likely that more land will be needed for food production. The Response Strategies Working Group of the Intergovernmental Panel on Climate Change has projected that the land in food production in developing countries will increase 50% by 2025 from the present level of about 700 million hectares (see Table 2) (IPCC, 1991). The demand can be compared to potential supply—that is, land physically capable of supporting economic crop production, within soil and water constraints. For 91 developing countries, potential cropland was estimated to be about 2000 million hectares—nearly three times present cropland (see Table 2) (FAO, 1991).

Looking to the year 2025 and assuming cropland requirements in developing countries increase 50% by then, there would still be a substantial surplus of potential cropland of nearly 1000 million hectares in these countries (see Table 2). There would be substantial regional differences, however, with major surpluses totaling more than 1100 million hectares in Latin America and Africa, and a 110 million hectare deficit in Asia. (China was not included in the U.N. Food and Agricultural Organization [FAO] analysis.) Thus it appears that substantial amounts of land suitable for energy plantations may be available in both Latin America and sub-Saharan Africa, even with major expansions of cropland to feed the growing population. But in Asia, with its high population density, conflicts with food production could become significant.

Table 2: *Present[1] and potential[2] cropland for 91 developing countries (million hectares)*

Region	Present Cropland	Potential Cropland						Total
		Low Rainfall	Uncertain Rainfall	Good Rainfall	Natural Flooded	Problem Land	Desert	
Central America	38	2.2	13	19	5.7	31	3.5	75
South America	142	26	38	150	106	493	2.8	815
Africa	179	73	97	149	71	358	3.8	753
Asia (excl. China)	348	60	67	67	81	118	20	413
Total	706	161	215	386	263	1000	30	2055

[1] From WRI, 1992.

[2] As estimated by the U.N. Food and Agriculture Organization (FAO) in 1990 (FAO, 1991). Potential cropland is defined by the FAO as all land that is physically capable of economic crop production, within soil and water constraints. It excludes land that is too steep or too dry or having unsuitable soils.

The extent of conflict with food production depends on future food crop productivities. Waggoner (1994) aruges that with feasible productivity gains a world population of 10 billion could be supported with no increase in cropland. Assessments are needed, country by country, to better understand the prospects for productivity gains and thereby the avoidance of food/fuel conflict.

Unfortunately, the FAO study does not clarify where new cropland would come from. To be sure, some forestlands are involved. Clearly, it would not be desirable to cut down virgin forests in favor of intensively managed biomass plantations. Cutting down virgin forests could be avoided, however, by targeting for biomass plantations lands that are deforested or otherwise degraded and that are suitable for reforestation. One estimate is that over 2000 million hectares of tropical lands have been degraded, of which about 600 million hectares are judged suitable for reforestation (see Table 3).

Outside of Asia the amount of degraded land suitable for reforestation (excluding degraded lands in the desertified drylands category), is substantial (see Table 3) —some 112 million hectares in Latin America and 62 million hectares in Africa. In Asia, such land areas are also large—some 115 million hectares; however, for Asia, country-by-country assessments are needed to determine the extent to which its degraded lands will be needed for food production or other purposes warranting higher priority than energy.

The main technical challenge of restoration is to find a sequence of plantings that can restore ground temperatures, organic and nutrient content, moisture levels, and other soil conditions to a point where crop yields are high and sustainable. Successful restoration strategies typically begin by establishing a hardy species with the aid of commercial fertilizers or local compost. Once erosion is stabilized and ground temperatures lowered, organic material can accumulate, microbes can

Table 3: *Geographical distribution of tropical degraded lands and potential areas for reforestation (million hectares)*

Region	Logged Forests	Forest Fallows	Deforested Watersheds	Desertified Drylands	Total
Latin America	44	85	27	162	318
Africa	39	59	3	741	842
Asia	54	59	56	748	917
Total	137	203	87	1650	2077
Area suitable for reforestation	–	203	87	331	621

From Grainger, 1988, 1990.

return, and moisture and nutrient properties can be steadily improved (OTA, 1992; Parham *et al.*, 1993).

If it is feasible to overcome this technical challenge and various other socioeconomic, political, and cultural challenges (Hall *et al.*, 1993), plantation biomass in developing regions could make substantial contributions to world energy without serious conflict with food production. In sub-Saharan Africa and Latin America, where potential land areas for plantations are especially large, biomass could be produced by the second quarter of the 21st century in quantities large enough to make these regions major exporters of biomass-derived liquid fuels, offering competition to oil exporters and bringing price stability to the global liquid fuels market (Johansson *et al.*, 1993). A possible interregional fuels trade scenario for 2050 is shown in Figure 4.

Converting such large areas of degraded lands to successful commercial plantations would be a formidable task. Research is needed to identify the most promising restoration techniques for all the different land types and conditions involved. Yet the fact that many of the successful plantations in developing countries have been established on degraded lands (Hall *et al.*, 1993) suggests that it may be feasible to deal with these challenges with adequate research and commitment. And interest in restoring tropical degraded lands is high, as indicated by the ambitious global net afforestation goal of 12 million hectares per year by the year 2000 set forth in the Noordwijk Declaration at the 1989 Ministerial Conference on Atmospheric and Climate Change in Noordwijk, The Netherlands (Ministerial Conference on Atmospheric Climatic Change, 1989).

Making Biomass Production for Energy Environmentally Attractive

Throughout the 19th and 20th centuries, there has been substantial deforestation worldwide, as a result of both land clearing for agriculture and nonsustainable mining of the forests for forest products. These cleared lands cannot support the diversity of species that once flourished there. Moreover, modern intensive agricultural management practices have created other serious environmental

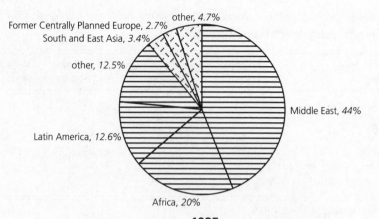

1985
18.6 million barrels per day of oil-equivalent

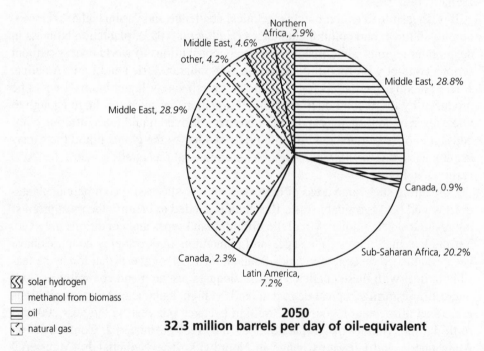

solar hydrogen
methanol from biomass
oil
natural gas

2050
32.3 million barrels per day of oil-equivalent

Figure 4. Fuel exports by region for a renewables-intensive global energy scenario. The importance of world energy commerce for the renewables-intensive global energy scenario developed in Johansson *et al*. (1993) and for which global primary energy consumption is shown in Figure 2 is illustrated here. The figures show that by the middle of the next century there would be comparable exports of oil, natural gas, and biomass-derived methanol, as well as small exports of hydrogen derived from renewable sources. This diversified export mix is in sharp contrast to the situation today, where oil dominates international commerce in liquid and gaseous fuels.

Most methanol exports would originate in sub-Saharan Africa and in Latin America, where there are vast degraded areas suitable for revegetation that will not be needed for cropland (see Tables 2 and 3). Growing biomass on such lands as feedstocks for producing methanol (or other biomass fuels) would provide a powerful economic driver for restoring these lands.

210

problems, including loss of soil quality, erosion, and contamination of runoff with nitrates and other chemicals arising from the use of fertilizers, herbicides, and pesticides. A major concern is that such problems would be aggravated by a major shift to biomass energy.

There is no doubt that biomass can be grown for energy purposes in ways that are environmentally undesirable. However, it is also possible to improve the land environmentally through the production of biomass for energy. The environmental outcome depends sensitively on how and where the biomass is produced.

Consider first the challenge of sustaining the productivity of the land. Since the harvesting of biomass removes nutrients from the site, care must be taken to ensure that these nutrients are restored. This challenge can be dealt with for energy plantations more easily than for agriculture, largely because the choice among plants is more flexible, so that choices can be better targeted to environmental objectives. This is especially true for biomass conversion technologies that begin with thermochemical gasification (which will often be the preferred approach for providing modern energy carriers from biomass [Johansson *et al.*, 1993]).

With thermochemical gasification it is feasible to recover all mineral nutrients as ash at the biomass conversion facility and to return the ash to the plantation site for use as fertilizer. By contrast, fixed nitrogen is lost to the atmosphere at the conversion facility, but it can be replenished in several environmentally acceptable ways. First, when trees are the harvested crop, the leaves, twigs, and small branches, in which nutrients are concentrated, can be left at the site to reduce nutrient loss. (So doing also helps maintain soil quality and reduce erosion through the addition of organic matter to the soil.) Also, nitrogen-fixing species can be selected for the plantation or for interplanting with the primary plantation species to eliminate or reduce to low levels the need for artificial fertilizer inputs. The promise of intercropping strategies is suggested by 10-year trials in Hawaii, where yields of 25 dry tonnes per hectare per year were achieved without nitrogen fertilizer when Eucalyptus was interplanted with nitrogen-fixing Albizia (DeBell *et al.*, 1989).

Energy crops also offer flexibility in dealing with erosion and with chemical pollution from herbicide use. These problems occur mainly at the time of crop establishment. Accordingly, if the energy crop is an annual crop (e.g., sweet sorghum), the erosion and herbicide pollution problems would be similar to those for annual row-crop agriculture. The cultivation of such crops should be avoided on erodable lands. However, the choices for biomass energy crops also include fast-growing trees that are harvested only every 5 to 8 years and replanted perhaps every 15 to 24 years, and perennial grasses that are harvested annually but replanted perhaps only once in a decade. In both cases erosion would be sharply reduced, on average, as would the need for herbicides.

A major concern about agriculture is water pollution from nitrate runoff associated with the excessive use of nitrate fertilizers. Where it would not be possible to deal with this problem by planting nitrogen-fixing plantation species as an alternative to chemical fertilizers, runoff could be controlled by planting fast-growing trees in riparian zones (see Schnoor and Thomas, this volume). In the future it will also be possible to use "designer fertilizers" whose release is timed to match the temporal variations of the plant's demand for fertilizer (Linder, 1989; Kimmins, 1990).

Another concern is chemical pollution from the use of pesticides to control the plantation crop against attack by pests and pathogens. While plantations in the tropics and subtropics tend to be more affected by disease and pest epidemics than those in temperate regions, experience with plantations in these regions shows that careful selection of species and good plantation design and management can be helpful in controlling pests and diseases, rendering the use of chemical pesticides unnecessary in all but extraordinary circumstances. A good plantation design, for example, will include: (1) areas set aside for native flora and fauna to harbor natural predators for plantation pest control, and perhaps (2) blocks of crops characterized by different clones and/or species. If a pest attack breaks out on one block, a now common practice in well-managed plantations is to let the attack run its course and to let predators from the set-aside areas help halt the pest outbreak (Hall *et al.*, 1993).

Biomass plantations are often criticized because the range of biological species they support is much narrower than for natural forests. But if biomass plantations were established on degraded lands, the result could be a net improvement in ecological diversity. Similarly, if biomass energy crops were to replace monocultural food crops, in many cases the shift would be to a more complex ecosystem (Beyea *et al.*, 1992).

Preserving biodiversity will require careful land-use planning. At the plantation level, as already noted, establishing and maintaining natural reserves can be helpful in controlling crop pests while providing local ecological benefits. At the regional level, natural forest patches could be connected via a network of undisturbed corridors (riparian buffer zones, shelterbelts, and hedgerows between fields), thus enabling species to migrate from one habitat to another (Beyea *et al.*, 1992).

Achieving Favorable Energy Balances in Biomass Production for Energy

For biomass energy systems to be viable, the net energy balance must be favorable—i.e., the useful energy produced must be greater than the fossil fuel energy inputs required to provide the biomass energy. Concerns about net energy balances have been widely voiced in the case of fuel ethanol from maize (corn), which is produced in the United States under subsidy at a rate of 4 billion liters per year. In this case the net energy balance is often marginal, and in some instances, the fossil fuel inputs to the system are greater than the alcohol energy produced (Wyman *et al.*, 1993). Maize, however, is a feedstock intended primarily for use as food, not fuel, and its production system is more energy-intensive than most other biomass crops.

212

Many biomass energy systems have favorable energy balances. For example, in the production of fuel ethanol from sugar cane in Brazil (where production at a level of 12 billion liters per year provides nearly one-fifth of total transport fuel requirements) the energy content of the alcohol is about six times the fossil fuel inputs required to grow, harvest, and transport the cane and convert it to alcohol (Goldemberg *et al.*, 1993). The energy balances are also favorable for many energy plantation crops that might be grown in temperate climates. Table 4 shows that with present plantation technology, the energy contents of hybrid poplar, sorghum, and switchgrass harvested and hauled 40 km range from 11 to 16 times the fossil fuel energy needed to provide the biomass (Turhollow and Perlack, 1991). This harvested biomass could be used with near-term technology to produce methanol at an overall efficiency of nearly 60% (Katofsky, 1993), so the net energy balance would be 6 to 9 times the fossil fuel input. With improved yields in the future, these ratios are likely to be still higher (Turhollow and Perlack, 1991).

Achieving Attractive Economics by Modernizing Biomass Energy

The planting, cultivation, and harvesting of biomass is generally more labor-intensive and costly than recovering coal or other fossil fuels from the ground. Thus, per unit of contained energy, biomass tends to be the more costly, especially where there are abundant indigenous fossil fuel resources. Nonetheless, biomass energy

Table 4: *Energy balances for biomass production on plantations, 1990 technology*

	Hybrid Poplar	Sorghum	Switchgrass
Energy inputs (GJ/hectare)			
Establishment	0.14	1.3	0.39
Fertilizers	3.3	8.9	5.3
Herbicides	0.41	1.8	—
Equipment	0.17	—	—
Harvesting	7.3	3.7	5.5
Hauling[1]	2.4	3.8	2.8
Total	13.8	19.5	13.9
Energy output[2] (GJ/hectare)	224	233	158
Energy ratio[3]	16	12	11

[1] The energy required to transport the biomass 40 km to a biomass processing plant.
[2] Yields net of harvesting and storage losses are assumed to be 11.3 tonnes per hectare per year for hybrid poplar (with a heating value of 19.8 GJ/tonne), 13.3 tonnes per hectare per year for sorghum (heating value of 17.5 GJ/tonne), and 9.0 tonnes per hectare per year for switchgrass (heating value of 17.5 GJ/tonne).
[3] The energy ratio = energy output/energy input.

systems can be cost-competitive with fossil fuel energy systems. A more meaningful measure of economic performance is the cost of the energy services, taking into account conversion into electricity and gaseous or liquid fuels, and the systems in which the energy is used. On a cost-of-service basis the economic outlook for biomass can be favorable if modern conversion and end-use technologies are used. This is illustrated below first for electricity generated from biomass, and then for biomass-derived gaseous and liquid fuels for transportation.

Biomass Electricity

Today biomass, mainly in the form of industrial and agricultural residues, is used to generate electricity with conventional steam-turbine power-generators. These biomass power systems can be cost-competitive where low-cost biomass fuels are available, in spite of the fact that steam-turbine technologies are comparatively inefficient and capital-intensive at the small sizes required for biomass electricity production. The United States currently has more than 8000 megawatts of electric generating capacity fueled with such feedstocks, most of which was developed in the 1980s. (For comparison, total U.S. generating capacity is about 700,000 megawatts.) Biomass electricity generation using steam turbines will not expand much in the future, because unused supplies of low-cost biomass residues are rapidly becoming unavailable.

Biomass power generation involving the use of more costly but more abundant feedstocks could be made cost-competitive by adapting advanced-gasification technologies originally developed for coal for use with gas turbine-based power systems (Williams and Larson, 1993). Its very low sulfur content gives biomass a marked advantage over coal for power generation applications. With currently available sulfur removal technology, coal must be gasified in oxygen; the sulfur is removed from the product gas in a "scrubber" prior to combustion of the gas in the gas turbine. Since biomass has negligible sulfur, it can be gasified in air, thereby saving the substantial cost of a plant that separates oxygen from air, and expensive sulfur removal equipment is not needed. Biomass gasifier/gas turbine power systems with efficiencies of 40% or more will be demonstrated in the mid-1990s and will probably be commercially available by 2000. These systems offer high efficiencies and low unit capital costs for baseload power generation at relatively modest capacity (below 100 megawatts). Electricity from such systems will probably be able to compete with coal-fired electricity in many circumstances—even with relatively costly biomass feedstocks. By 2025, gas turbines may give way to even more efficient high-temperature fuel cells (Williams and Larson, 1993).

The electric power industry is beginning to appreciate the importance of biomass for power generation. In an assessment by the Electric Power Research Institute of the potential for biomass-based power generation, it is projected that biomass could be used to support 50,000 megawatts of electric capacity in the United States by 2010 and probably twice that amount by 2030 (Turnbull, 1993).

Transport Fuels from Biomass

Unlike the auspicious near-term outlook for biomass-derived electricity, very large increases in the world oil price are required before biomass-derived transport fuels could compete in cost with gasoline on a cost-per-unit-of-fuel-energy basis. Nevertheless, ongoing changes in the transport sector could permit biomass fuels to compete in providing transport services at world oil prices near the present low level. This prospect will be illustrated here for methanol and gaseous hydrogen fuels derived from biomass via thermochemical gasification.

Based on the use of gasification technology that could be commercialized by the turn of the century, it should be feasible to provide the consumer with either methanol or hydrogen derived from biomass at a cost that is only 40 to 50% more than the price of gasoline expected at that time. Moreover, these biomass-derived fuels are expected to cost no more than the same fuels derived from coal—even though the biomass feedstock would be more costly than coal. This surprising result is due in part to the fact that costly sulfur cleanup technology is not needed for biomass and in part to the fact that biomass is more reactive than coal, which makes it possible to gasify the biomass at a lower temperature. With either feedstock the product of gasification is a nitrogen-free synthesis gas (mainly carbon monoxide and hydrogen, plus some methane) that is subsequently processed to either methanol or hydrogen. In the case of coal, this synthesis gas is produced by gasification in oxygen; the needed high gasification temperatures are provided by partial oxidation of some of the coal in the gasifier; as for power generation, this entails the high cost of a plant to separate oxygen from air. In the case of biomass, gasification can be carried out in steam, with the needed heat provided by an external combustor; the lower gasification temperatures realizable with such "indirectly heated" gasifiers are adequate for biomass gasification, thereby obviating the need for the costly oxygen separation plant (Williams, 1994; Katofsky, 1993).

Synthetic fuels will be needed for transportation, in part to avoid overdependence on insecure sources of foreign oil. U.S. domestic crude oil production peaked in 1970; by 1993 production had fallen to 70% of the peak level; by 2030 production is expected to be less than 30% of the peak level (Department of Energy, 1991). Since biomass-derived synthetic fuels are expected to be no more costly than those derived from coal, they would be preferred in light of the environmental advantages.

A shift to synthetic fuels will probably be accelerated by air quality concerns posed by the use of the internal combustion engine in transportation. While it has dominated road transportation since the automobile was introduced, the long-term outlook for this engine is clouded by growing perceptions that air quality goals cannot be met simply by mandating further incremental reductions in tail-pipe emissions of new vehicles, and that a shift to very-low- or zero-emission vehicles is needed. Already the state of California has mandated that 10% of new cars purchased in 2003 must be zero-emission vehicles, and 12 eastern states have agreed to adopt the same requirement (Wald, 1994).

The California air-quality initiative has led to a substantial industrial effort to commercialize the battery-powered electric car. While the battery-powered electric car is a zero-emission vehicle, its potential is limited, without major advances in battery technology that make it feasible to overcome the long (several-hour) recharging time (Johansson *et al.*, 1993; Williams, 1994).

An alternative is the fuel-cell car operated on compressed hydrogen. As in the battery-powered electric car, electric motors provide the mechanical power that drives the wheels. But the electricity to run the motors is provided not by a battery but rather by a fuel cell that converts energy stored in compressed hydrogen gas canisters directly into electricity. Unlike the battery-powered electric car, the hydrogen fuel-cell car can be refueled in several minutes. Moreover, the life-cycle cost, i.e., the total cost of owning and operating a fuel-cell car (in cents per km) operated on hydrogen derived from biomass is likely to be less than for a battery-powered electric car (Johansson *et al.*, 1993; Williams, 1994; Katofsky, 1993). The life-cycle cost could also be less than for a car with an internal combustion engine of comparable performance even though the fuel is expected to be more expensive than gasoline, mainly because the fuel-cell car is expected to be three times as energy efficient and because it is expected to have lower maintenance costs.

A drawback of the hydrogen fuel-cell option is the requirement for an infrastructure for gaseous hydrogen under pressure. An alternative is to use methanol at atmospheric pressure as a hydrogen carrier: the methanol would react with steam in a "reformer" under the hood of the car, producing a mixture of carbon dioxide and hydrogen, thereby providing the hydrogen needed to operate the fuel cell. The main advantages of the methanol option are that it is easier to establish a distribution infrastructure for a liquid fuel than for pressurized hydrogen, and it is easier to store methanol onboard the car than pressurized hydrogen. The main drawbacks are that a methanol fuel cell vehicle would be more complicated, and it would not qualify as a zero-emission vehicle because of the small amounts of air pollutants generated by the reformer. Lifecycle costs for the methanol and hydrogen fuel cell vehicle options would be comparable (Williams, 1994; Katofsky, 1993).

Fuels derived from biomass are expected to become competitive on a lifecycle cost basis because both hydrogen and methanol can be readily used in technologically superior fuel-cell vehicles, while gasoline and other hydrocarbon fuels cannot—at least for first-generation fuel-cell vehicles.[1] Thus biomass-derived methanol could become a major energy carrier in international commerce in a world that is sensitive to environmental values. The world trade pattern for liquid fuels, shown in Figure 4 for a renewables-intensive global energy scenario (Johansson *et al.*, 1993), shows equal levels of trade for oil and biomass-derived methanol in the period 2025–50, based on an assumed indifference between these fuels.

---

[1]The most likely first candidate fuel cell for automotive propulsion is the proton-exchange-membrane fuel cell, which operates at about 100 °C. At such a low temperature it is practical to reform methanol on board the car, but not other fuels. In the future, if high-temperature fuel cells (operating at about 1000°C) become practical for vehicles, it may be feasible to reform a wide range of hydrocarbon or alcohol fuels under the hood of the car.

Creating Major Energy Roles for Biomass with Limited Land Resources

Because the photosynthetic process is a relatively inefficient way of converting solar energy into chemical-fuel energy, large land areas are required if biomass is to make major contributions to energy supply. For example, displacing fossil fuels in the United States with biomass grown on plantations at the average productivity of U.S. forests (4 dry tonnes per hectare per year) would require an area approximately equal to the total U.S. land area. This suggests that biomass can never become a significant energy source. While there is not enough suitable land to enable biomass to provide all energy needs, the role of biomass can nevertheless be substantial, if modern technologies are used for biomass production and conversion.

The land constraints on biomass production can be reduced in part by intensively managing the biomass plantations. With modern production techniques, biomass productivities far in excess of natural forest yields can be realized. A reasonable goal for the average harvestable yield on large-scale plantations in the United States is 15 dry tonnes per hectare per year (Hall *et al.*, 1993), corresponding to a photosynthetic efficiency of about 0.5%. Assuming that by 2030 the amount of land in the United States committed to biomass plantations is 30 million hectares—approximately the amount of excess cropland in the United States at present—biomass production would be about 9 EJ per year, more than 10% of current U.S. primary energy use (about 80 EJ per year).

The land constraints on biomass production can also be eased by exploiting for energy purposes urban wastes and residues of the agricultural and forest-product industries that can be recovered in environmentally acceptable ways. It has been estimated that such residues in the United States could amount to about 6 EJ per year (Beyea *et al.*, 1992).

The biomass energy potentially available from these two sources, some 15 EJ per year, could probably be produced in the United States in environmentally acceptable ways without confronting significant land-use constraints. This is equivalent to about 20% of current U.S. primary energy use, exclusive of biomass. It does not follow, however, that these potential biomass supplies would displace 20% of conventional U.S. energy. The extent to which conventional energy would be displaced depends sensitively on the conversion technologies deployed.

Consider, for example, the two energy activities often targeted for replacement by biomass energy: the generation of electricity from coal and the running of light-duty vehicles (automobiles and light trucks) on gasoline. In the United States these activities accounted in 1987 for some 30 EJ of primary energy and about one-half of total carbon dioxide emissions from fossil fuel burning. If these two activities (at 1987 activity levels) could be replaced by biomass grown renewably, the result would therefore be a 50% reduction in U.S. carbon dioxide emissions. Three examples of replacement technologies based on biomass are considered here. The overall results are displayed in Figure 5.

217

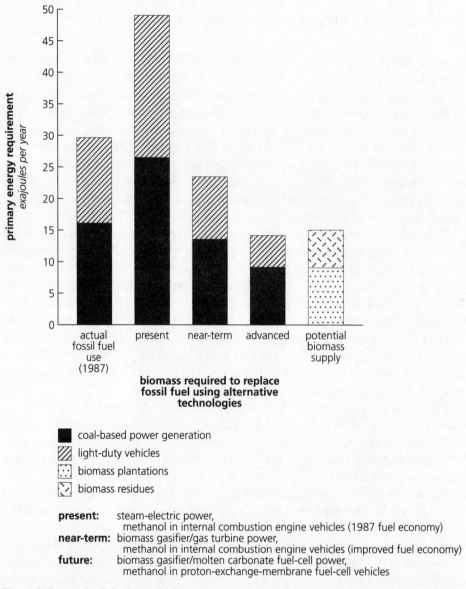

Figure 5. Energy for light-duty vehicles and power generation in the United States. Shown here are the biomass primary energy requirements for displacing all petroleum used by light-duty vehicles (automobiles and light trucks) and all coal-fired power generation in the United States, at the 1987 activity levels, with alternative biomass technologies, in relation to potential biomass supplies.

The bar on the left shows fuel actually consumed in 1987 by light-duty vehicles and by coal-fired power plants. The second bar shows the biomass primary energy requirements if light-duty vehicles and coal-fired power plants at 1987 activity levels were replaced by biomass energy systems that are commercially available today. The third bar shows biomass requirements if technologies likely to be available in the year 2000+ time frame were used to replace all oil used for light-duty vehicles and all coal-based power generation, at 1987 activity levels. The fourth bar shows the biomass requirements if technologies likely to be widely available in the 2020 time frame were used to replace all oil used for light-duty vehicles and all coal-based power generation, at 1987 activity levels. The bar on the

Suppose first that biomass were used with commercially available technologies: (1) replacing coal-based steam-electric power plants with biomass-based steam-electric power plants having a 20% average efficiency, and (2) replacing gasoline by methanol in internal-combustion-engine light-duty vehicles having 1987 average fuel economies, deriving the methanol from biomass (using commercially available technology designed to make methanol from coal but modified to accommodate biomass), and assuming no improvement in the fuel economy of the vehicles other than what would be inherent in a shift from gasoline to methanol. The amount of biomass needed annually for this conversion would be about 49 EJ per year, 60% more than the amount of fossil fuels now used for these purposes and more than three times the 15 EJ per year of biomass supplies estimated above to be potentially available.

Consider, instead, conversion technology likely to be available by the turn of the century. The first generation of biomass-integrated gasifier/gas turbine technology will probably be commercially available, making it possible to roughly double the efficiency of biomass power generation (Williams and Larson, 1993). In this time frame more energy-efficient biomass-to-methanol conversion technologies may also become available (Katofsky, 1993). Moreover, it will be feasible and cost-effective to introduce internal combustion engine vehicles operated on methanol having much higher fuel economies. Using these technologies the total biomass required to displace all coal-based power generation and all oil use by light-duty vehicles at 1987 activity levels would be reduced to 24 EJ per year.

During the second decade of next century, a third set of conversion technologies should be available, including energy-efficient fuel-cell technologies for both stationary electric power generation and motor vehicle applications. For stationary power applications one may anticipate 57% efficient biomass-integrated gasifier/fuel-cell systems employing molten-carbonate or solid-oxide fuel cells. For motor vehicle applications, one may see proton-exchange-membrane fuel-cell vehicles that are 2.5 times as energy-efficient as comparable gasoline-fired internal-combustion-engine vehicles. Using these technologies the total biomass required to displace all coal-based power generation and all oil use by light-duty vehicles at 1987 activity levels would be reduced to about 14 EJ per year, which is comparable to the above estimate of potential supplies from plantations and residues.

Thus with advanced technologies biomass can play major roles in the energy economy, despite the low efficiency of photosynthesis.

Figure 5 contd.

right shows potential biomass supplies from plantations on 30 million hectares of excess agricultural lands plus residues (urban refuse plus agricultural and forest product industry residues) that are recoverable under environmental constraints.

For details see Endnote for Figure 5, at the end of the chapter.

Conclusion

As a supply option for modern energy economies, biomass is unusual because its production and use would relate to a far wider range of human activities than any other energy source. This is in part because biomass energy is derived from photo-synthesis, without which human life would not be possible. Biomass for energy could emerge as a human use of photosynthesis that is comparable in scale to that for agriculture or forestry.

If biomass energy systems are poorly managed, the benefits may not outweigh the social costs. But if biomass is well managed, it is likely that the major concerns people have about biomass energy could be addressed. Successfully developed, biomass energy could provide a broad range of environmental, social, and economic benefits. Worldwide, people could enjoy:

- competitively priced modern energy carriers for a substantial fraction of human energy needs, if advanced conversion and end-use technologies are used;
- the opportunity to reduce CO_2 emissions at zero incremental cost, through the displacement of fossil fuels;[2]
- fuels that are compatible with zero-emission or near-zero emission fuel-cell vehicles, for combating urban air pollution problems.

In addition, the worldwide development of a biofuels industry could provide interfuel competition, energy price stability, and increased energy security in the world fuels market. In developing countries, biomass energy plantations could provide a strong basis for rural development; an opportunity to pay for the restoration of many subtropical degraded lands through their conversion to biomass plantations for energy; and stimulation for economic development in sub-Saharan Africa and Latin America as they become large-scale biofuels exporters. In industrialized countries the development of biomass energy could benefit the agricultural sector by providing a new livelihood for farmers. It would also provide industrialized countries an opportunity to phase out agricultural subsidies eventually, thereby strengthening the economies of these countries while leveling the playing field in world trade for farmers in developing countries.

But these benefits cannot be realized without innovative public policies, coordinated internationally, designed to launch a new industry. An obvious first step would be to eliminate the various national policies biased against biomass energy systems, such as subsidies, tax incentives, and regulations that are not neutral across fuels. A coordinated international research and development program will

---

[2] In addition, some carbon would be sequestered in the steady-state inventory of biomass plantations. Neglecting changes in soil carbon associated with establishing plantations and considering only the average inventory associated with biomass that will be harvested, the sequestering capacity would be about 9 billion tonnes of carbon (assuming an average rotation length of six years between cuts), corresponding to three years of buildup of carbon dioxide in the atmosphere.

also be necessary, with two parallel thrusts: (1) sustainable biomass production in a wide range of climates and soils, and (2) innovative, cost-effective energy conversion and end-use technologies. Agreement on environmental guidelines should be sought at an early stage, to give impetus to development and commercialization of options that are environmentally attractive, and to safeguard against undesirable and self-defeating approaches. Resources will have to be transferred from industrialized countries to developing countries to assure access to advanced biomass energy technologies in developing countries.

Five years ago, the required policy changes would have been deemed unrealistic, even unthinkable. But today this is no longer the case. A global consensus is emerging that the only acceptable development is sustainable development. This was the underlying theme of the United Nations Conference on Environment and Development in Rio de Janeiro, in June 1992. It is now realized that the only way to achieve sustainable development is to shift from one-dimensional policy-making to holistic approaches that deal with all direct and indirect impacts of a given economic activity, making concerted efforts to avoid adverse impacts before they occur.

Thus, though it is unfamiliar or at least not yet well understood by most people, biomass energy will receive focused attention in the forthcoming sustainable development debates—both because of potential disastrous consequences of ill-planned biomass energy developments and because of the enormous overall benefits in support of sustainable development goals that would arise from the proper development of biomass energy.

The timing of these debates could not be better for the biomass energy community. Because modernized biomass energy plays a negligible role in the world energy economy at present, the future shape of the biomass energy industry can be molded by the sustainable development debates, before the industry becomes well established.

It is a rare event in modern history that the big concerns about potential adverse impacts of technology are aired before the technology is implemented. And rarer still is the opportunity to address these concerns by timely changes of course.

Acknowledgments

This research was supported by the Office of Energy and Infrastructure of the U.S. Agency for International Development, the U.S. Environmental Protection Agency's Air and Energy Engineering Laboratory, the Geraldine R. Dodge Foundation, the Energy Foundation, the W. Alton Jones Foundation, the Merck Fund, the New Land Foundation, and the Rockefeller Foundation.

References

Beyea, J., J. Cook, D. O. Hall, R. H. Socolow, and R.H. Williams. 1992. *Toward Ecological Guidelines for Large-Scale Biomass Energy Development*. Report of a Workshop for Engineers,

Ecologists, and Policymakers Convened by the National Audubon Society and Princeton University, May 6, 1991, National Audubon Society, New York.

Davis, G. R. 1990. Energy for planet earth. *Scientific American 263(3)*, 54–63.

DeBell, D. S., C. D. Whitesell, and T. H. Schubert. 1989. Using N_2-fixing Albizia to increase growth of eucalyptus plantations in Hawaii. *Forest Science 35(1)*, 64–75.

Energy Information Administration. 1991. *State Energy Price and Expenditure Report 1989*. U.S. Department of Energy, DOE/EIA-0376(89).

FAO (U.N. Food and Agriculture Organization). 1991. *Agricultural Land Use: Inventory of Agroecological Zones Studies*. FAO, Rome, Italy.

Goldemberg, J., T. B. Johansson, A. K. N. Reddy, and R. H. Williams. 1987. *Energy for a Sustainable World*. World Resources Institute, Washington, D.C., 119 pp.

Goldemberg, J., T. B. Johansson, A. K. N. Reddy, and R. H. Williams. 1988. *Energy for a Sustainable World*. Wiley-Eastern, Delhi, 517 pp.

Goldemberg, J., L. C. Monaco, and I. Macedo. 1993. The Brazilian fuel-alcohol program. In *Renewable Energy: Sources for Fuel and Electricity* (T. B. Johansson, H. Kelly, A. K. N. Reddy, and R. H. Williams, eds.), Island Press, Washington, D.C., 841–864.

Grainger, A. 1988. Estimating areas of degraded tropical lands requiring replenishment of forest cover. *International Tree Crops Journal 5*, 31–61.

Grainger, A. 1990. Modelling the impact of alternative afforestation strategies to reduce carbon emissions. In *Proceedings of the Intergovernmental Panel on Climate Change Conference on Tropical Forestry Response Options to Climate Change*, Report No. 20P-2003, Office of Policy Analysis, U.S. Environmental Protection Agency, Washington, D.C.

Hall, D. O., F. Rosillo-Calle, R. H. Williams, and J. Woods. 1993. Biomass for energy: Supply prospects. In *Renewable Energy: Sources for Fuel and Electricity* (T.B. Johansson, H. Kelly, A. K. N. Reddy, and R. H. Williams, eds.), Island Press, Washington, D.C., 593–652.

Hummel, F. C. 1988. Biomass forestry: Implications for land-use policy in Europe. *Land Use Policy*, 375–384.

IPCC (Intergovernmental Panel on Climate Change). 1990. *Climate Change: The IPCC Scientific Assessment* (J.T Houghton, G. J. Jenkins, and J. J. Ephraums, eds.), Cambridge University Press, Cambridge, U.K.

IPCC. 1991. *Climate Change: The IPCC Response Strategies*. Island Press, Washington, D.C.

Johansson, T. B., H. Kelly, A. K. N. Reddy, and R. H. Williams. 1993. Renewable fuels and electricity for a growing world economy: Defining and achieving the potential. In *Renewable Energy: Sources for Fuel and Electricity* (T. B. Johansson, H. Kelly, A. K. N. Reddy, and R. H. Williams, eds.), Island Press, Washington, D.C., 1–72.

Katofsky, R. E. 1993. *Production of Fluid Fuels from Biomass*. PU/CEES Report No. 279, Center for Energy and Environmental Studies, Princeton University, Princeton, New Jersey.

Kimmins, J. P. 1990. Modelling the sustainability of forest production and yield for a changing and uncertain future. *The Forest Chronicle 6*, 271–280.

Linder, S. 1989. Nutrient control of forest yield. In *Nutrition of Trees*, Marcus Wallenberg Foundation Symposia Vol. 6, Falun, Sweden, 62–87.

Masters, C. D., D. H. Root, and E. D. Attanasi. 1990. World oil and gas resources—Future production realities. *Annual Review of Energy 15*, 23–31.

Ministerial Conference on Atmospheric and Climatic Change. 1989. *The Noordwijk Declaration on Atmospheric Pollution and Climatic Change*. Ministerial Conference on Atmospheric and Climatic Change, Noordwijk, Netherlands.

Netherlands Scientific Council for Government Policy. 1992. *Grounds for Choice: Four Perspectives for the Rural Areas in the European Community*. Netherlands Scientific Council for Government Policy, The Hague, Netherlands.

New Farm and Forest Products Task Force. 1987. *New Farm and Forest Products: Responses to the Challenges and Opportunities Facing American Agriculture*. A report to the Secretary of Agriculture, June 25, 1987.

OECD (Organization for Economic Cooperation and Development). 1992. *Agricultural Policies, Markets, and Trade: Monitoring and Outlook 1992*. OECD, Paris, France.

OTA (Office of Technology Assessment of the U.S. Congress). 1992. Technologies to Sustain Tropical Forest Resources and Biological Diversity. OTA-F-515, U.S. Government Printing Office, Washington, D.C.

Parham, W. E., P. J. Durana, and A. L. Hess (eds.). 1993. Improving degraded lands: Promising experiences from South China. In *Bishop Museum Bulletin in Botany 3*, The Bishop Museum, Honolulu, Hawaii.

Response Strategies Working Group of the Intergovernmental Panel on Climate Change. 1990. *Emissions Scenarios: Appendix of the Expert Group on Emissions Scenarios*. U.S. Environmental Protection Agency, Washington, D.C.

Smith, K. R., and S. A. Thorneloe. 1992. Household fuels in developing countries: Global warming, health, and energy implications. In *Proceedings from the 1992 Greenhouse Gas Emissions and Mitigation Research Symposium*, August 18–20, 1992, sponsored by the U.S. Environmental Protection Agency Air and Energy Engineering Laboratory, Washington D.C.

Soil Conservation Service. 1989. *The Second RCA Appraisal: Soil, Water, and Related Resources on Non-Federal Land in the United States—Analysis of Condition and Trends*. U.S. Department of Agriculture.

Turhollow, A. H., and R. D. Perlack. 1991. Emissions of CO_2 from energy crop production. *Biomass and Bioenergy 1*, 129–135.

Turnbull, J. 1993. *Strategies for Achieving a Sustainable, Clean, and Cost-Effective Biomass Resource*. Electric Power Research Institute, Palo Alto, California.

U.S. Department of Energy. 1991. *National Energy Strategy, Technical Annex 2. Integrated Analysis Supporting the National Energy Strategy: Methodology, Assumptions, and Results*. DOE/S-0086P.

Waggoner, P.E. 1994. *How Much Land Can Ten Billion People Spare for Native?* Council for Agricultural Science and Technology, Ames, Iowa.

Wald, M. L. 1992. California's pied piper of clean air. *The New York Times*, Business Section, September 13, 1992.

Wald, M. L. 1994. Harder auto emission rules agreed to by eastern states. *The New York Times, A1*, February 2, 1994.

Williams, M. 1989. *Americans and Their Forests: A Historical Geography*. Cambridge University Press, Cambridge, U.K.

Williams, R. H. 1994. The clean machine. *Technology Review 97(3)*, 20–30.

Williams, R. H., and E. D. Larson. 1993. Advanced gasification-based power generation. In *Renewable Energy: Sources for Fuel and Electricity* (T. B. Johansson, H. Kelly, A. K. N. Reddy, and R. H. Williams, eds.), Island Press, Washington, D.C., 729–786.

WRI (World Resources Institute). 1992. *World Resources 1992–93: A Guide to the Global Environment*. Oxford University Press.

Wyman, C. E., R. L. Bain, N. D. Hinman, and D. J. Stevens. 1993. Ethanol and methanol from cellulosic biomass. In *Renewable Energy: Sources for Fuel and Electricity* (T. B. Johansson, H. Kelly, A. K. N. Reddy, and R. H. Williams, eds.), Island Press, Washington, D.C., 865–924.

Endnote for Figure 5

The bar on the left represents fuel consumed in 1987 by light-duty vehicles and by coal-fired power plants. Automobiles and light trucks, with average fuel economies of 19.1 mpg and 12.9 mpg, respectively, consumed 103 billion gallons of gasoline. In 1987 coal-fired power plants, operated with an average efficiency of 32.9%, produced 1464 TWh of electricity.

The second bar shows the biomass primary energy requirements if light-duty vehicles and coal-fired power plants at 1987 activity levels were replaced by biomass energy systems that are commercially available today. With present biomass gasification technology (adapted directly from coal gasification) methanol can be produced from wood at 50% efficiency. Operated on methanol, cars and light trucks would have gasoline-equivalent fuel economies of 22.9 mpg and 15.5 mpg, respectively, some 20% higher than gasoline vehicles, because of the higher thermal efficiency of internal-combustion engines when operated on methanol (Wyman *et al.*, 1993). The net result is that the biomass feedstock requirements to support the 1987 level of light-duty vehicles would be $1/(0.5 \times 1.2) = 1.67$ times the amount of gasoline used by light-duty vehicles in 1987. The present average efficiency of biomass power plants operating in California is about 20%, so that the biomass plants would require $32.9/20 = 1.65$ times as much fuel to make electricity as the coal plants they would displace.

The third bar shows the biomass primary energy requirements if biomass technologies likely to be available in the year 2000+ time frame were used to replace all oil used for light-duty vehicles and all coal-based power generation, at 1987 activity levels. It is cost-effective to increase the average (on-the-road) fuel economy of new cars and light trucks to about 33.6 and 22.7 mpg (76% higher than in 1987), respectively, if operated on gasoline, and to 40.3 and 27.2 mpg of gasoline-equivalent energy (20% higher than on gasoline), respectively, if operated on methanol. A shift to such vehicles could be achieved over the next couple of decades. During this period it would be feasible to introduce methanol production technology involving indirect biomass gasification, for which the overall biomass-to-methanol conversion efficiency is 58% (Katofsky, 1993). With these technologies biomass fuel input requirements would be $(1/1.76)/(0.58 \times 1.2) = 0.82$ times as large as the petroleum required in 1987. If electricity were produced from biomass using biomass integrated gasifier/gas turbine technology that could be introduced commercially in this time frame, the efficiency of power generation would be 40%, nearly double that of existing biomass power plants, so that the biomass plants would require $32.9/40 = 0.82$ times as much fuel to make electricity as the coal plants they would displace.

The fourth bar shows the biomass primary energy requirements if biomass technologies likely to be available in the 2020 time frame were used to replace all oil used for light-duty vehicles and all coal-based power generation, at 1987 activity levels. By the end of the second decade of the 21st century, biomass-derived methanol could be routinely used in proton-exchange-membrane fuel-cell vehicles at gasoline-equivalent fuel economies that are 2.4 times the fuel economies of gasoline-powered internal-combustion-engine vehicles of comparable performance (Johansson *et al.*, 1993; Williams, 1993). With these technologies biomass fuel input requirements would be $(1/1.76)/(0.64 \times 2.4) = 0.408$ times as large as the petroleum required in 1987. Also, by that time, biomass integrated gasifier/fuel cell systems (using molten carbonate or solid oxide fuel cells) for stationary power generation could well be available with biomass-to-electricity conversion efficiencies of perhaps 57% (Williams and Larson, 1993), for which fuel requirements would be $32.9/57 = 0.577$ times coal requirements for power generation in 1987.

The fifth bar shows potential U.S. biomass supplies, consisting of (1) 9 EJ per year of plantation biomass grown on 30 million hectares of excess agricultural lands at an average productivity of 15 dry tonnes per hectare per year and a heating value of 20 GJ per dry tonne of biomass, plus (2) 6 EJ per year of those residues (urban refuse, agricultural residues, forest product industry residues) that are estimated to be recoverable under environmental constraints (Beyea *et al.*, 1992).

PART 3

TOXICS AND THE ENVIRONMENT

15

Introduction

The Editors

Industrial ecology is a comphrensive response to industrial systems damaging the environment. This damage has been amply demonstrated by numerous examples, usually involving specific industrial or consumer activities, specific pollutants, and localized ecosystems or groups of people. It is, however, much more difficult to make an integrated assessment of environmental damage, especially at a regional or global scale. Yet to understand the "industrial ecology" of our current industrial system, and especially to design an "eco-restructured" industrial system, integrated assessments are essential.

Part Three contains seven chapters on the environmental and human health consequences of pollutants emitted into the environment by industrialized societies. The emphasis is less on localized, intensely polluted sites, and more on the consequences of diffuse sources of widely dispersed pollutants.

There are two starting points from which the environmental effects of human activity can be approached. One is to focus on ecosystems, or other environmental entities such as "soils" or "biodiversity" and ask how they are being threatened by human activity. The first three chapters in this section take this approach in addressing global threats to ecosystems. Agriculture, while for the most part excluded from consideration in this volume, is nevertheless part of the industrial system, both as feedstocks for other industries and as an industrial activity itself, and it has become a significant cause of environmental damage. Agriculture is explicitly addressed in the chapter by Schnoor and Thomas, "Soils as a Vulnerable Environmental System." Major effects include not only habitat destruction from the clearing of land, but also erosion due to tillage, and environmental damage from the use of pesticides.

Schlesinger, in "The Vulnerability of Biotic Diversity" examines how industrial activity affects biodiversity. Sources of ecosystem stress including nitrogen inputs (both from the use of nitrogen fertilizers and from acid rain derived from automobile and power plant emissions), climate change, and stratospheric ozone depletion. While deposited nitrogen may act as a fertilizer, this fertilization can damage natural ecosystems, as can be seen in the eutrophication of lakes that receive high inputs of nitrogen. By altering the ecosystem dynamics, excessive nitrogen deposition is likely to result in a loss of species diversity. Similarly, species diversity may decrease as atmospheric carbon dioxide increases, because differences in plant responses to increased carbon dioxide will change competitive relationships. Moreover, the global and local climate change that may result from increased atmospheric carbon dioxide could have a major effect on ecosystems. Depletion of stratospheric ozone, and the resulting increase in exposure to ultraviolet UV-B radiation, can also be expected to stress some species and ecosystems and result in reduced species diversity.

Our knowledge of the effects of pollutants on ecosystems is still very limited. While lethal doses of pollutants for a number of species are known, we have very little knowledge of sublethal effects, or of mechanisms of tolerance and adaptation. The chapter by Anderson, "Global Ecotoxicity: Management and Science" provides an agenda for assessing the global ecological impact of toxic substances.

The other starting point in pollution assessment is to focus on an individual pollutant, class of pollutants, or activity, to understand its sources and its effects on ecosystems or human health. The last four chapters take this approach, a key element of which is the mapping of material flows through the industrial system and the environment.

Three of these chapters are about heavy metals. Because metals do not degrade or disperse rapidly, emissions of heavy metals may result in elevated environmental concentrations that are effectively permanent. Nriagu, in "Industrial Activity and Metal Emissions," quantifies the global emissions of heavy metals to air, water, and soil. He reports that while metals emissions in developed countries have decreased over the past decade, emissions in developing countries are increasing with the growth of industrial activity and transportation infrastructures. Worldwide, humans, and other species as well, are now exposed to concentrations of metals much higher than those with which they evolved.

Stigliani, Jaffé, and Anderberg in "Metals Loading of the Environment: Cadmium in the Rhine Basin," assess the emissions of cadmium in the Rhine River Basin over the past three decades, and estimate the resulting increase in human cadmium exposure. They find that emissions have decreased significantly since the 1960s, especially due to the decrease in industrial, point-source emissions. However, diffuse sources of pollution, from agricultural lands and urban runoff, have not been well controlled; they are now responsible for about one-half of aqueous emissions of cadmium. The focus of regulation is now shifting to these diffuse sources.

Thomas and Spiro, in "Emissions and Exposure to Metals: Cadmium and Lead," examine the global use and emissions of both lead and cadmium, with emphasis on the uses that cause the greatest environmental and human health damage. They show that the largest sources of emissions are typically very different from the most significant sources of human or ecosystem exposure. Exposure assessment suggests that cadmium has relatively minor impacts on the environment and human health. The largest potential effects appear to be due to trace amounts of cadmium in phosphate fertilizers and coal, rather than from cadmium products, such as nickel–cadmium batteries. Lead, in contrast, is recognized as one of the most significant pediatric health problems in the United States, despite the phase-out of leaded gasoline and lead-based household paint in the past 20 years. The ecosystem effects of lead are also significant: ingestion of lead shot is responsible for the deaths of millions of birds annually.

Berkhout's chapter, "Nuclear Power: An Industrial Ecology That Failed?" underscores the importance of social and institutional factors in determining the acceptability of a technology. The nuclear industry took the industrial ecology approach, in the sense that accounting for and containing all of its materials was of fundamental importance for the industry. Large investments were made in materials control, thereby eliminating certain risks. Yet this approach seems to have failed; the public still does not embrace nuclear power, and the industry has not flourished.

The nuclear fuel example also challenges the assumption that recycling is always better than disposal. For the nuclear fuel cycle, it is widely believed that the recycling of nuclear fuels increases the risks of nuclear weapons proliferation and human exposure to radiation, so that disposal after one use is the preferred alternative. Fuel management which does not include recycling may therefore be the most environmentally sound, at least in the short term.

16

Soil as a Vulnerable Environmental System

Jerald Schnoor and Valerie Thomas

Abstract

Since 1945, soil degradation has affected nearly 20% of the vegetated land area of the earth, with agriculture, overgrazing, and deforestation the primary causes on all continents. Industrial emissions appear as a relatively minor cause of soil degradation, because the effects of pollutants are typically measured as acute effects on a few species rather than in wholesale depletion of the soil. Soils also play a critical role in regulating the carbon dioxide, methane, and nitrogen oxide concentrations in the atmosphere. In view of the major impacts of agricultural activities and deforestation on soils and terrestrial ecosystems, these topics should become a priority for industrial ecology.

Introduction

Soil is the compartment of terrestrial ecosystems where the lithosphere, biosphere, and hydrosphere most actively interact, at several spatial and temporal scales. Soil genesis is a complex process which takes place on time scales of millennia and is intimately linked with the subsoil (parent material), the local relief, the climatic history, and, more recently, human activities.

Soil supports the entire terrestrial ecosystem. It makes possible the development of a variety of types of vegetation, and the maintenance of our agroecosystems. During soil genesis, rocks and till are weathered by chemical and physical geologic processes, and soil properties, both chemical and physical, respond slowly to changes in inputs. Vegetation may provide the first clue to long-term changes in soil quality because plants respond more quickly than soils to changes in inputs. Soils recover slowly once they have become contaminated or infertile.

Soil modifies precipitation before water enters surface and groundwater. Microbial processes in soils transform trace gases and exchange them with the atmosphere. Transfers of solutes and gases between the atmosphere, biota, soil, rivers, and oceans are so important that the soil must be considered whenever global change in the atmosphere, biota or any other compartment is assessed.

Of the total land area of the earth (15 billion hectares), about one-third is cropland or permanent pasture, 25% is forest, 25% is unvegetated, and the remaining

15% includes wetlands, non-pasture grasslands, and built-on areas. "Wilderness," defined as large blocks of land (at least 4000 km^2) with no roads, buildings, or other structures, covers about a third of the earth's land area. One-half of the wilderness area is in Antarctica, the former Soviet Union, and Canada, and much of the rest is in desert areas. Only 5% of the U.S. land area is wilderness, and the only European countries with any wilderness area at all are Finland, Iceland, Norway, and Sweden (McCloskey and Spalding, 1989; World Resources Institute, 1992).

Very broadly speaking, environmental threats to the soil system are connected with effects on or from agriculture, or with the destruction of wilderness areas and biodiversity. Threats to the soil system include the direct degradation of the capacity of soil to support life, primarily through erosion. A second type of environmental problem involves pollutants used in or deposited on soil that, while they may not destroy the soil's capacity to support life, do harm specific species or parts of the ecosystem. This includes the effects on birds of the pesticide DDT, or the human health threat from nuclear radiation absorbed into crops. A third type of effect is change in global cycles, such as global warming. These global system changes will affect soils, and changes in soils and soil processes will in turn affect the atmospheric cycling of nutrients. Each type of effect will be addressed in turn.

Although the general processes which threaten soils and their ecosystems are found worldwide, the causes, extent, and significance of these processes differ from region to region. For example, industrial emissions are more of a problem in Europe than elsewhere, while deforestation is more important in tropical countries.

Soil Degradation

The best study to date of the global extent of soil degradation is a compilation of expert estimates, in a study sponsored by the United Nations Environment Program (Oldeman *et al.*, 1990). The study concluded that since 1945, human activities have "degraded" 2 billion hectares of land, about 20% of the total vegetated area of the earth. One-half of this land is classified as "moderately" or "strongly" degraded. Moderately degraded land still permits agricultural use, but with greatly reduced productivity; on strongly degraded land agriculture is no longer possible and most biotic functions have been destroyed.

As shown in Figure 1, the main causes of soil degradation are overgrazing, agricultural activities, and deforestation, each accounting for roughly one-third of the degradation. Overexploitation, stripping land of vegetation for fuelwood, is significant in some regions, accounting for about 7% of worldwide soil degradation. Industrial activity, including acidification by airborne pollutants, and contamination with industrial and urban wastes, pesticides and other pollutants, is a significant factor only in Europe, where it is responsible for 9% of the degraded land; on all other continents it accounts for 1% or less.

From the industrial ecology perspective, what is most striking is the minor role of industrial emissions, as compared with deforestation and agricultural and

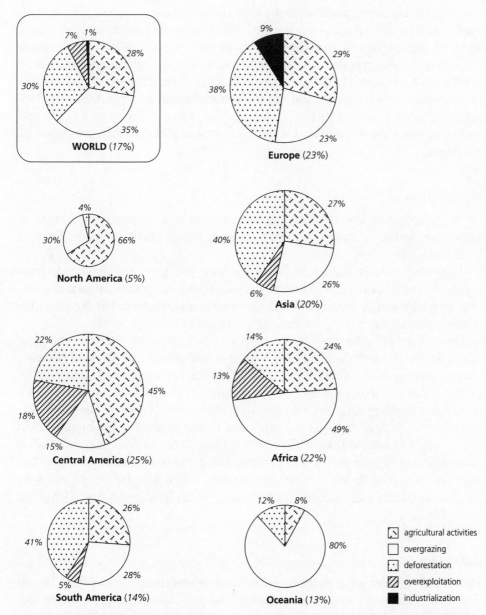

Figure 1. Causes of soil degradation, by fraction of vegetated land affected in each region. The area of each circle is proportional to the fraction of vegetated land in that region that has been degraded since 1945, including light, moderate, strong, and extreme degradation. The circle for North America is smaller than the others because Canada and Alaska have been subject to relatively little soil degradation. Within each circle, the shaded areas indicate the causes of degradation. Categories not shown represent less than 1%. (Data from Oldeman et al., 1990; figure adapted from World Resources Institute, 1992).

livestock management practices, in the degradation of land, in industrialized countries as well as developing countries. A similar story is told by the relative importance of the mechanisms of degradation: wind erosion (56%), water erosion (28%), chemical degradation (12%), and physical degradation from compaction or flooding (4%). Industrial emissions are included in chemical degradation, but this category also includes loss of soil nutrients and salinization. Only in Europe does pollution account for the majority of chemical degradation; elsewhere nutrient loss and salinization are responsible for far more land degradation than are industrial emissions.

Deforestation

While deforestation can be found the world over, it is especially significant in tropical countries. In many countries the annual loss is about 1% of the forested land (FAO, 1993).

When land is cleared, animal habitats are destroyed. Part of the habitat destruction is caused by changes in soil quality or loss of soil via erosion. Habitat destruction is not limited to the area that is converted to agriculture, but includes fish spawning areas of receiving streams, covered by silt.

The effects of deforestation on atmospheric carbon dioxide are complex, depending on whether the biomass is burned or stored, and how the land is subsequently managed, as well as the specific characteristics of the soil. Deforestation releases carbon to the atmosphere, not only through the combustion or degradation of biomass above ground, but also through the loss of soil carbon. Globally, soils contains two to three times as much organic carbon as does living biomass, and after deforestation soil typically loses about 20% of its organic carbon (Schlesinger, 1991). In recent years, deforestation has contributed one to two billion tons of carbon to the atmosphere annually, which may be compared to the roughly six billion tons currently contributed by fossil fuel combustion (Houghton *et al.*, 1987).

Agriculture

Crop yields in most areas of the world have increased since World War II, due to new, high-yield crops, irrigation, fertilizers, and pesticides. Over the same period, soil degradation has advanced at an alarming rate. Soil erosion decreases crop yields, and only in some cases can its effects be compensated by fertilizers (OTA, 1982). To some extent modern agricultural technology is masking the consequences of soil degradation. Can agricultural technology stay ahead of the curve?

If agricultural technology falls behind, global food production will be dependent on "spare" arable land, allowing agricultural production to move away from degraded land, and giving it time to recover. As is discussed in the chapter by Williams (this volume), there is currently more agricultural land than is needed in the United States and Europe. Specifically, Williams cites estimates that in the United States and Europe, one-third to one-half of current cropland will be idle by the second decade of the next century. In developing countries, however, growing populations and standards of living are expected to result in demand for about 50% more cropland by the year 2025. In Africa and Latin America, enough high-quality potential cropland can be found, but in Asia, a 50% expansion of cropland would require use of land with low or uncertain rainfall and other problems. In summary, in the United States and Europe, cropland is plentiful enough for the agricultural system to withstand significant soil degradation, but on a global basis, there is a much smaller safety margin.

Still more land would be required if a global biomass fuel system is developed. As discussed by Williams, such a system would require the equivalent of about a fourth of current world cropland. Although the ability to feed the world's population does not seem to be at risk, maintenance of agricultural productivity will require both soil quality and soil quantity to be preserved.

Agriculture, including both livestock and crops, is the main source of soil degradation worldwide. Overgrazing alone is estimated to account for one-third of the world's soil degradation. As of 1980, according to a U.S. Department of Agriculture estimate, half of the rangelands in the contiguous United States were seriously degraded, and only 15% were rated in good condition (OTA, 1982). Second to overgrazing, tillage is probably the most damaging agricultural activity. Tilling the soil once or more each year makes soils vulnerable to wind and water erosion.

In the United States, annual average soil loss from wind and water erosion averages about 10 tons/hectare/year (OTA, 1982). In some regions of the world, soil losses are on the order of 100 tons/hectare/year (Pimentel *et al.*, 1993).

Soil regeneration rates are not known with any certainty. In the United States, soil loss tolerances estimated by the Soil Conservation Service of the U.S. Department of Agriculture range from 12 tons/hectare/year for deep, permeable, well-drained, productive soils, down to 2 tons/hectare/year for shallow soils having unfavorable subsoils and parent materials that severely restrict root penetration and development. However, there is little science to support these numbers (OTA, 1982). Another estimate of a typical topsoil renewal rate under tropical and temperate agricultural conditions is 1 ton/hectare/year (Pimentel *et al.*, 1993).

Pollutants

The amount of land degraded by industrial sources is small compared to degradation caused by agriculture and deforestation. Industrial effects are best understood not in terms of soil degradation, but rather in terms of the effects of pollutants on particular species or ecosystems. Of the great variety of pollutants, we will address here four general categories: acid deposition, agricultural chemicals, metals, and radionuclides.

Acid Deposition

Acid deposition is due to oxides of nitrogen and sulfur, that are emitted into the atmosphere by combustion, fertilizer use, and other industrial processes. Acid deposition can have severe effects on forests, and is believed to have been responsible for large-scale forest decline in Europe. In Czechoslovakia, for example, 100,000 acres (40,000 ha) of Norway spruce have died; one-third to one-half of the forests are estimated to have suffered moderate to severe damage (Moldan and Schnoor, 1992). Similar damage has occurred in Germany and in other eastern European countries.

Agricultural Chemicals

Fertilizer inputs to row crop agriculture have increased dramatically since 1950 throughout the world (for nitrogen, see Figure 10 in Ayres, Schlesinger, and Socolow, this volume). A continuing increase in agricultural use of fertilizers is widely believed to be necessary in order to continue to feed the world's growing and more prosperous population. Nitrogen and phosphorus applied to cropland affect the chemistry of rivers, streams, lakes, bays, and groundwater. Among the effects are eutrophication of the receiving waters and contamination of drinking water. Figure 2 shows a rise in nitrogen fertilizer application in Iowa in the 1960s, and a subsequent increase in nitrate concentrations in a representative Iowa river (McDonald, 1992).

Pesticide usage has also increased dramatically since World War II. The better understood detrimental consequences of pesticides are their effects on species high on the food chain, such as birds and humans. The organochlorine pesticides aldrin, dieldrin, endrin, camphechlor, chlordane, heptachlor, HCH, lindane, DDT, and mirex are persistent, fat-soluble substances whose concentrations increase with each link up the food chain. These substances and some of their metabolites have been found in human fatty tissues and human milk in countries all over the world. Some pesticides can cause gross birth defects, especially in birds. Exposure to DDT by predatory birds such as eagles, ospreys and falcons has led to high reproductive losses in these species (Carson, 1962). While the acute health effects of some pesticides are well established, the effects of long-term exposure at low doses are not well understood. Pesticides can affect soil microbes and invertebrates (OTA, 1982). Many of the organochlorine pesticides are thought to be human carcinogens (WHO, 1990). Some biologists and toxicologists hypothesize that organochlorine pesticides suppress the immune and endocrine systems of animals (Colborn and Clement, 1992).

Many persistent pesticides have been banned or restricted in industrialized countries, but they are still manufactured and sold to the developing world, where they continue to be a problem reaching beyond national borders (IASA, 1987). These chemicals have a long environmental lifetime due to the toxic nature of the chlorine atoms that renders attack by aerobic soil microbes difficult. They can be transported thousands of miles after being volatilized from soils or vegetation.

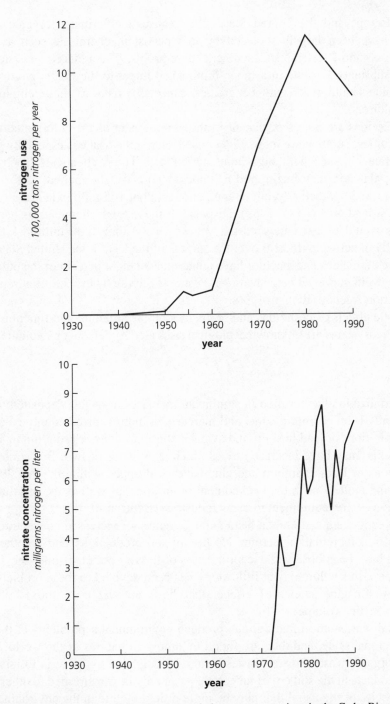

Figure 2. Fertilizer use in Iowa, 1930–90 (top); and nitrate concentrations in the Cedar River, Iowa, 1972–91 (bottom). The upper figure shows total nitrate fertilizer use in Iowa; the area of Iowa is about 15 million hectares (56,000 square miles), of which about 50% is under intensive agricultural use. The lower figures shows nitrate concentrations in Cedar River, Iowa, measured 1 meter below the surface. (Source: McDonald, 1992.)

Toxics and the Environment

In Europe and the United States the persistent chlorinated organic insecticides have been largely replaced by less persistent chemicals such as carbamates, organo-P esters, and synthetic pyrethroids. The reduced persistence of these substances results in a much diminished threat to the global environment, but some health risks may be greater, especially risks to those employed in agriculture.

Herbicides are not as persistent in the environment as the chlorinated organic insecticides, but they are more widely used. Herbicide usage is especially great in temperate climates with high-input agriculture. These chemicals (such as triazines, alachlor, metribuzin, and metalochlor) are widely applied. In the United States, 1 to 3 kg/hectare/year are applied over about 165 million hectares, 17% of the U.S. land area (CEQ, 1989). Disposal of these toxic chemicals as hazardous wastes would be extremely costly, yet we freely apply 200 million kilograms annually of active ingredient over the surface of the land in the United States. The triazine herbicides and alachlor have contaminated eight percent of the public and private wells surveyed in the National Pesticide Survey by the U.S. Environmental Protection Agency (EPA) in 1990.

Subtle effects of pesticides and herbicides on the structure and function of off-target ecosystems are unknown ecological risks (see Anderson, this volume).

Metals

Industrialization has resulted in significant increases in metals deposited to soils, especially lead, cadmium, zinc, and mercury. In industrialized countries, average concentrations of lead in soils, and average concentrations of cadmium in agricultural soils, may have increased by as much as a factor of two. But atmospheric emissions of lead, cadmium and zinc have all dropped within the last 20 years, following reductions in the lead content of gasoline, as well as the installation of pollution control equipment in many industries (Boutron, 1991).

The ecosystem and human health consequences of increased concentrations of metals in soils remain uncertain. No loss of soil productivity and no other overt effects have been observed except in cases of heavily polluted soils (see Thomas and Spiro, this volume). But little study has been devoted to these problems, and one can imagine numerous effects that have not yet been looked for (see Anderson, this volume).

Metal wastes and mine tailings produce environmental problems at the local and regional scale, and they are found in almost all nations of the world. Unlike toxic organic wastes, metal wastes cannot be degraded by bacteria to innocuous end-products in the soil environment; they can only be transformed to other chemical species of the metal that may be more or less mobile in the environment and more or less toxic. At hazardous waste sites containing mixed wastes, it is usually cadmium, lead, mercury, and arsenic that are considered to pose the greatest risk to human health. Often, the pathway of greatest exposure to humans is from inhalation of dust or ingestion of wind-blown particles in homes. Revegetating the sites can help to decrease wind-blown dust, but the vegetation can become a pathway for the metal to enter the food chain.

Radionuclides

Radionuclide hazardous wastes are a problem in countries with uranium mine wastes and nuclear power or nuclear weapons industries. The nuclear weapons program of the United States, for example, involved manufacturing, testing and waste disposal at 15 government facilities; major environmental problems are now being confronted at most of these sites, as they begin the process of decommissioning. A similar situation exists in the former Soviet Union, former East Germany, Czechoslovakia, Romania, China, and probably other countries.

During the 1960s when the United States, Soviet Union, China, and France were all testing nuclear weapons above ground, soils began to accumulate radionuclides at low levels worldwide. Lead-210, krypton-85, cesium-137, iodine-131, and strontium-90 were all monitored and were bioaccumulating in pasteurized milk. Due to the accumulation in grass and milk, above-ground bomb testing was eventually halted.

The signal of radioisotopes from nuclear testing remains in lake sediments around the world, and is used to date recent sediment cores. The persistence of radioactivity can be read as a warning of the widespread soil contamination that would result from even a limited nuclear war and the vulnerability of our soil resource to planetary atmospheric pollution.

Radionuclides were transported to soils from the Chernobyl nuclear reactor explosion in 1986. Iodine-131, with a short half-life (8 days), was assimilated in human thyroid glands; the source was contaminated milk from cows grazing in contaminated areas. Cesium-137, with a half-life of 30 years, continues to contaminate soils hundreds of miles from Chernobyl in two countries, Ukraine and Belarus.

Climate Change and Soils

Soils will both affect and be affected by climate change. Climate change can directly affect soil development, and it can also affect the transfer of trace gases between the soil and the atmosphere. Radiatively important trace gases that emanate from and return to soils include carbon dioxide, methane, and nitrogen oxides (especially nitrous oxide, N_2O). Soil moisture affects all of these gas exchanges, and ammonia concentrations in soils control nitrogen balances.

Methane

Methane accounts for 15% of the total greenhouse radiative trapping effect of anthropogenic trace gases. Methane gas concentrations in the atmosphere are increasing at a rate of about 1% annually. The increase in methane is probably due both to increases in methane sources and losses of methane sinks (Ayres, Schlesinger, and Socolow, this volume).

Wet soils are important sources of atmospheric methane. The largest methane sources are believed to be rice paddies and natural wetlands, and thus increased rice production can lead to increased methane emissions, while draining wetlands has the opposite effect. Dry soils can be methane sinks, but this is a much smaller effect: soil microbes in dry soils are believed to be sinks for only about 2% of annual methane emissions (Schlesinger, 1991).

Carbon Dioxide

Soils play a critical role in regulating the carbon dioxide concentration of the atmosphere. Soil supports vegetation and the primary production and respiration/decomposition that occur there. When soil organic matter and litter decompose on the forest floor, carbon dioxide is emitted into the atmosphere. The higher the soil temperature, the faster the decomposition of soil organic matter, and the lower the quantity of soil organic carbon. Significantly warmer surface temperatures would increase the decomposition rate of soil organic matter and would cause a large input of carbon dioxide to the atmosphere—a positive feedback effect, if this additional carbon dioxide resulted in a further increase in surface temperatures.

The Intergovernmental Panel on Climate Change (IPCC) reports an expected 0.03°C/yr increase in average air temperature at the surface, based on current trends in the increase of atmospheric trace gases (IPCC, 1990). Jenkinson *et al.* (1991) project a net flux of carbon dioxide from soils of about 1 gigaton/yr carbon during the next 60 years if global soils are warmed according to the IPCC projections. For comparison, about 6 gigatons of carbon are emitted from current anthropogenic fossil fuel combustion each year.

Countering this effect, increasing concentrations of carbon dioxide in the atmosphere and increasing surface temperatures may "fertilize" plants and increase global primary production. This negative feedback would slow the rise in carbon dioxide concentrations.

Implications for Industrial Ecology

The boundaries of industrial ecology are not well-defined. Industrial pollution, heavily mechanized agriculture, and the use of fertilizers, pesticides, and herbicides are usually included. Deforestation, overgrazing, and soil erosion are generally excluded.

But it appears that deforestation and soil degradation due to overgrazing and agriculture have a greater impact than "industrial effects," both on the productive capacity of the soils and on the ecosystems that soils support. This appears to be true worldwide. For the new enterprise of industrial ecology to be coherent, land use and soil and land degradation must be included.

Fortunately, a number of strategies can slow the rate of degradation. Revegetation and reforestation can renew degraded land and slow erosion, and at the same time sequester carbon. Revegetation can take up excess nitrogen in runoff, can assimilate and metabolize some organic chemicals, and can enhance pesticide transformations by soil microorganisms. Conservation tillage can improve soil quality by returning more crop residues to the soil, by disturbing the soil less, and by using less fossil fuel for farm tractors.

Vegetation in riparian zones, that border streams and lakes, is especially important for water quality because it is the last soil through which water is transported before it enters the stream (Licht, 1990; Nair and Schnoor, 1992; Paterson and Schnoor, 1993).

However, protecting land from current agricultural practices will require much more than reforestation and stabilization of riparian zones. In many areas, achieving sustainable agriculture will require radical changes from current practices. Although much current soil degradation can be attributed to growing pressures for food production, the global need for food security may also encourage improved land management practices.

Damage to soil and methods for its repair have been studied only on a piecemeal basis. On a global scale, the sustainability of current land-use practices is not well understood, much less the prospects for improvement. Can livestock production be compatible with sustained land quality? Can crop production proceed with no net loss of soil? With overgrazing and agriculture responsible for so much damage to soils, these questions are central to the welfare of human society and to biological diversity.

Acknowledgments

This research would not have been possible without the help of Jerald Schnoor's colleagues and students at the University of Iowa including Louis Licht, Todd Rees, and Kurtis Paterson. Support for the work was provided by the United States Agency for International Development (USAID) funded cooperative agreement DHR-5555-A-00-1086. Valerie Thomas would like to thank R. Socolow and L. Solarzano for helpful comments on the manuscript. The views, interpretations, and any errors are those of the authors and should not be attributed to USAID, the United States Government, or anyone acting on their behalf.

References

Boutron, C. F., U. Gorlach, and J. P. Candelone. 1991. Decrease in anthropogenic lead, cadmium, and zinc in Greenland snows since the late 1960s. *Nature 353*, 153–156.

Carson, R. 1962. *Silent Spring*. Houghton-Mifflin, New York.

CEQ (Council on Environmental Quality). 1989. *Environmental Quality*. Twentieth annual report, Executive Office of the President, Washington, D.C., 494 pp.

Colborn, T., and C. Clement (eds.). 1992. *Chemically Induced Alterations in Sexual and Functional*

Development: The Wildlife/Human Connection. Princeton Scientific Publishing Co., Princeton, New Jersey.

Driscoll, C. T., C. E. Johnson, T. G. Siccama, and G. E. Likens. 1992. Biogeochemistry of trace metals at the Hubbard Brook Experimental Forest, New Hampshire. In *Proceedings of the Air and Waste Management Association*, Pergamon Press, New York.

FAO (U.N. Food and Agricultural Organization). 1993. *The 1990 Tropical Forest Resources Assessment*. K. D. Singh, 174, Vol. 55, 10–19.

Houghton, R. A. *et al.* 1987. The flux of carbon from terrestrial ecosystems to the atmosphere in 1980 due to changes in land use: Geographic distribution of the global flux. *Tellus 39B*, 122–139.

IASA (International Alliance for Sustainable Agriculture). 1987. *Breaking the Pesticide Habit*. International Alliance for Sustainable Agriculture, Minneapolis, Minnesota.

IPCC (Intergovernmental Panel on Climate Change). 1990. *Climate Change: The IPCC Scientific Assessment*. Intergovernmental Panel on Climate Change (J. T. Houghton, G. J. Jenkins, and J. J. Ephraums, eds.), Cambridge University Press, Cambridge, United Kingdom.

Jenkinson, D. S., D. E. Adams, and A. Wild. 1991. Model estimates of CO_2 emissions from soil in response to global warming. *Nature 351*, 304–309.

Licht, L. A. 1990. Ph.D. dissertation. The University of Iowa, Iowa City, Iowa, 249 pp.

McCloskey, J. M. and H. Spalding. 1989. A reconnaissance-level inventory of the amount of wilderness remaining in the world. *Ambio 18(4)*, 221–227.

McDonald, D. B. 1992. Cedar river operational ecological study. *Annual Report 1991*, Duane Arnold Energy Center, Palo, Iowa, 52 pp.

Moldan, B., and J. L. Schnoor. 1992. Czechoslovakia: Examining a critically ill environment. *Environmental Science and Technology 26(1)*, 14–21.

Nair, D. R., and J. L. Schnoor. 1992. Effect of two electron acceptors on atrazine mineralization. *Environmental Science and Technology 26(11)*, 2298–2300.

Oldeman, L. R., V. W. P. van Engelen, and J. H. M. Pulles. 1990. The extent of human-induced soil degradation. Annex 5 of *World Map of the Status of Human-Induced Soil Degradation: An Explanatory Note*, rev. 2nd edn., International Soil Reference and Information Centre, Wageningen, Netherlands.

OTA (Office of Technology Assessment). 1982. *Impacts of Technology on U.S. Cropland and Rangeland Productivity*. OTA-F-166, Office of Technology Assessment, U.S. Government Printing Office.

OTA. 1991. *Changing By Degrees: Steps to Reduce Greenhouse Gases*. Office of Technology Assessment, U.S. Government Printing Office.

Paterson, K. G., and J. L. Schnoor. 1993. Vegetative alteration of nitrate fate in unsaturated zone. *Journal Environmental Engineering 119(5)*, 986–993.

Pimentel, D., *et al.* 1993. Soil erosion and agricultural productivity. In *World Soil Erosion and Conservation* (D. Pimentel, ed.), Cambridge University Press.

Schlesinger, W. H. 1991. *Biogeochemistry: An Analysis of Global Change*. Academic Press.

WHO (World Health Organization). 1990. *Public Health Impact of Pesticides Used in Agriculture*. World Health Organization and U.N. Environment Programme, Geneva, Switzerland, p. 59.

World Resources Institute. 1992. *World Resources 1992–3*. Oxford University Press, New York.

17

The Vulnerability of Biotic Diversity

William H. Schlesinger

Abstract

The presence or absence of a single species can cause a dramatic change in ecosystems, but our ability to predict which species can cause such change is limited. The disappearance of species can indicate changes in the ability of an ecosystem to sustain life over a long period of time. While there is an enormous literature on the losses of species associated with various industrial activities, the focus here is on changes in the earth's atmosphere that may cause biotic impoverishment. These include nitrogen deposition from fossil fuel combustion, increases in atmospheric carbon dioxide, and increased exposure to ultraviolet light due to a decrease in stratospheric ozone, all of which can be expected to result in changes in ecosystem function and species diversity.

Introduction

There are probably close to ten million species on earth. They are the products of long-term evolution in a variety of natural ecosystems, or biomes, that extend from the Arctic tundra to the tropics and from high mountains to the deepest sea. The presence of life on earth has profoundly affected the basic environmental conditions on the surface of this planet. Preservation of the biosphere is essential for the preservation of those conditions in which the human species has evolved and flourished on earth.

Humans must now assume a stewardship role for the biotic diversity of the planet. Barring a catastrophic geologic event, we have the power to determine our own success in this role. That is, will humans allow the resources of the planet to be fully usurped to support our own growing population and its material desires? Or will we allow a diversity of other living things to persist in nature?

Species are disappearing at an alarming rate—perhaps at the greatest rate ever seen during the history of life on earth. Although devastating examples of local industrial pollution are well known, we may rightly ask if global industrialization is responsible for a general loss of biotic diversity on earth and if it will contribute to a continuing loss of species in the future. Industrial activity has the capacity to affect environmental conditions far removed from its actual location. Many of its effects are spread through the atmosphere and by changes in global climate. My objective here is to examine how global industrial activity may affect biotic diversity. I do not focus on acute exposures of natural ecosystems to air and water pollution or on species extinctions that are due to direct land disturbance, such as tropical deforestation. Rather, my goal is to document ways in which global industrialization may affect the biotic diversity of the planet by changes wrought at the global level.

Do Species Matter?

It is legitimate to begin by asking: does biotic diversity really matter? Certainly a diverse world will be a more interesting habitat for future generations of our own species. But human society might be able to persist in the absence of national parks and other spots of natural beauty (Dubos, 1976). Already we may be usurping up to 40% of the world's plant productivity (Vitousek *et al.*, 1986), and with our current rate of population growth we may soon usurp it all. A world where all the land is devoted to the needs of humans would certainly be a less refreshing and stimulating world, equivalent to the loss of all the world's great museums of art. Far more urgent, however, is the possibility that a disruption and simplification of the natural biosphere and its species diversity could so alter the environmental conditions of the planet as to lead to the demise of human society and perhaps even the human species. Our motivation in stewardship extends beyond the ethics and aesthetics of preservation to primary self-interest.

Throughout much of North America, populations of migratory songbirds have declined over the last several decades (Robbins *et al.*, 1989; Terborgh, 1989; Figure 1). The potential causes are manifold: loss and fragmentation of habitat in breeding and wintering grounds, pesticide use, predation, and parasitism (Holmes *et al.*, 1986; Sherry and Holmes, 1992). Just as, in coal mines, the demise of canaries once signaled unhealthful conditions, the slow demise of songbirds may be a global indication of unhealthful conditions in the biosphere. Warblers and other songbirds help keep forest insect populations in check (Holmes *et al.*, 1979). Should we be alarmed that the disappearance of songbirds may lead to outbreaks of insects and to the destruction and loss of forests over much of North America?

A century ago flocks of passenger pigeons darkened the skies of the eastern United States, and the American chestnut was a dominant species in the mixed hardwood forests of the Appalachians. Both species are now gone: the former extinct and the latter reduced to an insignificant level by a blight that infects any sizable individual. These species must have played a major role in biogeochemical cycling and primary production over the majority of the eastern half of this continent. What we now regard as "natural" forest is not natural at all. Yet despite some missing species, normal levels of ecosystem function seem to persist in this region. In this case, the ecosystem seems to have carried a certain amount of redundancy that has allowed adjustment to change.

Not so in southern New Mexico. Here, a century of overgrazing, periodic drought, and perhaps higher concentrations of atmospheric carbon dioxide have

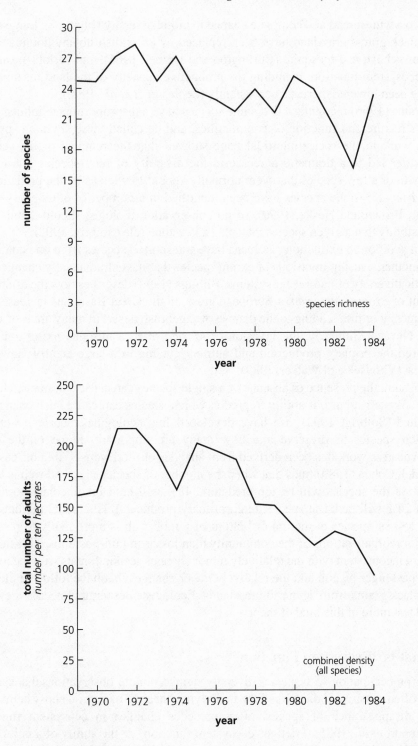

Figure 1. Decline in the density of birds at Hubbard Brook Experimental Forest, New Hampshire, from 1969 to 1984 (from Holmes et al., 1986, reprinted by permission).

led to a widespread and complete change in biotic diversity (Figure 2). Large areas of black grama grassland have been replaced by an arid shrubland dominated by creosotebush and mesquite (Buffington and Herbel, 1965), and various measures of ecosystem function, including the productive capacity of the land for humans, have been lowered, perhaps permanently (Schlesinger *et al.*, 1990).

Paine (1966) recognized that some species have a disproportionate influence on the structure and function of communities, and he called these *keystone* species. His work in the rocky intertidal zone showed that the removal of the starfish *Pisaster* led to a dramatic decrease in the diversity of prey species due to the growth of a few species that were normally kept at low levels by the predation of *Pisaster*. Keystone species have been identified in a number of other ecosystems (e.g., Brown and Heske, 1990), so no conservation biologist should dismiss the possibility that a given species may play a keystone role (Solbrig, 1992).

In addition to extinctions, humans have transported species from one continent to another, causing invasions of exotic species that have dramatically changed the biotic diversity of natural ecosystems. Billings (1990) describes how the establishment of *Bromus tectorum*, a European grass, in the Great Basin has increased the frequency of fire, leading to the demise of sagebrush desert in many areas of western Nevada. Similarly, the invasion of *Myrica faya*, a nitrogen-fixing tree, has altered the primary production and nutrient cycling in a large area of Hawaiian forest (Vitousek and Walker, 1989).

In sum, the presence or absence of a single species can cause a dramatic change in ecosystems, but our ability to predict which species can cause such changes is limited (Solbrig, 1992). We have developed few ecological theories to tell us which species to preserve and how many are important. Some of the most provocative work has been derived from analysis of food webs. Using his cascade model, Cohen (1989) finds that when the number of species in a food web is large, 26% of the species will be top predators, 48% will hold intermediate positions, and 26% will be basal species (e.g., primary producers). His model predicts that the loss of species at the top or bottom of a food web is more likely to result in major reorganizations of the community than losses in mid-positions. This finding is not inconsistent with the relatively minor changes seen in forests with the loss of the passenger pigeon and the relatively major changes that have followed the loss of black grama from semiarid grasslands. Ecologists desperately need to develop and test more of this kind of theory.

What Is "Ecosystem Function"?

Our preoccupation with species diversity stems from our observations that species are often sensitive indicators of the underlying "health" of natural ecosystems, and the disappearance of species often precedes changes in ecosystem function (Rapport *et al.*, 1985). I define "ecosystem function" as the ability of a unit of the earth's surface to sustain life over a long period of time. As the human population grows and its technology both links and affects the entire planet, we must expand

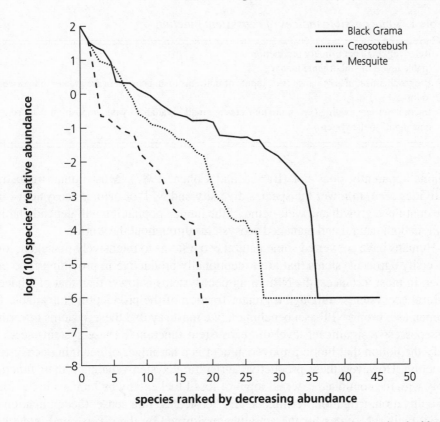

Figure 2. Plant species diversity in black grama grasslands, creosotebush, and mesquite shrublands in southern New Mexico (Huenneke, unpublished data).[†]

this concept to the global scale. The current emphasis on earth system science reflects our new perception that the entire planet acts as a single ecosystem that must continue to function if life is to persist on earth.

There are a variety of possible indices of ecosystem function (Table 1), but I suggest that the single best index is net primary production (NPP), which determines the amount of sunlight energy that is captured by plants to fuel life on earth. In any ecosystem, NPP is constrained by climate and geology, so its use as an index of ecosystem function must be relative, i.e., the NPP currently available to sustain life as a proportion of the long-term average NPP that has supported the persistence of species in the ecosystem in question.

Among natural ecosystems, the relationship of NPP to species diversity is not direct. The biomass and productivity of herbivores seem to increase with NPP (McNaughton et al., 1989; Oesterheld et al., 1992), but the overall length of food

---

[†] The points along each curve represent the relative abundance of the species encountered in each community, plotted as a declining function from the most abundant species to the least. This forms a dominance-diversity curve (see Whittaker and Woodwell, 1973).

Table 1: *Some possible indices of ecosystem function*

1. Relatively high net primary production
2. Stable levels of soil organic matter
3. Biogeochemical mass balance (i.e., inputs of nutrients to an ecosystem are greater than or equal to losses)
4. Internal nutrient cycling (i.e., more nutrients are made available by decomposition than from new inputs to the system).

chains apparently does not (Briand and Cohen, 1987). Most human industrial activities seem to lower the species diversity and NPP of natural ecosystems, but the continued growth and well-being of the human population will depend on high NPP in both natural and managed ecosystems throughout the world.

Humans have converted some natural ecosystems to intensively managed, low-diversity agroecosystems that are exceptionally productive in providing food and fiber. In most instances, the NPP of agroecosystems is lower than that of adjacent natural ecosystems, although a greater fraction of the production is available for human use. Using NPP as a benchmark, we must say that these systems neverthe-less possess a significant level of "ecosystem function." These systems seem to defy the notion that biotic impoverishment is a harbinger of declining ecosystem function. However, these productive agricultural systems maintain their function only with the continued external input of fossil fuel energy by humans in the form of cultivation, irrigation, fertilizers, and herbicides. In a sense, the application of fossil fuels substitutes for the activities performed by the diverse and integrated food web of a natural ecosystem. We substitute pesticides for insect predators and fertilizers for the efficient turnover of soil nutrients by decomposers. The true net productivity of managed, agricultural systems should be calculated by subtracting the fossil net production that has been used to sustain them (Pimentel *et al.*, 1973).

Potential Causes of Biotic Impoverishment

An enormous literature describes losses of species that are associated with various industrial activities. Here, I concentrate on three cases in which industrialization may cause biotic impoverishment by global changes in the earth's atmosphere.

Nitrogen Deposition

Vegetation on much of the earth's land surface exists in a state of nitrogen defi-ciency, due in part to the low natural rate of nitrogen fixation (Gutschick, 1981; Vitousek and Howarth, 1991) and persistent losses of available nitrogen to denitri-fication and nitrate leaching (Schlesinger, 1991). Through industrial activities, humans have roughly doubled the supply of fixed nitrogen on land, and most of it is distributed rather haphazardly on the earth's surface (Lyons *et al.*, 1990; Muller 1992; Schlesinger and Hartley, 1992). Recent layers of the Antarctic snowpack show increased levels of nitrate, as a result of the global dispersion of industrial fixed nitrogen (Mayewski and Legrand, 1990). Thus, we might anticipate that an excess atmospheric deposition of fixed nitrogen in acid rain could act as a fertilizer, stimulating plant growth in many regions. Indeed, recent work suggests that the forests of central Europe are now growing much faster than they were several decades ago, in part due to excess nitrogen deposition from the atmosphere (Kauppi et al., 1992). However, we know from a large ecological literature that the fertilization of natural ecosystems, perhaps first noted in the eutrophication of lakes, is likely to result in a loss of species diversity (e.g., Bakelaar and Odum, 1978; Tilman, 1987; Huenneke et al., 1990).

Studies of natural and experimental populations of plants often show that an increase in plant growth in fertile conditions is associated with intense competition. A long-time fertilized pasture in Great Britain showed a remarkable reduction in plant diversity from 1856 to the present, as a large number of rare species were replaced by a small number of aggressive weedy species that assumed extreme dominance in the community (Figure 3). In experimental mixtures of plants, Austin and Austin (1980) also showed that species diversity is lowest and the concentration of dominance is greatest in fertile conditions. Ecologists have ample theories to explain these observations; the maximum species diversity in natural communities will persist when resources are low, so that each species is limited by a different resource and no one species can outcompete the rest (Levin, 1970; Rosenzweig, 1971; Tilman, 1982). Any addition of a resource to such a community will lead to the dominance of the species that can use that resource most efficiently.

The decline in species diversity with the addition of unusual levels of essential resources resembles the decline that results from exposure to an overt ecosystem stress, such as the addition of a toxic pollutant (Odum et al., 1979). Woodwell (1970) proposed the unifying concept *retrogression* to emphasize that the changes in species diversity as a result of ecosystem stress or subsidy are opposite to the trends seen during the recovery of ecosystems after disturbance (e.g., Bazzaz, 1975). During retrogression, communities are likely to lose species first, followed by changes in ecosystem function, such as declining net primary production and increasing nutrient loss (Rapport et al., 1985; Whittaker and Woodwell, 1973).

Atmospheric deposition of nitrogen in the Netherlands (50 kg/ha/yr) is somewhat smaller than that applied as fertilizer to the pastureland in Britain (144 kg/ha/yr), but heathlands and pastures of the Netherlands show recent trends in species diversity that are similar to the changes seen in fertilization experiments. Over large areas, formerly diverse pastures are now dominated by *Brachypodium*, which increases upon experimental applications of fertilizer (Bobbink, 1991). In other areas *Calluna* heathlands have been invaded by grasslands of *Deschampsia* and *Molinia* (Heil and Diemont, 1983). The changes in these plant communities appear to be driven by changes in the competitive relations among species, and changes in basic ecosystem properties have followed (Table 2). Unlike the losses of species that are seen as a result of local, toxic industrial emissions, these changes in plant community structure represent a regional loss of biotic diversity,

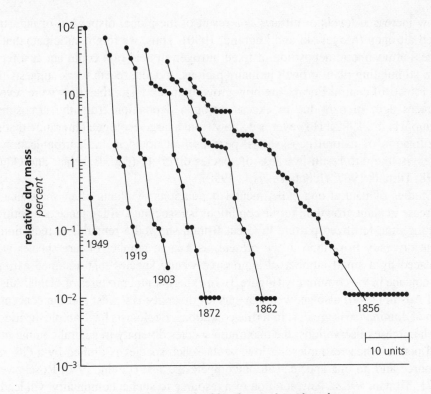

species ranked by decreasing abundance

Figure 3. Plant species diversity in pasture in the Parkgrass Experiment, Rothamstead, United Kingdom, following the continuous application of nitrogen fertilizer: 1856 to 1949 (reproduced from Kempton, 1979, with permission from the Biometric Society).[*]

as a result of persistent exposure to an effluent of the industrialized world that would superficially seem to be beneficial to plant communities. The ecosystem shows signs of eutrophication, i.e., it is more productive but has lower species diversity and a greater potential for nitrate loss. Persistent atmospheric deposition of fixed nitrogen to forest ecosystems may eventually lead to retrogressive changes in ecosystem function, including greater losses of nitrous oxide (N_2O), a greenhouse gas, to the atmosphere and greater leaching of nitrate in runoff (Aber *et al.*, 1989). Nitrate loss reduces the water quality for human use and increases the eutrophication of downstream ecosystems.

Atmospheric Carbon Dioxide

The atmosphere now contains about 360 ppm of carbon dioxide. Prior to the Industrial Revolution, the concentration was about 270 ppm. This increase means

---

[*] The plotted line is a dominance-diversity curve as in Figure 2. Species abundance < 0.01% are plotted as 0.01%.

Table 2: *Changes in ecosystem properties associated with vegetation change in Dutch heathlands*

Vegetation	Net Primary Production (g/m^2/yr)	Net N Mineralization (gN/m^2/yr)	Percent Nitrified
Calluna (native)	730	6.2	4.8
Molinia (invader)	2050	10.9	33.0
Deschampsia (invader)	430	12.6	42.9

Note: As species composition has changed, the potential for nitrogen loss as nitrate has increased. (Nitrogen mineralization is a measure of the rate at which nitrogen is made available to plants; the percent nitrified is the percentage of the mineralized nitrogen that is converted to nitrate, which has the potential for being leached from the soil. Measurements were made in areas dominated by Calluna, Molinia, and Deschampsia, respectively.)

From Van Vuuren et al., 1992. Reprinted by permission of Klumer Academic Publishers.

that no natural ecosystem, no matter how remote or protected, remains unaffected by the human species and its industrial activities. Changes in atmospheric carbon dioxide have a variety of well-documented direct effects on plants, including increased rates of photosynthesis and greater water-use efficiency. However, different species have different rates of response to high carbon dioxide, and the extent to which the concentration of carbon dioxide in today's atmosphere has altered the competitive relationships among plants could well lead to a loss of species in some ecosystems (e.g., Miao *et al.*, 1992). Thus, industrial activity may be responsible for biotic impoverishment far removed from the source of carbon dioxide release.

Of particular interest is the relation between C3 plants[1], which make up the bulk of the earth's vegetation, and C4 plants, which include a variety of well-known and economically important grasses that are often superior competitors in hot, dry conditions. C3 plants are much more responsive to high concentrations of atmospheric CO_2, which increases their water-use efficiency and competitive ability in dry climates (Bazzaz and Carlson, 1984; Bazzaz and Fajer, 1992). Johnson (1992) suggests that the invasion of black grama grasslands (C4) by creosotebush and mesquite during the last century may be due to an increased competitive ability of these C3 shrubs as atmospheric CO_2 has risen. If he is correct, then rising CO_2 has

[1] Plants are classified as C3 plants when the first product of their photosynthetic reaction is a sugar containing three carbon atoms; in C4 plants this product is a four-carbon sugar. This difference in biochemistry leads to other physiological differences, such as the differential response to carbon dioxide.

led to the biotic impoverishment of much of the southwestern United States (cf. Figure 2).

The indirect effects of greater atmospheric carbon dioxide, i.e., global climate changes, are potentially even more severe. Most general circulation models (GCMs) of global climate suggest that a global warming of 1.5 to 4.5°C is likely within the next century. Long-recognized correlations between the distribution of land vegetation and temperature suggest that a directional climatic change of this magnitude would cause major changes in the distribution of the earth's vegetation (Emanuel *et al.*, 1985), especially where temperature now determines the range of the dominant species (e.g., Arris and Eagleson, 1989; Black and Bliss, 1980). One climate model predicts that the future extent of tundra and boreal forest ecosystems will be 32–37% lower than present-day values (Emanuel *et al.*, 1985), while another suggests a loss of 63–66% of the current area of these ecosystems (Prentice and Fung, 1990).

Although broad-scale reorganizations in terrestrial plant communities are not unknown in the geologic record, the rate of anticipated climatic change is unprecedented (Woodwell, 1989). Climatic changes equivalent to those that brought jack pine from the coastal plain of North Carolina to the forests of southern Canada during several thousand years at the end of the last glacial would occur in only several hundred years. In eastern North America, the predicted range of some forest species is 500 to 1000 km north of their present distribution (Overpeck *et al.*, 1991). Those species whose dispersal and establishment cannot accommodate this rate of change within 100 years will be likely to go extinct unless their populations are directly managed by humans.

Climatic warming may have particularly severe effects on the fauna of lakes, from which dispersal to new habitats is problematic. Schindler *et al.* (1990) show that the length of the ice-free season has already increased by 3 weeks/yr in lakes of northwestern Ontario. Deepening of the thermocline has reduced the habitat for coldwater species, such as lake trout, which may disappear from shallow lakes if the warming continues. Although we do not know whether this recent warming has been caused by the global greenhouse effect, its effects are indicative of the kinds of changes that we might expect in lake ecosystems during the anticipated greenhouse warming of the future. Lake morphometry may strongly influence these effects. Using a model for the thermal structure of lakes, Magnuson (1991) suggests that global warming might actually *increase* the favorable habitat for coldwater species in large, deep lakes, such as Lake Superior.

Most models of global climate change predict drier conditions for the interior portion of continents (e.g., Manabe and Wetherald, 1987), especially as the temperature of the land and its rate of evapotranspiration increase more rapidly than that of the sea surface (Rind *et al.*, 1990). Although not yet statistically significant, current trends in climate for central North America are consistent with these predictions (Karl *et al.*, 1991). Hot, dry conditions in the central United States may lead to the expansion of areas of desert shrubland at the expense of semiarid grasslands, despite the superior growth of C4 grasses in dry environments (Neilson,

1986; Schlesinger *et al.*, 1990). The prairie pothole wetlands, so important as breeding grounds for waterfowl, may dry up (Poiani and Johnson, 1991).

All of these examples show the potential for human use of fossil fuels to affect global biotic diversity by altering global climate through the greenhouse effect. Moreover, the expected changes in terrestrial ecosystems are not likely to increase the role of vegetation and soils as a sink for CO_2 that might slow the rate of greenhouse warming in the future (Schlesinger, 1993; but see also Prentice and Fung, 1990). Of course, none of these predictions is proven, but the magnitude of the potential global change is sufficient to merit our deepest concern. Feedbacks in the global system will complicate our ability to develop effective international policy. For example, an onset of global warming may stimulate decomposition in cold, northern ecosystems, releasing CO_2 from soils that will further contribute to the greenhouse effect (Jenkinson *et al.*, 1991; Raich and Schlesinger, 1992).

Changes in Ultraviolet (UV-B) Light

No environmental skeptic should now doubt that there has been a dramatic decline in the ozone content of the polar stratosphere, and recent reports suggest that the depletion periodically may extend into the North Temperate Zone (Stolarski *et al.*, 1992; Correll *et al.*, 1992). Atmospheric chemists are in fair agreement that a variety of chemicals of the human industrialized world are the cause of ozone destruction (Rowland, 1989; Solomon, 1990). Physicists tell us that ozone is an effective absorber of ultraviolet radiation in the short mutagenic wavelengths of 280–320 nm, so ozone depletion should lead to a greater flux of such radiation to the earth's surface. As a form of high-energy electromagnetic radiation, ultraviolet radiation should have effects on ecosystems that are similar to, though less severe than, the effects of gamma radiation, documented by Woodwell (1967) and Woodwell and Houghton (1990).

We know that UV-B is associated with risk of skin cancer and cataracts in humans, but we know surprisingly little about the effect of UV-B radiation on most other organisms. Depletion of stratospheric ozone should lead to a greater period of exposure of the plants and animals of high-latitude ecosystems to UV-B radiation (Frederick and Snell, 1988). Unfortunately, these species may be the most susceptible, having evolved in an environment of relatively low natural exposure to these wavelengths (Robberecht *et al.*, 1980).

Wide variation exists in the tolerance of plant species to UV-B radiation. Field studies suggest that the dominant effect of UV-B radiation may be to alter competitive relations among species, allowing different species to achieve dominance (Gold and Caldwell, 1983; Barnes *et al.*, 1988). As a result of changes in plant biochemistry, a wide variety of species may become more susceptible to insect pests (Berenbaum, 1988; Caldwell *et al.*, 1989). The parallels to Woodwell's early work are strong; current depletion of stratospheric ozone suggests an alarming potential for a reduction in the diversity and activity of the biosphere. Other signs of retrogression may follow. Already, Smith *et al.* (1992) find that phytoplankton production in the Antarctic ocean may be reduced by 6–12% as a result of a greater flux of ultraviolet radiation.

Ozone depletion in temperate regions is less severe, and the anticipated increase in the flux of UV-B to the earth's surface may be somewhat attenuated by increases in tropospheric ozone from air pollution. Scotto *et al.* (1988) report no measurable increase in UV-B at eight locations in the United States between 1974 and 1985. Our ability to detect the severity of the global change in UV-B is hampered, however, by the absence of a network of stations to measure it throughout the world.

Conclusions

In each case—nitrogen deposition, CO_2 fertilization, and ozone depletion—global industrialization is likely to cause a loss of species from the planet. While higher supplies of nitrogen and CO_2 may increase the growth of some crops, agricultural yields are also likely to suffer from global climate change and greater ultraviolet-B radiation (Adams *et al.*, 1990; Caldwell *et al.*, 1989). Impoverished ecosystems, both natural and agricultural, are vulnerable to instability from greater pestilence, creating an upward spiral in the need for the human industrialization of nature to provide food, fiber and shelter for our exponentially increasing population. Those species that survive are likely to be a group of hardy generalists that are superior competitors in disturbed environments—in short, weeds. Cause and effect will be difficult to prove, but a loss of species often portends other changes in ecosystems that reduce their carrying capacity for life (Rapport *et al.*, 1985). The trends of retrogression, clearly demonstrated in the response of ecosystems to acute exposures to stress, will be played out at the global level. Unfortunately, we know relatively little about how biotic diversity relates to earth system function. It is as if we are slowly dismantling an intricate watch, all the while expecting that it will continue to function, without knowing when we will throw out the mainspring. Can we fully usurp the biotic productivity of this planet for our own species? Is there time to stop?

Acknowledgments

This manuscript has benefited from critical reviews by Anne Cross, Inge Horkeby, Laura F. Huenneke, Matt Leibold, Jim McNeal, and Lisa D. Schlesinger; I thank them all. The work is a contribution to the Long-Term Ecological Research (LTER) program at the Jornada Experimental Range, New Mexico, supported by the National Science Foundation.

References

Aber, J. D., K. J. Nadelhoffer, P. Steudler, and J. M. Melillo. 1989. Nitrogen saturation in northern forest ecosystems. *BioScience 39*, 378–386.

Adams, R. M., C. Rosenzweig, R. M. Peart, J. T. Ritchie, B. A. McCarl, J. D. Glyer, R. B. Curry, J. J. Jones, K. J. Boote, and L. H. Allen. 1990. Global climate change and US agriculture. *Nature 345*, 219–224.

Arris, L. L., and P. S. Eagleson. 1989. Evidence for a physiological basis for the boreal-deciduous forest ecotone in North America. *Vegetatio 82*, 55–58.

Austin, M. P., and B. O. Austin. 1980. Behaviour of experimental plant communities along a nutrient gradient. *Journal of Ecology 68*, 891–918.

Bakelaar, R. G., and E. P. Odum. 1978. Community and population level responses to fertilization in an old-field ecosystem. *Ecology 59*, 660–665.

Barnes, P. W., P. W. Jordan, W. G. Gold, S. D. Flint, and M. M. Caldwell. 1988. Competition, morphology, and canopy structure in wheat (*Triticum aestivum* L.) and wild oat (*Avena fatua* L.) exposed to enhanced ultraviolet-B radiation. *Functional Ecology 2*, 319–330.

Bazzaz, F. A. 1975. Plant species diversity in old-field successional ecosystems in southern Illinois. *Ecology 56*, 485–488.

Bazzaz, F. A., and R. W. Carlson. 1984. The response of plants to elevated CO_2. I. Competition among an assemblage of annuals at two levels of soil moisture. *Oecologia 62*, 196–198.

Bazzaz, F. A., and E. D. Fajer. 1992. Plant life in a CO_2-rich world. *Scientific American 266(1)*, 68–74.

Berenbaum, M. 1988. Effects of electromagnetic radiation on insect–plant interactions. In *Plant Stress–Insect Interactions* (E.A. Heinrichs, ed.), John Wiley and Sons, New York, 167–185.

Billings, W. D. 1990. *Bromus tectorum*, a biotic cause of ecosystem impoverishment in the Great Basin. In *The Earth in Transition: Patterns and Processes of Biotic Impoverishment* (G.M. Woodwell, ed.), Cambridge University Press, Cambridge, United Kingdom, 301–322.

Black, R. A., and L. C. Bliss. 1980. Reproductive ecology of *Picea mariana* (Mill.). BSP at tree line near Inuvik, Northwest Territories, Canada. *Ecological Monographs 50*, 331–354.

Bobbink, R. 1991. Effects of nutrient enrichment in Dutch chalk grassland. *Journal of Applied Ecology 28*, 28–41.

Briand, F., and J. E. Cohen. 1987. Environmental correlates of food chain length. *Science 238*, 956–960.

Brown, J. H., and E. J. Heske. 1990. Control of a desert–grassland transition by a keystone rodent guild. *Science 250*, 1705–1707.

Buffington, L. C., and C. H. Herbel. 1965. Vegetational change on a semidesert grassland range from 1858 to 1963. *Ecological Monographs 35*, 139–164.

Caldwell, M. M., A. H. Teramura, and M. Tevini. 1989. The changing solar ultraviolet climate and the ecological consequences for higher plants. T*rends in Ecology and Evolution 4*, 363–367.

Cohen, J. E. 1989. Food webs and community structure. In *Perspectives in Ecological Theory* (J. Roughgarden, R.M. May, and S.A. Levin, eds.), Princeton University Press, Princeton, New Jersey, 181–202.

Correll, D. L., C. O. Clark, B. Goldberg, V. R. Goodrich, D. R. Hayes, W. H. Klein, and W. D. Schecher. 1992. Spectral ultraviolet-B radiation fluxes at the Earth's surface: Long-term variations at 39°N, 77°W. *Journal of Geophysical Research 97*, 7579–7592.

Dubos, R. 1976. Symbiosis between the earth and humankind. *Science 193*, 459–462.

Emanuel, W. R., H. H. Shugart, and M. P. Stevenson. 1985. Climatic change and the broad-scale ldistribution of terrestrial ecosystem complexes. *Climatic Change 7*, 29–43, 457–460.

Frederick, J. E., and H. E. Snell. 1988. Ultraviolet radiation levels during the Antarctic spring.

Science 241, 438–440.

Gold, W. G., and M. M. Caldwell. 1983. The effects of ultraviolet-B radiation on plant competition in terrestrial ecosystems. *Physiologica Plantarum 58*, 435–444.

Gutschick, V. P. 1981. Evolved strategies in nitrogen acquisition by plants. *American Naturalist 118*, 607–637.

Heil, G. W., and W. H. Diemont. 1983. Raised nutrient levels change heathland into grassland. *Vegetatio 53*, 113–120.

Holmes, R. T., J. C. Schultz, and P. Nothnagle. 1979. Bird predation on forest insects: An exclosure experiment. *Science 206*, 462–463.

Holmes, R. T., T. W. Sherry, and F. W. Sturges. 1986. Bird community dynamics in a temperate deciduous forest; Long-term trends at Hubbard Brook. *Ecological Monographs 56*, 201–220.

Huenneke, L. F., S. P. Hamburg, R. Koide, H. A. Mooney, and P. M. Vitousek. 1990. Effects of soil resources on plant invasion and community structure in Californian serpentine grassland. *Ecology 71*, 478–491.

Jenkinson, D. S., D. E. Adams, and A. Wild. 1991. Model estimates of CO_2 emissions from soil in response to global warming. *Nature 351*, 304–306.

Johnson, H. 1992. Increasing CO_2 and plant–plant interactions; Effects on natural vegetation. In *CO_2 and the Biosphere* (J. Rozema, H. Lambers, S.C. van de Geijn, and M.L. Cambridge, eds.), Kluwer Academic Publishers, Dordrecht, Netherlands.

Karl, T. R., R. R. Heim, and R. G. Quayle. 1991. The greenhouse effect in central North America: If not now, when? *Science 251*, 1058–1061.

Kauppi, P. E., K. Mielikainen, and K. Kuusela. 1992. Biomass and carbon budget of European forests, 1971 to 1990. *Science 256*, 70–74.

Kempton, R. A. 1979. The structure of species diversity and measurement of diversity. *Biometrics 35*, 307–321.

Levin, S. A. 1970. Community equilibria and stability, and an extension of the competitive exclusion principle. *American Naturalist 104*, 413–423.

Lyons, W. B., P. A. Mayewski, M. J. Spencer, and M. S. Twickler. 1990. Nitrate concentrations in snow from remote areas: Implications for the global NO_x flux. *Biogeochemistry 9*, 211–222.

Magnuson, J. J. 1991. Fish and fisheries ecology. *Ecological Applications 1*, 13–26.

Manabe, S., and R. T. Wetherald. 1987. Large-scale changes in soil wetness induced by an increase in atmospheric carbon dioxide. *Journal of Atmospheric Science 44*, 1211–1253.

Mayewski, P. A., and M. R. Legrand. 1990. Recent increase in nitrate concentration of Antarctic snow. *Nature 346*, 258–260.

McNaughton, S. J., M. Oesterheld, D. A. Frank, and K. J. Williams. 1989. Ecosystem-level patterns of primary productivity and herbivory in terrestrial habitats. *Nature 341*, 142–144.

Miao, S. L., P. M. Wayne, and F. A. Bazzaz. 1992. Elevated CO_2 differentially alters the responses of cooccurring birch and maple seedlings to a moisture gradient. *Oecologia 90*, 300–304.

Muller, J.-F. 1992. Geographic distribution and seasonal variation of surface emissions and deposition velocities of atmospheric trace gases. *Journal of Geophysical Research 97*, 3787–3804.

Neilson, R. P. 1986. High-resolution climatic analysis and Southwest biogeography. *Science 232*, 27–34.

Odum, E. P., J. T. Finn, and E. H. Franz. 1979. Perturbation theory and the subsidy-stress gradient. *BioScience 29*, 349–352.

Oesterheld, M., O. E. Sala, and S. J. McNaughton. 1992. Effect of animal husbandry on herbivore-carrying capacity at a regional scale. *Nature 356*, 234–236.

Overpeck, J. T., P. J. Bartlein, and T. Webb. 1991. Potential magnitude of future vegetation change in eastern North America; Comparisons with the past. *Science 254*, 692–695.

Paine, R. T. 1966. Food web complexity and species diversity. *American Naturalist 100*, 65–76.

Pimentel, D., *et al.* 1973. Food production and the energy crisis. *Science 182*, 443–448.

Poiani, K. A., and W. C. Johnson. 1991. Global warming and prairie wetlands. *BioScience 41*, 611–618.

Prentice, K. C., and I. Y. Fung. 1990. The sensitivity of terrestrial carbon storage to climate change. *Nature 346*, 48–51.

Raich, J. W., and W. H. Schlesinger. 1992. The global carbon dioxide flux in soil respiration and its relationship to vegetation and climate. *Tellus 44B*, 81–89.

Rapport, D. J., H. A. Regier, and T. C. Hutchinson. 1985. Ecosystem behavior under stress. *American Naturalist 125*, 617–640.

Rind, D., R. Goldberg, J. Hansen, C. Rosenzweig, and R. Ruedy. 1990. Potential evapotranspiration and the likelihood of future drought. *Journal of Geophysical Research 95*, 9983–10004.

Robberecht, R., M. M. Caldwell, and W. D. Billings. 1980. Leaf ultraviolet optical properties along a latitudinal gradient in the Arctic-Alpine life zone. *Ecology 61*, 612–619.

Robbins, C. S., J. R. Sauer, R. S. Greenberg, and S. Droege. 1989. Population declines in North American birds that migrate to the neotropics. *Proceedings of the National Academy of Sciences US 86*, 7658–7662.

Rosenzweig, M. L. 1971. Paradox of enrichment: Destabilization of exploitation ecosystems in ecological time. *Science 171*, 385–387.

Rowland, F. S. 1989. Chlorofluorocarbons and the depletion of stratospheric ozone. *American Scientist 77*, 36–45.

Schindler, D. W., K. G. Beaty, E. J. Fee, D. R. Cruikshank, E. R. DeBruyn, D. L. Findlay, G. A. Linsey, J. A. Shearer, M. P. Stainton, and M. A. Turner. 1990. Effects of climatic warming on lakes of the central boreal forest. *Science 250*, 967–970.

Schlesinger, W. H. 1991. *Biogeochemistry: An Analysis of Global Change*, Academic Press, San Diego, California.

Schlesinger, W. H. 1993. Response of the terrestrial biosphere to global climate change and human perturbation. *Vegetatio 104*, 295.

Schlesinger, W. H., and A. E. Hartley. 1992. A global budget for atmospheric NH_3. *Biogeochemistry 15*, 191–211.

Schlesinger, W. H., J. F. Reynolds, G. L. Cunningham, L. F. Huenneke, W. M. Jarrell, R. A. Virginia, and W. G. Whitford. 1990. Biological feedbacks in global desertification. *Science 247*, 1043–1048.

Scotto, J., G. Cotton, F. Urbach, D. Berger, and T. Fears. 1988. Biologically effective ultraviolet radiation: Surface measurements in the United States, 1974 to 1985. *Science 239*, 762–764.

Sherry, T. W., and R. T. Holmes. 1992. Population fluctuations in a long-distance neotropical migrant: Demographic evidence for the importance of breeding season events in the American redstart. In *Ecology and Conservation of Neotropical Migrant Landbirds* (Hagan and Johnston, eds.), Smithsonian Institution, Washington, D.C.

Smith, R. C., B. B. Prezelin, K. S. Baker, R. R. Bidigare, N. P. Boucher, T. Coley, K Karentz, S. MacIntyre, H. A. Matlick, D. Menzies, M. Ondrusek, Z. Wan, and K. J. Waters. 1992. Ozone depletion: Ultraviolet radiation and phytoplankton biology in Antarctic waters. *Science 255,* 952–959.

Solbrig, O. T. 1992. The IUBS-SCOPE-UNESCO program of research in biodiversity. *Ecological Applications 2,* 131–138.

Solomon, S. 1990. Progress towards a quantitative understanding of Antarctic ozone depletion. *Nature 347,* 347–354.

Stolarski, R., R. Bojkov, L. Bishop, C. Zerefos, J. Staehelin, and J. Zawodny. 1992. Measured trends in stratospheric ozone. *Science 256,* 342–349.

Terborgh, J. 1989. *Where Have All The Birds Gone?* Princeton University Press, Princeton, New Jersey.

Tilman, D. 1982. *Resource Competition and Community Structure,* Princeton University Press, Princeton, New Jersey.

Tilman, D. 1987. Secondary succession and the pattern of plant dominance along experimental nitrogen gradients. *Ecological Monographs 57,* 189–214.

Van Vuuren, M. M. I., R. Aerts, and F. Berendse. 1992. Nitrogen mineralization in heatherland ecosystems dominated by different plant species. *Biogeochemistry 16(3),* 151–166.

Vitousek, P. M., and R. W. Howarth. 1991. Nitrogen limitation on land and in the sea: How can it occur? *Biogeochemistry 13,* 87–115.

Vitousek, P. M., and L. R. Walker. 1989. Biological invasion by *Myrica faya* in Hawaii: Plant demography, nitrogen fixation, and ecosystem effects. *Ecological Monographs 59,* 247–265.

Vitousek, P. M., P. R. Ehrlich, A. H. Ehrlich, and P. A. Matson. 1986. Human appropriation of the products of photosynthesis. *BioScience 36,* 368–373.

Whittaker, R. H., and G. M. Woodwell. 1973. Retrogression and coenocline distance. In *Ordination and Classification of Communities* (R.H. Whittaker, ed.), Dr. W. Junk Publishers, The Hague, Netherlands, 55–73.

Woodwell, G. M. 1967. Radiation and patterns of nature. *Science 156,* 461–470.

Woodwell, G. M. 1970. Effects of pollution on the structure and physiology of ecosystems. *Science 168,* 429–433.

Woodwell, G. M. 1989. The warming of the industrialized middle latitudes 1985–2050: Causes and consequences. *Climatic Change 15,* 31–50.

Woodwell, G. M., and R. A. Houghton. 1990. The experimental impoverishment of natural communities: Effects of ionizing radiation on plant communities, 1961–1976. In *The Earth in Transition: Patterns and Processes of Biotic Impoverishment* (G.M. Woodwell, ed.), Cambridge University Press, Cambridge, United Kingdom.

18

Global Ecotoxicology: Management and Science

Susan Anderson

Abstract

Global environmental pollution is here defined to include widespread, low-level increases in environmental concentrations of toxic substances; the net effects of a patchwork of regional pollution problems; and the increase in ultraviolet radiation (UV-B) due to decreases in stratospheric ozone. Current understanding of the effects of global pollution on ecosystems is poor, especially for low-level, widespread contamination by toxic substances. A research agenda is proposed to focus on understanding the sublethal effects of toxic substances, the mechanisms of tolerance and adaptation, the relationship of elevated tissue levels of pollutants to health consequences for the organism, and the behavior and effects of toxic substances in complex media, such as in sediments and soils.

Introduction

It is well known that toxic substances have the potential to harm ecological systems. However, the role of such substances as agents of global, rather than regional, change is poorly understood. Do toxic substances (and the practices that introduce them) simply cause a patchwork of regional insults, or do they harm the biosphere in ways that have profound global implications? Are subtle global impacts more important than acute but localized ones? Are the effects of toxic substances significant in comparison to effects attributable to other agents of global change? These questions cannot be answered in depth until advances are made in the science of ecotoxicology.

This chapter examines the most critical scientific barriers to answering the questions posed above. First, characteristics of global pollution are developed. Second, scientific questions that are vital for an understanding of both global and regional pollution problems are discussed and linkages described. Finally, priorities for global ecotoxicology are proposed. Examples are taken primarily from the field of aquatic toxicology, and an emphasis is placed on the effects of contaminants rather than their fate in ecosystems.

For over two decades, scientists in the United States and much of the world have focused primarily on local and regional pollution assessment and control.

Largely, they have found that point-source inputs and acute effects of environmental pollution are relatively easy to study and are amenable to technological control (Stumm, 1986). However, it has also been recognized that evaluation and control of low, sublethal levels of environmental pollution emanating from diffuse sources are far more difficult tasks—a true curse upon the scientists and engineers who helped to solve so many pollution problems of the 1970s. This chapter forwards the proposal that the key scientific barriers to more sophisticated regional ecotoxicological assessments are largely the same as those confronting the newer area of global ecotoxicology. This viewpoint leads to the formulation of research priorities in global ecotoxicology and suggests that many key problems are truly at the frontiers of science.

Characteristics of Global Pollution

What are the key characteristics of global environmental pollution, and how should they be classified? One approach is to classify global pollution problems according to spatial and temporal characteristics. For example, localized acute emissions are quite distinct from diffuse, low-level emissions. Below, I have identified four key characteristics of global pollution problems based on this spatio-temporal approach (Figure 1).

Characteristic 1: Global Pollution Can Be Partially Characterized as a Patchwork of Regional Acute and Chronic Effects

Activities introducing such pollution include mining, oil development and processing, agriculture, pulp and paper production, municipal sewage disposal, industrial hazardous waste disposal, and many direct and indirect effects of urbanization, especially urban runoff. Current worldwide concerns include widespread mercury contamination of the Amazon Basin as a consequence of gold processing (Malm *et al.*, 1990; Martinelli *et al.*, 1988), acute pesticide poisoning in Central America (Whelan, 1988), fisheries impacts attributable to the *Exxon Valdez* oil spill in Alaska, and degradation of near coastal waters of the East Asian seas by untreated sewage (Gomez, 1988). These incidents represent only the breadth of the types of problems; they do not adequately represent the temporal and spatial distribution of such insults on a global level.

There has been no serious global assessment of regional impacts of pollution by toxic substances. Such an assessment would ideally characterize the nature of the polluting activity, the distribution of the substances or of their toxic effects, the duration of the exposure, and relevant information on the biological resources of the receptor site. This information, of course, is rarely all available; it is questionable whether ecological risk assessments can be performed using less complete data. Nevertheless, an understanding of the degree and extent of insults worldwide is necessary to determine whether acute regional insults confer greater harm to the global environment than do lower-level, more widespread effects. One also must

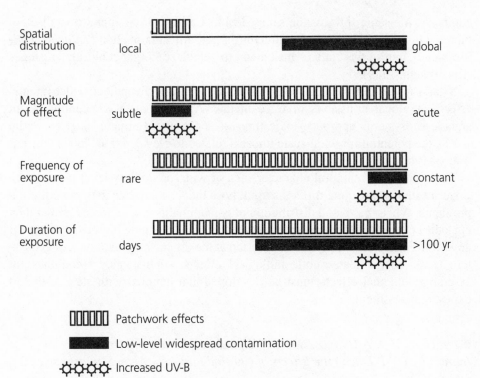

Figure 1. Spatial and temporal characteristics of global pollution.

seriously consider whether the detrimental effects of toxic substances are as significant as other agents of global change, such as tropical deforestation, war,
eutrophication, international water development, soil erosion, or potentially,
global warming.

*Characteristic 2: Global Pollution Can Be Partially Characterized by Low-Level
but Widespread Increases in the Occurrence of (or Background Concentrations
of) Numerous Toxic Substances*

The distribution of toxic substances in waterways of the world has been monitored
in international programs such as the Global Mussel Watch Program (Risebrough,
1989). Nevertheless, our understanding of the fate, fluxes, and effects of low-level
environmental pollution is not sufficient to make predictions of the potential for
broad-scale detrimental ecological effects. Regional examples, however, document that sublethal effects of environmental pollution can have significant impacts
on organisms in marine ecosystems. One of the most remarkable examples
includes the decline of brown pelican (*Pelecanus occidentalis*) populations in
California. This decline was attributed to the accumulation of DDT in body fat or
egg lipids of the birds; elevated DDT concentrations caused eggshell thinning
which resulted in unhatched eggs being crushed by the weight of incubating

adults. At the same time, at the same sites in California, an approximately 25% increase in the incidence of premature births of California sea lion (*Zalophus californianus*) pups was also attributed to elevated organochlorine residues (Risebrough, 1989).

Nriagu (this volume) has raised concern regarding the potential detrimental effects of global emissions of trace metals. As trace metal concentrations in aquatic ecosystems approach toxic thresholds for many organisms (e.g., Patin, 1985), the potential for wide-scale impacts related to elevated trace metal concentrations is becoming a serious issue.

The spatial and temporal characteristics of widespread, low-level increases in toxic substances are quite different from those that characterize regional pollution problems. When the spatial distribution of contaminants is widespread, rather than episodic or localized, the contaminants tend to produce subtle biological effects, if any at all. Such effects could be manifested at the organism, population, or ecosystem level. To elucidate subtle biological effects, sophisticated techniques for assessing sublethal effects must be developed and coupled with state-of-the-art exposure assessment.

Characteristic 3: A Unique Aspect of Global Pollution is the Occurrence of Increasing UV-B Due to the Effects of Chlorofluorocarbons on Stratospheric Ozone

Many of the challenges related to assessing the effects of low-level increases in ultraviolet light (UV-B) are similar to those described for low-level increases in toxic substances, but there are also some significant differences. First, in contrast to toxic contaminants the effects of solar UV on living cells are reasonably well characterized (Jagger, 1985). The mechanisms include a broad spectrum of DNA damage as well as effects on biological membranes. Designing programs to assess and monitor these selected mechanisms is a different task than is doing so for mixtures of contaminants for which mechanisms of effect may vary or may not be known. There is the *possibility* that, if mechanistic approaches are used, the effects of UV-B can be distinguished from other potentially confounding factors. This opportunity to establish cause and effect relationships is somewhat unusual within global ecotoxicology.

The spatio-temporal scale of the global UV problem is also unique (Figure 1). Predictions of levels of ozone depletion in the Antarctic and in temperate latitudes are rapidly changing. However, estimates of ozone depletion up to 50% during the Antarctic spring and of 3–8% at selected seasons in temperate latitudes are now widely supported (Blumthaler and Ambach, 1990; Solomon, 1990; Stamnes *et al.*, 1992). These phenomena have seasonal patterns which have been thoroughly described (e.g., Solomon, 1990). Although it is not known how quickly ozone depleting chemicals will be phased out worldwide, recent predictions are that the problem of ozone depletion is certainly a major concern for at least 50–100 years (Hammitt *et al.*, 1987; Manzer, 1990). The exposure characteristics of the UV-B problem are also different from those broadly described for chemical contaminants, because there is no concern about bioaccumulation, and because there is no need to predict fate and fluxes from varied media.

Controversy exists as to whether detrimental effects of increasing solar UV have already been documented in natural ecosystems. However, strong evidence has recently been presented that UV-B inhibition of photosynthesis is occurring in Antarctic phytoplankton communities during the spring ozone depletion (Smith *et al.*, 1992). These recent findings highlight the need for rapid action to evaluate potential ecological impacts and to develop global monitoring strategies.

Characteristic 4: Regional Activities May Have Global Ecological Effects

Ecological parameters and the life history traits of individual species may give regional pollution problems global significance. Particularly vulnerable are critical habitats such as wetlands and migratory corridors (major rivers and important habitat on flyways). For example, hydroelectric dams have decimated fisheries in many parts of the world, largely because one acute action at a single spot can deter migration and spawning of fish throughout the system. It is not implausible that acute environmental pollution in one area along a river could create "chemical dams" that would result in similar impacts throughout the system. Foe and coworkers[1] have documented stretches of toxicity dozens of miles long (using acute toxicity tests on indicator species) in the Sacramento and San Joaquin rivers in California that are attributable to agricultural chemicals. It is not yet known whether these toxic stretches have created chemical dams for the already depleted salmon and striped bass populations, but this possibility is under investigation.

Regional pollution may also affect rare and declining species (Lubchenco *et al.*, 1991). Although numerous factors may contribute to loss of biological diversity worldwide, the relative significance of pollution, as a factor in declining biological diversity, has not been systematically evaluated. Citations abound referring to the potential threat of toxic substances on biological diversity, but hard evidence is lacking. As was implied above, pollution may cause direct toxic effects on the species of interest or indirect effects on its prey and competitors. In addition, pollution may alter genetic diversity in exposed populations (Bishop and Cook, 1981; Gillespie and Guttman, 1989). Populations with low genetic plasticity (such as rare and declining species) may be particularly vulnerable to the effects of toxic substances.

Scientific Priorities

Today, assessments of the effects of toxic substances are widely invoked in management decisions about the integrity of ecosystems; yet, research is needed

---

[1] Data presented in reports to the Central Valley Regional Water Quality Control Board, Sacramento, California.

to develop more sensitive and robust techniques. For aquatic ecosystems, effluent toxicity assessments typically rely on the use of short-term toxicity tests to predict the effects of mixed wastes as well as analytical chemical analyses to evaluate compliance with water quality objectives[2] (Bascietto *et al.* 1990; Cairns and Mount, 1990; Anderson *et al.*, 1991). Managers do not have reliable techniques to predict many types of sublethal effects[3] of environmental contamination; the assessment of sublethal effects in complex media such as soil and sediment is a particularly thorny problem. The latter problem is crucial to better management of Superfund site cleanups and disposal of dredged materials. Managers also lack tools to assess the significance of bioaccumulation of toxic substances on the health of fish and wildlife. Such problems, and the overarching need for a better understanding of the population and ecosystem-level effects of pollution, are common barriers to improved assessments at the local, regional, and global levels.

Below are described four key research topics that are central to a better understanding of regional and global pollution problems, with a focus on responses to xenobiotic substances.

Assessing Sublethal Effects

When the potential impacts of global pollution problems are evaluated, sublethal as well as lethal effects of toxic substances must be considered. The oldest rationale for this statement is perhaps the best one: we should be able to detect the "disease" before the patient is dead. In addition, diseased or even dead organisms may not be immediately apparent in the environment, particularly in their early life stages. Thus, many losses go undetected, and natural phenomena modify the population-level consequences of such occurrences.

Sublethal effects assessments are frequently more sensitive than are assessments of lethality. Unfortunately, a major failure of ecotoxicology in the past two decades has been the inability of researchers to determine the ecological significance of these sensitive techniques or to describe the potential utility of such responses as early warning signals. Even if some responses may not be appropriate triggers for regulatory action, early warning signals may help to prioritize pollution problems. The potential ecological significance of any given technique may vary dramatically depending on where it fits within a spectrum encompassing strictly compensatory responses to noncompensatory responses that reflect "disease."

There is an urgent need to determine the hazards of low-level exposures to toxic substances and UV and to develop appropriate applications for existing sensitive assessment tools. These problems can only be addressed by further focused and prioritized research into the sublethal effects of contaminants and their potential

---

[2] Water quality objectives are protection levels that are based on toxicity test results. Consequently, they are an estimate of levels that are protective of aquatic life. They are not technology-based standards.

[3] Sublethal effects may include effects on growth, development, reproduction, histopathologic alterations, DNA damage, or enzyme and cellular responses.

ecological significance. This area of ecotoxicology is significant in both global and regional studies.

In general terms, sublethal effects research is composed of two broad components. One component involves the development of early life history and life cycle tests in indicator organisms. The second and burgeoning component is the development of biomarkers, that is, biochemical, physiological, or histopathologic indicators of either effects of or exposure to toxic substances.

Early life stage and life cycle tests have been widely developed and standardized using fish, amphibia, and aquatic invertebrates (e.g., Birge *et al.*, 1985; DeGraeve *et al.*, 1985; Bantle *et al.*, 1989). They typically assess developmental abnormalities, growth, and reproductive success in organisms exposed over either long time periods (60–90 days) or an entire life cycle. Many such tests require only minor amounts of further research and validation before they can be applied to the decision-making process (e.g., Anderson and Harrison, 1990b). The short-term effluent toxicity tests that have become so widely used in managing aquatic ecosystems (EPA, 1985a, 1985b; Anderson *et al.*, 1991) are the shortest and simplest incarnations of the early life stage tests. Although this area of research has recently received less attention than the popular topic of biomarkers, further work is warranted. Detrimental effects on growth, reproduction, and development are more clearly related to effects on populations and communities than are many biomarker responses. The key disadvantages of this type of test are that the long-term assays are not easily applied to field monitoring and that mechanisms of effect are not elucidated.

Biomarkers of sublethal effects have been researched for two decades, although they have only recently attained a very high profile. The pitch of research has heightened as a critical mass of data have been attained (catalogued in Huggett *et al.*, 1992) and as decision-makers have come to an understanding that the tools they have to assess pollution effects are not adequate. To some extent, research in this area has also been accelerated by the development of new technologies such as monoclonal antibody technology and advances in molecular genetic techniques.

The most common classes of biomarkers (Huggett *et al.*, 1992) are DNA alterations, metabolic product indicators, immunologic responses, histopathologic measurements, and enzyme and protein synthesis responses. These types of indicators hold enormous potential for evaluating low-level effects of global pollution *in situ*, for determining their potential cause, for discerning long-term or latent effects using short-term indicators, and for examining the effects of specific toxic agents. In addition, these techniques can be excellent indicators of exposure to rapidly metabolized toxic substances that do not bioaccumulate. Although the ultimate promise of such techniques is enormous, they have only been suggested for use in a limited number of decision-making applications (e.g., Landner, 1988; Anderson and Harrison, 1990b); and, broadly speaking, there is limited consensus as to which techniques are most useful. The level of development of the techniques varies widely. Strengths and weaknesses of any given test are dependent upon sensitivity, specificity, inherent variability, applicability to field conditions, ecological relevance, methodologic utility, and other considerations.

McCarthy (1990) has recently concluded that improved integration and prioritization of biomarker research is crucial for its full promise to be realized. The prioritization of further research on the application of biomarkers is a task beyond the scope of this chapter, but a few simplifying principles should be discussed. It is crucial to emphasize research that will lead to indicators that are:

• More sensitive than more overt responses but that can be linked to significant detrimental effects.
• Applicable to reliable field monitoring of the effects of and exposure to a wide range of chemicals with high specificity in target organs. This ideal trait of "something that will detect everything" will never be realized, but limited batteries of tests with broad applicability and low inherent variability are a potential reality.
• Applicable to field monitoring of effects of and exposure to specific chemicals with high specificity in target organs (such as specific DNA adducts or cDNA probes to P450 enzymes). These could be very useful in monitoring biologically effective doses and exposures in complex media when single contaminants or classes of contaminants are of particular concern.
• Related to the basic mechanisms of effect. If programs are to be prioritized, specific mechanisms could be targeted for chemicals of concern. This avoids the application of inappropriate biomarker techniques and the potential risks of false negatives.

The evaluation of genotoxic responses fulfils many of these criteria. Biological dosimeters of effective doses, such as specific DNA adducts, are being developed; sensitive responses are being linked to ecologically significant endpoints such as reproduction (Liguori and Landolt, 1985; Anderson *et al.*, 1990; Anderson and Harrison, 1990a) and carcinogenesis (reviewed in Shugart *et al.*, 1992); and numerous techniques are being adapted from the human health literature (e.g., Hose and Puffer, 1983; Shugart, 1988; Singh *et al.*, 1988) that are useful in evaluating mechanisms of effect. In addition, field validations of selected techniques have been performed (Long and Markel, 1992). Finally, the potential for using molecular genetic techniques to evaluate specific mechanisms of effect (e.g., Cariello *et al.*, 1990) may enable researchers to "fingerprint" the spectrum of DNA damage produced by classes of contaminants.

Understanding Mechanisms of Tolerance and Adaptation

Animal populations may be altered by mutations that confer tolerance to toxic substances. Although this topic has been treated critically in only a limited number of ecotoxicological studies (Bishop and Cook, 1981), it may have unique significance in global ecotoxicology. This is because the widespread, long-term, low-level, and constant exposures that typify global pollution problems are more likely to produce stable population changes than are the intermittent or localized exposures that typify regional pollution problems.

Bishop and Cook (1981) have provided a general synthesis of this topic. They summarize key examples on insecticide resistance, industrial melanism, and the tolerance of higher plants to selected toxic materials. Shugart *et al.* (1992) have also recognized the need for further work on genetic adaptation attributable to pollutant exposure, and they suggest that modern molecular techniques used in evolutionary genetics be applied to this problem.

Globally, metals are mobilized into the air and water at rates that are greatly accelerated by human activity (Nriagu, this volume). We do not know at what point narrow differences between natural (and sometimes essential) levels and toxic levels of heavy metals will have eroded too far. Tolerance to metals has been evaluated in numerous aquatic organisms (Luoma, 1977; Kuwabara and Leland, 1986). However, as gradual increases occur in the concentrations of trace metals in aquatic ecosystems, it becomes increasingly difficult to determine how control sites and control organisms should be selected for experiments on adaptation. It is possible that for some organisms, no true controls exist. The limits of genetic adaptation and their relation to thresholds of toxicity should be fully explored.

Assessing the Significance of an Elevated Contaminant Tissue Concentration on the Health of an Organism

One of the most frequently assessed parameters in pollution studies is the concentration of toxic substances in the tissues of varied organisms. Despite their widespread use, such measurements usually provide only a very limited understanding of potential for detrimental biological effects. Only rarely are sufficient laboratory studies conducted to correlate elevated tissue concentrations with detrimental effects on an entire organism. This is a significant problem in both aquatic toxicology (Cairns and Mount, 1990) and wildlife toxicology (Hoffman *et al.*, 1990), because there is a widespread need to predict and assess how much chemical in specific tissues of an animal is too much. Currently, then, managers generally lack sufficient information to develop criteria for fish and wildlife protection for bioaccumulative substances. Further laboratory research is needed to link elevated tissue concentrations to detrimental biological effects for specific substances. Research is also needed to link biomarker responses to elevated tissue concentrations for bioaccumulative substances.

The need to determine the biological significance of elevated tissue concentrations is exemplified by the following recent incident. At the Kesterson Reservoir in California, field observations of deformed birds were linked to trace metals, particularly selenium, that had leached from subsurface agricultural drainwater (Ohlendorf *et al.*, 1986a; 1986b). Previous testing of Kesterson waters, using standard bioassay tests, had not indicated that there would be deleterious effects. In global ecotoxicology, there will be a similar urgent need to assess the significance of low-level, widespread increases of toxic substances in diverse organisms and tissues.

A concerted effort is needed to link research into the overt detrimental effects of elevated tissue concentrations with research on biomarkers. Biomarkers should be incorporated into laboratory studies of overt effects, such as developmental abnor- malities and decreased growth, to elucidate mechanisms of effect and to validate the use of selected biomarkers for field monitoring. Of course, for many rapidly metabolized substances, correlations between tissue concentration and detrimental effect are not expected. Similarly, such relationships for toxic heavy metals are known to be complex because of the detoxification action attributable to metal binding proteins.

Assessing Effects in Complex Media

The difficulty of determining detrimental effects of toxic substances in complex media is a problem that affects both global and regional assessments, and it over- laps with many aspects of sublethal effects research. On a regional level, there are needs to predict the bioavailability and potential effects of toxic substances in soils and sediments at hazardous waste (Anderson, 1992) and dredge disposal sites. This information is needed to make accurate predictions of the hazards associated with specific cleanup procedures, to determine appropriate disposal options for excavated wastes, and to determine targets for remediation activities. Unfortunately, even simple toxicity test techniques may be fraught with variability and interferences when applied to soils and sediments. Positive interferences are attributable to such factors as sulfides in anaerobic sediments and variations in sediment grain sizes. Selection of reference soils and sediments can also be a severe problem for many sites. Some reliable acute toxicity tests do exist, but many fewer tests assess sublethal responses (Burton, 1991).

In global ecotoxicology (with the exception of solar UV effects), there will also be a need to evaluate effects of toxic substances in complex media such as soil and sediment. Whether deposited by atmospheric distribution or ocean circulation, aquatic sediments are an important sink for contaminants. How can decision-mak- ers determine the effects of low-level but widespread increases of toxic substances on sediment and soil communities? This can only be accomplished if there are adequate methods to detect the sublethal effects and bioavailability of contami- nants in complex media. Both early life stage tests and biomarker responses are needed. Although field comparisons of such responses have been made at a lim- ited number of sites (Chapman *et al.*, 1992; Long and Markel, 1992), the relative utility of various techniques has not been thoroughly characterized.

The complex exposure scenarios occurring in sediment and soil make it an extremely complex task to develop criteria for sediment and soil protection. Nevertheless, there are numerous regulatory efforts to establish sediment and soil quality criteria. These criteria would be numerical limits for specific chemicals that would indicate "safe" exposure levels. Alternatively, integrated chemical monitoring and toxicity criteria could be applied to regulation (e.g., Ginn and Pastorok, 1992; Southerland *et al.*, 1992). Research in predicting the bioavailabil- ity of sediment-sorbed chemicals (summarized in Knezovich *et al.*, 1987) is essen- tial for regulatory efforts to result in meaningful improvements in water quality.

Although such research is widely pursued and incorporated into management, it remains the source of extensive controversy.

Other current topics in sediment and soil toxicity assessment include: (1) problems associated with sample handling, (2) modeling and assessment of bioaccumulation (Lee, 1992), and (3) assessment of population- and community-level effects (Burton, 1991; Cairns *et al.*, 1992). All of these topics are important to producing more sophisticated regional and global assessments.

Conclusions and Recommendations

Research priorities for global ecotoxicology must be formulated. Below are listed key recommendations for further research.

Low-Level, Widespread Contamination by Toxic Substances

This is the most intractable of the problems, because extensive research is needed to determine the potential significance of low-level contamination. Nevertheless, key actions are identifiable. First, research in the four key areas discussed above should be a high priority in global ecotoxicology. Secondly, toxic substances of greatest concern should be identified based on existing data, with special emphasis on heavy metals. Third, crosscutting programs should be devised to accelerate the progress of biomarker research with respect to specific toxic substances. Combined field and laboratory programs to evaluate both methods for *in situ* monitoring and mechanisms of action of priority substances would be extremely valuable. These interdisciplinary programs would include assessments in complex media and assessments of the potential significance of elevated tissue concentrations in sentinel species.

Regional Effects

To better determine the magnitude, duration, and frequency of regional pollution problems, the feasibility of a global toxicity assessment should be evaluated. The feasibility analysis would first consider available information from existing international programs. Secondly, a strategic assessment should be devised to survey selected land-use categories, selected habitats, or selected industrial processes from a representation of nations. The assessment could be followed with field validation. Partnerships could be developed between ecologists and toxicologists to identify critical habitats, and biomarkers could be tested in varied environments on an experimental basis. In addition, social scientists could collaborate in devising the assessment strategy.

UV-B Effects

Monitoring and assessment of UV-B effects must be accelerated. Despite methodological difficulties associated with solar UV research, the UV-B problem is still much more focused than the determination of the potential broad-scale effects of low-level contamination by toxic substances. Consequently, the UV-B problem is more amenable to immediate "mission-oriented" solutions coupling both research and monitoring.

Ecological Boundaries

There is an urgent need for increased collaboration between ecotoxicologists and ecological theorists, on topics such as the effects of pollutants on global biodiversity, the frequency of identification of pollution-tolerant species and their ecological significance, the identification of key habitats (e.g., wetlands and major migratory pathways) for protection and research, the identification of life history strategies that may confer vulnerability to toxic substances, and the study of rare and declining species. Recent efforts in the ecological research community (Lubchenco *et al.*, 1991) encouraging ecologists to be more responsive to pollution problems undoubtedly signal that the time is ripe for change.

I propose that scientists in global ecotoxicology should work toward the goal of making a first estimate of the potential significance of contaminants (and UV-B) as agents of global change within ten years. This goal may seem unambitious to some and overly ambitious to others. Although this chapter has not discussed the potential institutional aspects of the proposed research, one thing is clear: this problem will require clear goals, strategy, and partnerships.

Acknowledgments

John Harte, Jerome Nriagu, Bruce Paton, Robert Socolow, and Valerie Thomas provided insightful comments on drafts of this chapter. This work was supported by the NIEHS Superfund Basic Research Program, University of California Berkeley Program Project (NIH P42 ES04705-03), through the U.S. Department of Energy under contract No. DE-AC03-76SF00098. This work was also partially supported by the Pew Charitable Trusts. Gillian Wild, Walter Sadinski, and Hee-Sung Kang provided invaluable assistance at various stages of manuscript preparation.

References

Anderson, S. L. 1992. Monitoring genetic damage to ecosystems from hazardous waste. In *Proceedings of the Pacific Basin Conference on Hazardous Waste Management,* Pacific Basin Consortium for Hazardous Waste Research, Honolulu, Hawaii.

Anderson, S. L., and F. L. Harrison. 1990a. Predicting the ecological significance of exposure to genotoxic substances in aquatic organisms. In *In Situ Evaluations of Biological Hazards of Environmental Pollutants* (S.S. Sandhu *et al.*, eds.), Plenum Press, New York, 81–93.

Anderson, S. L., and F. L. Harrison. 1990b. Prepermit Bioassay Testing Guidance Document for Marine Low Radioactive Waste Disposal. EPA/520/1-88-005, U.S. Environmental Protection Agency, Washington, D.C.

Anderson, S. L., M. P. Carlin, and A. L. Suer. 1991. Effluent and ambient toxicity programs in the San Francisco Bay region. *Canadian Technical Report of Fisheries and Aquatic Sciences 1774 (1),* 152–164.

Anderson, S. L., F. L. Harrison, G. Chan, and D. H. Moore II. 1990. Comparison of cellular and whole-animal bioassays for estimation of radiation effects in the polychaete worm *Neanthes arenaceodentata* (Polychaete). *Archives of Environmental Contamination and Toxicology 19,* 164–174.

Bantle, J. A., D. J. Fort, and B. L. James. 1989. Identification of developmental toxicants using the frog embryo teratogenesis Assay-*Xenopus* (FETAX). *Hydrobiologia 188/189,* 577–585.

Bascietto, J., D. Hinckley, J. Plafkin, and M. Slimak. 1990. Ecotoxicity and ecological risk assessment. *Environmental Science and Technology 24(1),* 10–15.

Birge, W., J. Black, and A. Westerman. 1985. Short-term fish and amphibian embryo-larval tests for determining the effects of toxicant stress on early life stages and estimating chronic values for single compounds and complex effluents. *Environmental Toxicology and Chemistry 4,* 807.

Bishop, J. A., and L. M. Cook (eds.). 1981. *Genetic Consequences of Man Made Change,* Academic Press, New York, 409 pp.

Blumthaler, M., and W. Ambach. 1990. Indication of increasing solar ultraviolet-b radiation flux in Alpine regions. *Science 248,* 206–208.

Burton, G. A., Jr. 1991. Assessing the toxicity of freshwater sediments. *Environmental Toxicology and Chemistry 10,* 1585–1627.

Cairns, J., Jr., and D. I. Mount. 1990. Aquatic toxicology. *Environmental Science and Technology 24(2),* 154–161.

Cairns, J. Jr., B. R. Neiderlehner, and E.P. Smith. 1992. The emergence of functional attributes as endpoints in ecotoxicology. In *Sediment Toxicity Assessment,* Lewis Publishers, Boca Raton, Florida, 111–125.

Cariello, N. F., P. Keohavong, A. G. Kat, and W. G. Thilly. 1990. Molecular analysis of complex human cell populations: Mutational spectra of MNNG and ICR-191. *Mutation Research 231,* 165–176.

Chapman, P. M., E. A. Power, and G. A. Burton, Jr. 1992. Integrative assessments in aquatic ecosystems. In *Sediment Toxicity Assessment,* Lewis Publishers, Boca Raton, Florida, 313–335.

DeGraeve, G. M., J. D. Cooney, R. S. Carr, and J. M. Neff. 1985. Identification, comparison and evaluation of existing freshwater, estuarine, and marine rapid chronic bioassays. American Petroleum Institute, Washington, D.C.

EPA (U.S. Environmental Protection Agency). 1985a. *Short-term Methods for Estimating the Chronic Toxicity of Effluents and Receiving Waters to Freshwater Organisms* (W. Horning and C. Weber, eds.), EPA/600/4-85-014, Environmental Protection Agency, Washington, D.C.

EPA. 1985b. *Technical Support Document for Water Quality-based Toxics Control,* EPA/440/4-85-032, Environmental Protection Agency, Washington, D.C.

Gillespie, R. B., and S. I. Guttman. 1989. Effects of contaminants on the frequencies of allozymes in populations of the central stoneroller. *Environmental Toxicology and Chemistry 8,* 309–317.

Ginn, T. C., and R. A. Pastorok. 1992. Assessment and management of contaminated sediments in Puget Sound. In *Sediment Toxicity Assessment,* Lewis Publishers, Boca Raton, Florida, 371–397.

Gomez, E. D. 1988. Overview of environmental problems in the East Asian Seas region. *Ambio 17(3),* 166–169.

Hammitt, J. K., F. Camm, P. S. Connell, W. E. Mooz, K. A. Wolf, D. J. Wuebbles, and A. Bamezai.

1987. Future emission scenarios for chemicals that may deplete stratospheric ozone. *Nature 330(24/31),* 711–716.

Hoffman, D. J., B. A. Rattner, and R. J. Hall. 1990. Wildlife toxicology. *Environmental Science and Technology 24(3),* 276–283.

Hose, J. E., and H. W. Puffer. 1983. Cytologic and cytogenetic anomalies induced in purple sea urchin embryos (*Strongylocentrotus purpuratus S.*) by parental exposure to benzo[a]pyrene. *Marine Biology Letters 4,* 87–95.

Huggett, R. J., R. A. Kimerle, P. M. Mehrle, Jr., and H. L. Bergman (eds.). 1992. *Biomarkers: Biochemical, Physiological, and Histological Markers of Anthropogenic Stress,* Lewis Publishers, Boca Raton, Florida, 347 pp.

Jagger, J. 1985. *Solar-UV Actions on Living Cells,* Praeger Publishers, New York, 202 pp.

Knezovich, J. P., F. L. Harrison, and R. Wilhelm. 1987. The bioavailability of sediment-sorbed organic chemicals: A review. *Water, Air, and Soil Pollution 32,* 233–245.

Kuwabara, J. S., and H. V. Leland. 1986. Adaptation of *Selenastrum capricornutum* (Chlorophycea) to copper. *Environmental Toxicology and Chemistry 5,* 197–203.

Landner, L. 1988. Hazardous chemicals in the environment—some new approaches to advanced assessment. *Ambio 17(6),* 360–366.

Lee, H., II. 1992. Models, muddles and mud: Predicting bioaccumulation of sediment-associated pollutants. In *Sediment Toxicity Assessment,* Lewis Publishers, Boca Raton, Florida, 267–289.

Liguori, V. M., and M. L. Landolt. 1985. Anaphase aberrations: An *in vivo* measure of genotoxicity. In *Short Term Bioassay in Analysis of Complex Environmental Mixtures IV* (M. D. Waters, S. S. Sandhu, J. Lewtas, L. Claxton, G. Strauss and S. Nesnow, eds.), Plenum Press, New York.

Long, E. R., and R. Markel. 1992. *An Evaluation of the Extent and Magnitude of Biological Effects Associated with Chemical Contaminants in San Francisco Bay, California,* National Oceanic and Atmospheric Administration, Seattle, Washington, 86 pp.

Lubchenco, J., A. M. Olson, L. B. Brubaker, S. R. Carpenter, M. M. Holland, S. P. Hubbell, S. A. Levin, J. A. MacMahon, P. A. Matson, J. M. Melillo, H. A. Mooney, C. H. Peterson, H. R. Pulliam, L. A. Real, P. J. Regal, and P. G. Risser. 1991. The sustainable biosphere initiative: An ecological research agenda. *Ecology 72(2),* 371–412.

Luoma, S. N. 1977. Detection of trace contaminant effects of aquatic ecosystems. *Journal of the Fisheries Research Board of Canada 34(3),* 436–439.

Malm, O., W. C. Pfeiffer, C. M. M. Souza, R. Reuther. 1990. Mercury pollution due to gold mining in the Madeira River Basin, Brazil. *Ambio 19(1),* 11–15.

Manzer, L.E. 1990. The CFC-ozone issue: Progress on the development of alternatives to CFCs. *Science 249,* 31–35.

Martinelli, L. A., J. R. Ferreira, B. R. Forsberg, and R. L. Victoria. 1988. Mercury contamination in the Amazon: A gold rush consequence. *Ambio 17(4),* 252–254.

McCarthy, J. M. 1990. Concluding remarks: Implementation of a biomarker-based environmental monitoring program. In *Biomarkers of Environmental Contamination,* (J. F. McCarthy, and L. R. Shugart, eds.), Lewis Publishers, Boca Raton, Florida, 457 pp.

Ohlendorf, H. M., D. J. Hoffman, M. K. Saiki, and T. W. Aldrich. 1986a. Embryonic mortality and abnormalities of aquatic birds: Apparent impacts of selenium from irrigation drainwater. *Science of the Total Environment 52,* 49–63.

Ohlendorf, H. M., R. W. Lowe, P. R. Kelly, and T. E. Harvey. 1986b. Selenium and heavy metals in San Francisco diving ducks. *Journal of Wildlife Management 50,* 64–71.

Patin, S. A. 1985. Biological consequences of global pollution of the marine environment. In *Global Ecology* (C.H. Southwick, ed.), Sinauer Associates, Sunderland, Massachusetts.

Risebrough, R. W. 1989. Monitoring and surveillance systems. In *Marine Pollution* (J. Albaiges, ed.), Hemisphere Publishing Corp., New York.

Shugart, L. 1988. An alkaline unwinding assay for the detection of DNA damage in aquatic organisms. *Marine Environmental Research 24,* 321–325.

Shugart, L. R., J. Bickham, G. Jackim, G. McMahon, W. Ridley, J. Stein, and S. Steinert. 1992. DNA alterations. In *Biomarkers: Biochemical, Physiological, and Histological Markers of Anthropogenic Stress* (R. J. Huggett, R. A. Kimerle, P. M. Mehrle, Jr., and H. L. Bergman, eds.), Lewis Publishers, Boca Raton, Florida, 347 pp.

Singh, N. P., M. T. McCoy, R. R. Tice, and E. L. Schneider. 1988. A simple technique for quantitation of low levels of DNA damage in individual cells. *Experimental Cell Research 175,* 184–191.

Smith, R. C., B. B. Prezelin, K. S. Baker, R. R. Bidgare, N. P. Boucher, T. Coley, D. Karentz, S. MacIntyre, H. A. Matlick, D. Menzies, M. Ondrusek, Z. Wan, and K. J. Waters. 1992. Ozone depletion: Ultraviolet radiation and phytoplankton biology in Antarctic waters. *Science 255,* 952–958.

Solomon, S. 1990. Progress towards a quantitative understanding of Antarctic ozone depletion. *Nature 347,* 347–354.

Southerland, E., M. Kravitz, and T. Wall. 1992. Management framework for contaminated sediments (The U.S. EPA Sediment Management Strategy). In *Sediment Toxicity Assessment,* Lewis Publishers, Boca Raton, Florida, 341–369.

Stamnes, K., Z. Jin, J. Slusser, C. Booth, and T. Lucas. 1992. Several-fold enhancement of biologically effective ultraviolet radiation levels at McMurdo Station Antarctica during the 1990 ozone "hole." *Geophysical Research Letters 19(10),* 1013–1016.

Stumm, W. 1986. Water, an endangered ecosystem. *Ambio 15(4),* 201–207.

Whelan, T. 1988. Central American environmentalists call for action on the environment. *Ambio 17,* 72–75.

275

19

Industrial Activity and Metal Emissions

Jerome Nriagu

Abstract

Worldwide industrial emissions of toxic elements to the atmosphere, soils, and water are inventoried. Sources of metal emissions include fossil fuel combustion, mining, metal smelting, industrial processes, and waste incineration. While much of the wastes result in localized or regional contamination problems, the atmospheric emissions of metals have resulted in global contamination. Soils are the largest sink for metal pollution, but aquatic ecosystems are more vulnerable to metals, and may be more affected by the emissions. While there has been a decrease in global atmospheric emissions since 1980, due to pollution control in developed countries, metals pollution in developing countries is believed to be increasing.

Introduction

Metals (and metalloids) and their compounds have been exploited by mankind for many millennia. Brightly colored ochres were put to several uses, including painting, by the cave people. Documented use of galena (lead sulfide, PbS) as magical charms, body paint, or ceremonial powder by pre-colonial inhabitants of North America dates back to the Early Archaic Period (over 8000 years BP; Walthall, 1981). Techniques for the recovery of metals from their ores were discovered around 9000 BP, and by the Roman Empire period, mine production of lead was in excess of 100,000 metric tonnes per year (Nriagu, 1983). It was, however, the dawn of the Industrial Revolution that brought about the unmitigated demand for things metallic. By the present time, over a million metal-containing substances have been synthesized, and at least 10,000 of them are in common use to maintain human health and welfare, drive the industrial economy, and protect national security. A basic human instinct has always been to transform and tame nature in pursuit of a better lifestyle, and the development and continued existence of a modern technological society would certainly be impossible without metals.

While the benefits of metals and their compounds are indisputable, the price of their utilization is only beginning to be appreciated. Each year, large quantities of toxic metal wastes are discharged into the environment from the ever-increasing production and manufacture of goods and services and from the burning of fossil

277

fuels to generate the energy needed to sustain the industrial and domestic activities. The number of toxic metal compounds as well as the quantity of the industrial discharges continue to grow, and the inference has been made that a large portion of the earth's resources is being consumed in order to generate metalliferous wastes (Nriagu, 1990). Metal pollution has slowly worked its way to the most remote areas of the atmosphere and the furthest reaches of the globe and unavoidably into the human food chain. The ultimate environmental and human health effects of global metal pollution are unknown.

The effects of metal production have been of concern for a long time. In his classic 16th-century book about mining and metallurgy, Georgius Agricola (1556) included a lengthy defense of metals in response to widespread concern that metals kindled greed and were disruptive of people's morals and lifestyles: "Several good men have been so perturbed by the tragedies that they conceive an intense hatred towards metals, and they wish absolutely that metals had never been created, or being created, that no one had ever dug them out." Little did they know the dangers they were being exposed to from the dominant industrial activity, mining, in Europe in their time. With the Industrial Revolution came blackened skies and foul air, and the following petition by the residents of the cradle of industrial development in Britain was symptomatic of the price exacted by the smoke of progress: "The gas from these manufactories is of such a deleterious nature as to blight everything within its influence, and is alike baneful to health and property. The herbage of the fields in their vicinity is scorched, and gardens neither yield fruit nor vegetables; many flourishing trees have lately become rotten naked sticks. Cattle and poultry droop and pine away. It tarnishes the furniture in our houses, and when we are exposed to it, which is of frequent occurrence, we are afflicted with cough and pains in the head" (*Proceedings of the Town Council of Newcastle-upon-Tyne*, January 9, 1839; cited by Davis, 1984, p. 40). Although measures were introduced much later to control gross emissions of noxious gases, little was done to regulate the toxic metals which were invariably discharged as an adjunct of the noxious gases. The benign neglect was to be forcefully emblazoned in the public consciousness by the discovery of the Minamata and Itai-itai diseases in the 1950s and 1960s (Eisenbud, 1978).

Industrial Emissions of Toxic Metals

The use of metals has become so pervasive that few industrial operations currently do not discharge metal-containing wastes into the air, water, or soil. An inventory of these industrial emissions is a critical first step in developing an appropriate control strategy, in modeling the regional and global cycling of trace metals, for determining source/receptor relationships, and for interpreting historical records of past environmental pollution. The following discussions are based on global inventories of emissions published by Nriagu and Pacyna (1988) and Nriagu (1989; 1990). The emphasis is on industrial emissions into the atmosphere since this is the key medium in the global distribution of metal pollution.

Inventories of worldwide industrial inputs of toxic metals into the atmosphere, soils, and aquatic environments are compared in Table 1. Major sources of anthro-pogenerated metals in the aquatic ecosystems include domestic and industrial wastewaters, sewage discharges, urban runoff and atmospheric fallouts. Major metal-containing wastes discharged on land include coal and wood ash, sewage sludge and solid wastes, fertilizers, metalliferous pesticides, and corrosion of metallic installations. It should be noted that the flow of pollutant metals into the atmosphere is less than the discharges into water (Pb is the only exception) and is small when compared to the loadings into soils. Basically, soils represent the ulti-mate sink for pollutant metals in the continental areas while the atmosphere serves as a medium for the transfer of trace metals from industrial and natural sources to the target soil or water body. The disruption of the natural biogeochemical cycle of lead in the ocean by eolian inputs is a well-documented phenomenon (see Schaule and Patterson, 1981; Boyle *et al.*, 1986; Bruland *et al.*, 1991).

The sustained inputs of pollutant metals into aquatic environments with very low baseline concentrations of trace metals have resulted in a range of environ-mental problems on a local, regional, and even global scale. On the basis of the relative sizes of the various environmental compartments, one can surmise that freshwater ecosystems are particularly at risk in terms of toxic metal pollution. If it is assumed that 10–20% of the global discharge into aquatic environments goes into lakes and rivers (the remaining 80–90% of the loadings go directly into the seas and coastal marine waters), the calculated pollution load would, in fact, dwarf the natural background concentrations of trace metals in such ecosystems. The

Table 1: *Worldwide industrial discharges of trace metals into the air, water, and soils (×1000 tonnes/yr)*

Element	Atmosphere	Aquatic Ecosystems	Soils
As	19	42	82
Cd	7.6	9.1	22
Cr	31	143	900
Cu	35	110	970
Hg	3.6	6.5	8.3
Mn	38	260	1700
Mo		11	87
Ni	52	110	290
Pb	330	140	760
Sb	3.5	18	26
Se	6.3	41	41
Zn	130	240	1300

Discharge to the atmosphere are for 1983 (Nriagu and Pacyna, 1988); discharges into soils and aquatic environments are for the 1980s (Nriagu, 1990).

impacts of the elevated levels are likely to be severe since aquatic organisms in general tend to be more sensitive to toxic metals in their habitat than are the terrestrial biota (Bowen, 1979). Furthermore, the propensity to biomagnify pollutant metals up the food chain is a feature of aquatic environments exemplified by the highly elevated concentrations of Hg in fish in many lakes in the northeastern and north central United States, eastern Canada, and Scandinavia (Fitzgerald and Clarkson, 1991). In recent years, at least 26 states in the United States have issued fish consumption advisories to restrict or eliminate the consumption of mercury-contaminated fish caught in tens of thousands of lakes. In the Florida Everglades, piscivorous alligators, bald eagles, and raccoons have acquired highly concentrated levels of Hg, and some panther deaths have been blamed on eating the raccoons (EPRI, 1991). In the Great Lakes region, reproductive problems in fish-eating eagles, otters, minks, and other animals have been attributed to excessive ingestion of Hg (EPRI, 1991). As far as can be determined, most of the Hg in the problematic lakes is derived from anthropogenic sources, and there is some evidence to suggest that the Hg levels in fish in northern Minnesota lakes have increased by 3–5% per year during the past 30 years (Glass *et al.*, 1991).

Soils are the major sink for pollutant metals, and the data in Table 1 bear this fact out. Undue human influence is now reflected in the highly elevated levels of trace metals that characterize soils in many urban areas and around major industrial installations (Purves, 1985). During the 1970s, the average atmospheric deposition rates for metals in urban areas of North America were estimated to be, in g/ha/yr, 18 for Cd, 160 for Cu, 910 for Pb, and 3200 for Zn; the fallout rates in European cities were 15 for Cd, 310 for Ni, 320 for Cu, and 400 for Pb (Jeffries and Snyder, 1981). Such deposition rates would result in the levels of trace metals in soils being doubled in roughly a decade, depending on the baseline metal contents. And the atmosphere is just one of the sources of metal pollution in urban soils. Since the trace metals are generally immobile in soils (unless under the influence of acid rain), the pollutant metals tend to accumulate in the surface layers from where they are more liable to get into the human food chain.

The dominant sources of trace metals in the atmosphere are grouped into five categories (Table 2). As of 1983, energy production by burning of fossil fuels accounted for about 80% of the Ni, 60% of the Hg and Se, and 30–50% of the Sb, Cr, and Mn released annually by industries into the global atmosphere. Smelters represented the leading sources of As (63%), Cu (66%), and Zn (54%), while the production of iron and steel was responsible for the largest fractions of Cr (55%) and Mn (39%) released to the atmosphere by industries. Although the use of leaded gasoline had already been reduced in a number of countries, automotive tailpipes still accounted for about three-quarters of the anthropogenerated lead in the atmosphere (Table 2).

Smelters, waste incinerators, and combustion-related activities of energy production and transportation are also the principal sources of oxides of sulfur (SO_x) and nitrogen (NO_x) in the atmosphere (Logan, 1983; Eisenbud, 1978). As the metals, SO_x, and NO_x cycle through the atmosphere, they may react to

Table 2: *Worldwide emissions of trace metals from the principal industrial sources to the atmosphere (×1000 tonnes/yr)*

Process	As	Cd	Cr	Cu	Hg	Mn	Ni	Pb	Sb	Se	Zn
Energy production	2.2	0.79	13	8.0	2.3	12	42	13	1.3	3.8	17
Mining	0.06	–	–	0.42	–	0.62	0.80	2.6	0.10	0.16	0.46
Smelting &refining	12	5.4	17	23	0.13	2.6	4.0	46	1.4	2.2	72
Industrial processes & applications	2.0	0.60	0.84	2.0	–	15	4.5	21	–	–	36
Waste incineration	0.31	0.75	–	1.6	1.2	8.3	0.35	2.4	0.67	0.11	5.9
Leaded gasoline	–	–	–	–	–	–	–	250	–	–	–
Total (industrial)	19	7.6	31	35	3.6	38	52	330	3.5	6.3	130
Total (natural sources)[1]	12	1.4	43	28	2.5	320	29	12	2.6	10	45

[1] Data for natural sources (windborne soil, volcanoes, sea spray, forest fires) taken from Nriagu, 1990.

Data for 1983 from Nriagu and Pacyna, 1988. Reprinted with permission from *Nature*, copyright © 1988 Macmillan Magazines Limited.

form "toxic" rain. The common origin of these critical pollutants needs to be considered in making an environmental assessment. The genetic affinity should also be of interest in terms of environmental controls since programs designed to fight acid rain may also result in the lowering of the burden of pollutant metals in the atmosphere.

Two divergent trends in the flow of metal pollution into the environment have become evident. As a result of various control measures, there has been a substantial reduction in the rate of discharge and hence in ambient concentrations of trace metals in the developed countries (see Stigliani, Jaffé, and Anderberg, this volume). The downward trend in the atmospheric metal burden is nicely demonstrated in Greenland ice layers, where the concentrations of Pb, Zn, and Cu have declined by factors of 7.5, 2.5, and 2.5 respectively since the 1960s (Boutron *et al.*, 1991). By contrast, the emissions of toxic metals in the developing countries continue to rise sharply because (1) inadequate financial resources force the countries to rely on technologies that are often obsolete and pollution prone, (2) environmental regulations are either lax or ineffectual, (3) industrial operations that cannot meet the strict environmental regulations in the developed countries are now being relocated to the developing countries, and (4) the need for foreign exchange encourages the exploitation of natural resources for short-term benefits with little concern for the long-term environmental consequences. One should therefore not be surprised that some of the highest concentrations of toxic metals in the environment being reported in the recent scientific literature come from the developing countries.

The growing disparity in the atmospheric emissions of toxic metals in the developed and developing countries can be adduced from the data in Table 3. By the late 1980s although the United States and Canada produce 26% of the world's energy and employ about 22% of the Pb, 26% of the Cd, 29% of the Hg, and 47% of the As consumed worldwide each year (UNEP, 1991), they only contributed about 4% of the Cd, 8% of the Hg, and 8% of the As released annually to the global atmosphere (from Tables 2 and 3). Automotive Pb emissions in North America had declined from about 175,000 tonnes/yr (or over 90% of the total industrial emissions) during the early 1970s to <7000 tonnes in 1986. By 1989, although the US and Canada were consuming 46% of the world's gasoline, they contributed only 1% of the worldwide automative lead emissions (see Thomas and Spiro, this volume). The effect of lead removal from gasoline on environmental levels has been dramatic; the average concentrations of atmospheric lead in the urban areas of North America have declined from about 1.0 μg/m^3 during the early 1970s to <0.1 μg/m^3 in recent years (EPA, 1986).

A global inventory of worldwide emissions of trace metals from natural sources to the atmosphere is also indicated in Table 2. From these data, the ratios of industrial to natural emission rates are estimated to vary from about 21 for Pb to 0.63 for Se and 0.12 for Mn. Thus, with the exception of Mn, the outputs from anthropogenic sources either exceed or are comparable to fluxes from natural sources, suggesting that industrial activities have become a key factor in the local, regional,

Table 3: *Industrial atmospheric emissions (in metric tonnes/yr) of selected toxic metals to the atmosphere in the United States and Canada, for the mid-1980s*

Element	Fossil Fuel Combustion	Non-ferrous Metal Smelting	Industrial Processes	Waste Incineration	Others	Total
Lead						
Canada	25	1500	720	200	1800[1]	4200
USA	800	1400	2300	2800	5100[1]	12,000
Total	830	2900	3000	3000	6900[1]	17,000
Cadmium						
Canada	28	69	6	5	–	110
USA	120	58	12	20	–	210
Total	150	130	18	25		320
Mercury						
Canada	10	13	2	6	8[2]	39
USA	105	10	16	49	72[2]	250
Total	120	23	18	54	81[2]	290
Aarsenic						
Canada	21	330	26	2	5[3]	390
USA	470	370	53	3	150[3]	1000
Total	490	700	79	5	160[3]	1400

[1] Mostly from gasoline.

[2] Primarily from latex painted buildings, lamp breakage, etc.

[3] From spray of arsenical pesticides.

Adapted from Voldner and Smith, 1989.

and global atmospheric cycling of the trace elements. The dominating influence of industrial sources is particularly manifested in air in the urban areas where the trace metal levels are now many times higher than the concentrations in the fairly pristine Antarctic (Nriagu and Davidson, 1986; Suttie and Wolff, 1991).

Accumulation in the Environment

It has been estimated that between 1850 and 1900, the worldwide industrial emissions of Ni, Cd, Cu, Zn, and Pb to the atmosphere averaged approximately 240, 380, 1800, 17,000, and 22,000 tonnes/yr respectively (Nriagu, 1979). From the turn of the century until the 1980s, there was a roughly exponential increase in the industrial emissions of toxic metals (Table 4), a pattern that parallels the rates of industrial and population growths during the same period. Between 1900 and 1980, the atmospheric emission rates for Cu, Zn, Cd, Pb and Ni increased by 6-, 8-, 8-, 9-, and 51-fold respectively (Table 4). The decline in emission rates during

Table 4: *World industrial emissions of trace metals to the atmosphere during this century ($\times 1000$ tonnes/10 yr period)*

Period	Cd	Cu	Pb	Ni	Zn
1901–1910	9	53	470	8	390
1911–1920	11	80	490	21	490
1921–1930	14	96	1100	21	620
1931–1940	17	120	1700	49	750
1941–1950	22	170	1700	80	960
1951–1960	34	230	2700	140	1500
1961–1970	54	440	3700	260	2400
1971–1980	74	590	4300	420	3300
1981–1990	59	470	3400	330	2600
Total	290	2200	20,000	1300	13,000

Updated from Nriagu, 1979.

1980–90 reflects the effects of pollution control programs in the developed countries. The cumulative (all-time) emissions from industrial sources have been estimated to be 0.37, 2.6, 23, 1.3, and 17 million tonnes respectively for Cd, Cu, Pb, Ni, and Zn (Table 4).

The ongoing contamination of the air, soil, and water with nondegradable toxic metals needs to be viewed with some concern. In fact, the problem is no longer limited to the developed countries but has become a global phenomenon (Nriagu, 1990). Each environmental compartment has a limited carrying capacity for toxic metals. At the current industrial input rates, the abilities of many ecosystems to cope with toxic metal pollution are being rapidly depreciated, and the existing margin of environmental safety for many organisms is disappearing rapidly.

References

Agricola, G. 1556. *De re Metallica* (Translated and edited by H. C. Hoover and L. H. Hoover), Dover Publications, New York, 1950.

Boutron, C., U. Gorlach, J. P. Candelone, M. A. Bolshov, and R. J. Delmas. 1991. Decrease in anthropogenic lead, cadmium and zinc in Greenland snows since the late 1960s. *Nature 353*, 153–156.

Bowen, H. J. M. 1979. *Environmental Chemistry of the Elements*, Academic Press, New York.

Boyle, E. A., S. D. Chapnick, and G. T. Chen. 1986. Temporal variability of lead in the Western North Atlantic. *Journal of Geophysical Research 91*, 8573–8593.

Bruland, K. W., J. R. Donat, and D. A. Hutchins. 1991. Interactive influence of bioactive trace metals on biological production in oceanic waters. *Limnology and Oceanography 36*, 1555–1577.

Davis, L. N. 1984. *The Corporate Alchemists*, William Morrow and Co., New York, 329 pp.

Eisenbud, M. 1978. *Environment, Technology and Health*, New York University Press, New York, 384 pp.

EPA (U.S. Environmental Protection Agency). 1986. *Air Quality Criteria Document for Lead*, U.S. Environmental Protection Agency, Environmental Criteria and Assessment Office, Research Triangle Park, North Carolina.

EPRI (Electric Power Research Institute). 1991. *Mercury in the Environment*, Electric Power Research Institute (EPRI Journal), Palo Alto, California, 11 pp.

Fitzgerald, W. F., and T. W. Clarkson. 1991. Mercury and monomethylmercury: Present and future concerns. *Environmental Health Perspectives 96*, 159–166.

Glass, G., K. W. Schmidt, J. Sorenson, J. K. Huber, and G. Rapp, Jr. 1991. *Mercury Sources and Distribution in Minnesota Aquatic Resources*, Final Report, Minnesota Pollution Control Agency, St. Paul, Minnesota.

Jeffries, D. S., and W. R. Snyder. 1981. Atmospheric deposition of heavy metals in Central Ontario. *Water, Air and Soil Pollution 15*, 127–152.

Logan, J. A. 1983. Nitrogen oxides in the troposphere: global and regional budgets. *Journal of Geophysical Research 88*, 10,785–10,807.

Nriagu, J. O. 1979. Global inventory of natural and anthropogenic emissions of trace metals to the atmosphere. *Nature 279*, 409–411.

Nriagu, J. O. 1983. *Lead and Lead Poisoning in Antiquity*, Wiley, New York.

Nriagu, J. O. 1989. A global assessment of natural sources of atmospheric trace metals. *Nature 338*, 47–49.

Nriagu, J. O. 1990. Global metal pollution. *Environment 32*, 7–11, 28–33.

Nriagu, J. O., and C. I. Davidson, eds. 1986. *Toxic Metals in the Atmosphere*, Wiley, New York.

Nriagu, J. O., and J. M. Pacyna. 1988. Quantitative assessment of worldwide contamination of the air, water and soils with trace metals. *Nature 333*, 134–139.

Purves, D. 1985. *Trace-Element Contamination of the Environment*, Elsevier, Amsterdam.

Schaule, B. K., and C. C. Patterson. 1981. Lead concentrations in the North Pacific: Evidence for global anthropogenic perturbations. *Earth and Planetary Science Letters 54*, 97–116.

Suttie, E. D., and E. W. Wolff. 1991. Seasonal input of heavy metals to Antarctic snow. In *Proceedings of the International Conference on Heavy Metals in the Environment* (J.G. Farmer, ed.), CEP Consultants, Edinburgh, Vol. 1, 78–81.

UNEP. 1991. *Environmental Data Report*, United Nations Environment Program, Basil Blackwell Ltd., Oxford.

Voldner, E. C., and L. Smith. 1989. *Production, Usage and Atmospheric Emissions of 14 Priority Toxic Chemicals*, Water Quality Board, International Joint Commission on the Great Lakes, Windsor, Ontario, Canada.

Walthall, J. A. 1981. *Galena and Aboriginal Trade in Eastern North America*, Scientific Papers, Vol. 17, Illinois State Museum, Springfield, Illinois, 66 pp.

20

Metals Loading of the Environment: Cadmium in the Rhine Basin

William Stigliani, Peter Jaffé, and Stefan Anderberg

Abstract

Air emissions of cadmium in the Rhine Basin are estimated for the years 1955 to 1988. Sources include iron and steel production, coal and oil combustion, nonferrous metal production, cement manufacturing, and solid waste incineration. Emissions have decreased by an order of magnitude since the 1950s. Cadmium in solid wastes are also estimated to have decreased, by about a factor of two. Aqueous emissions have decreased by an order of magnitude, with about one-half of the remaining emissions from nonpoint sources, such as urban and agricultural runoff. The topsoil cadmium concentration is estimated to have increased by about 10% between the mid-1970s and the late 1980s. Assuming that the soil pH is maintained in agricultural soils, cadmium intake is estimated to remain below World Health Organization recommended thresholds for many decades.

Introduction

The Rhine Basin encompasses most of Switzerland, southwestern Germany, northeastern France, Luxembourg, and most of the Netherlands. Its total area, about 200,000 km^2, is about the size of the state of Georgia, with a current population of approximately 50 million. It has been one of the most industrialized and polluted river basins in the world for many decades. Beginning in the early to mid-1970s, however, there has been a marked decrease in pollution for most chemicals, even though the level of industrial activity has not declined appreciably since that time. Thus, the cleanup of the basin may be considered a success story which hopefully may be emulated in river basins still trapped in the production–pollution syndrome.

On the other hand, consideration must be given to the fact that thousands of tons of pollutants, including heavy metals, pesticides, and toxic organic compounds, have been deposited in the basin from atmospheric deposition, application of agrochemicals, tipping of wastes in industrial and municipal landfills, etc. What has happened to these wastes, and can we assume that they pose no potential problems now or in the future? To gain some insight into this question, this chapter examines the history of cadmium pollution in the basin since the 1950s. Sources of pollution are identified, and the potential impact on human health of the cumulative loading of cadmium in agricultural soils since the 1950s is assessed.

Although there is a large uncertainty associated with the reconstruction of historical pollution trends over so long a time period and so broad a spatial scale, we feel that the analysis provided below raises important scientific questions worthy of further investigation.

Historical Perspective

As shown in Figure 1, there has been a dramatic decrease in the loading of cadmium in the Rhine Basin from atmospheric deposition since the 1950s. Deposition peaked in the early to mid-1960s at more than 200 tons per year, and was more than two times higher in the mid-1950s than it was in the mid-1970s. By the late 1980s, deposition had declined to less than 30 tons per year—less than 15% of the load in the mid-1960s. Also shown are the contributions to the total deposition from atmospheric emissions originating in the countries in which the Rhine Basin is located, the Western European countries outside of the basin (Belgium, United Kingdom, and Italy), and the Eastern European countries (the former German Democratic Republic, Poland, Czechoslovakia, and the former Soviet Union).

Another trend evident from Figure 1 is the increasing share of deposition originating from Eastern Europe, accounting for over 25% of the total by the late 1980s compared to less than 5% in the mid-1950s. The current share is surprisingly high, considering that the prevailing winds blow from west to east and most of the emissions in Eastern Europe are deposited far east of the Rhine Basin. The observed trend is not due to increasing emissions from Eastern Europe, but rather to regional differences in efforts to limit air pollution. While Western Europe vigorously implemented measures for reducing air emissions, in Eastern Europe emissions remained largely unabated.

The main sources of air emissions in the Rhine Basin from the 1950s to the late 1980s are provided in Table 1. One reason for the decline in emissions was the implementation of emission control technologies in the basin nations. Additional factors, however, also were important. Emissions from coal and oil combustion declined also because of the adoption of energy conservation measures and the increased use of nuclear power. Emissions from iron and steel production declined in part because of the stagnation of production in the basin. The very large reductions in nonferrous metal production (zinc, copper, and others) were in part the result of the closing down of large pyrometallurgical smelters.

Figure 2 shows the trend for available cadmium in solid wastes generated in the Rhine Basin since 1970. These values refer not to the total amount of cadmium in the wastes, but rather to the maximum fraction estimated to be mobilizable over a 30- to 50-year period under rather severe environmental conditions (Stigliani and Anderberg, in press). The total annual cadmium load declined by nearly 60% between 1970 and the late 1980s. As shown in the figure, the largest sources of cadmium in solid wastes historically have been municipal waste disposal, iron and

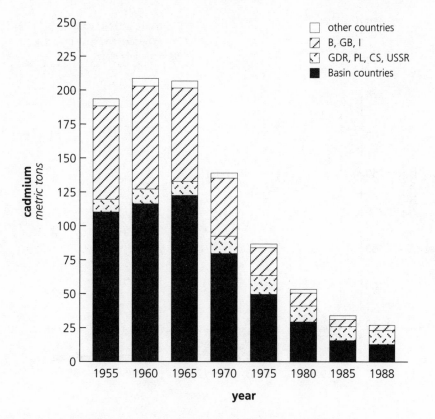

Figure 1. Atmospheric deposition of cadmium in the Rhine Basin, 1955–88.

Table 1: *Sources and trends in air emissions of cadmium in the Rhine Basin (tons per year)*

Source	1955	1960	1965	1970	1975	1980	1985	1988
Iron and steel production[1]	26	33	38	49	32	12	7	7
Coal and oil combustion	60	51	49	45	34	23	18	16
Non-ferrous metal production	120	130	140	61	39	14	5	5
Cement manufacturing	4	5	7	8	4	2	1	1
Solid waste incineration	1	3	5	8	10	11	8	5
Total	210	220	240	170	120	62	39	34

[1] Includes coke production.

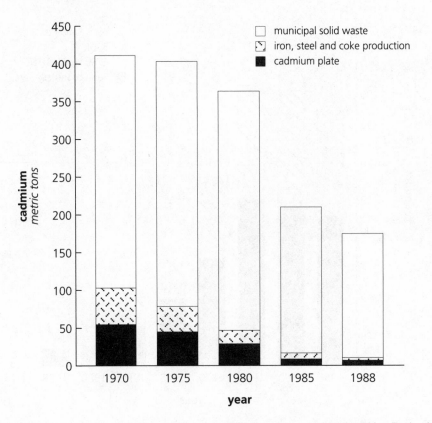

Figure 2. Environmentally available cadmium in solid wastes generated in the Rhine Basin since 1970.

steel production, and cadmium-product manufacturing. Reductions in solid cadmium wastes in iron and steel production occurred for two reasons. One was the extensive recycling of these wastes as feedstocks to zinc–cadmium smelters beginning in the late 1970s. Another reason was the decreasing concentration of cadmium in recycled scrap steel. The major use of cadmium in the 1950s and 1960s was as a protective coating on steel and other metals. Basin-wide, approximately 600 tons of cadmium per year were used for this purpose in the 1960s, and the recycling of Cd-coated steel scrap constituted the largest source of cadmium contamination in iron and steel production. By the late 1980s, cadmium use as a coating for steel was nearly phased out, with a corresponding reduction of cadmium wastes in iron and steel production.

There are four major cadmium-containing products: cadmium plate, cadmium pigments, cadmium as a stabilizer in polyvinyl chloride (PVC), and nickel–cadmium batteries. Of these, only the manufacture of cadmium plate results in appreciable mobilizable solid wastes. Thus, the reduced production of cadmium plate during the 1970s and 1980s led to reduced amounts of solid manufacturing wastes. The reduction of cadmium in municipal wastes is the result of the reduced use of all cadmium products except for nickel–cadmium batteries.

Figure 3 depicts the aqueous emissions of cadmium in the Rhine Basin since 1970. One may observe two trends. One is that the total emissions declined markedly over this period, from over 200 tons in 1970 to under 30 tons in the late 1980s. Secondly, the share of emissions from diffuse sources relative to point sources has steadily increased, from about 20% of the total in 1970 to about 50% in the late 1980s.

As shown in Table 2, the major point sources of aqueous emissions have been nonferrous metal production (particularly zinc production), iron and steel production, cadmium-product manufacturing, the production of phosphate fertilizer, and activities of small industry (almost entirely from small scale cadmium plating firms). A large part of these emissions prior to the late 1970s resulted from indiscriminate dumping of cadmium wastes to the river. One reason for this practice was that air emissions began to be regulated in the early to mid-1970s, before there were stringent regulations on water emissions. Thus, it was relatively easy to make large reductions in aqueous emissions by simple "good housekeeping" practices. By the 1980s, regulation of aqueous emissions became increasingly more stringent, and further reductions were achieved by installing in-house waste-water treatment facilities.

The major sources of diffuse aqueous emissions, accounting for about 35 tons in 1970 and about 13 tons in the late 1980s, were runoff from urban storms, and runoff and erosion from agricultural areas. The origins of diffuse cadmium in urban areas were cadmium as a component of atmospheric deposition (from both short and long range transport), the corrosion of zinc-containing building materials in which cadmium is present as an unwanted impurity, and leaching from landfills. In agricultural areas, the main sources of cadmium were cadmium in atmospheric deposition (from long range transport) and applied phosphate fertilizer in which cadmium is present as an unwanted impurity.

Cadmium Accumulation in Soils and Its Relevance

The main pathway for cadmium deposition on urban and forest soils is atmospheric deposition, and the main removal processes are surface runoff, erosion, and infiltration. On agricultural soils, on the other hand, cadmium is also deposited through the application of phosphorus fertilizers, which contain cadmium as an impurity, and through the application of sewage sludge. Removal of cadmium from agricultural soils is also affected by plant uptake followed by harvesting.

Based on these processes, Stigliani and Anderberg (in press) estimate net inputs of cadmium to soils since 1950, using available information on historical phosphate use (Behrendt, 1988) and calculating atmospheric deposition. Historical atmospheric deposition was calculated from the TRACE 2 model (Alcamo *et al.*, 1992), using historical emissions as an input. Emissions were estimated from historical levels of production for the major industrial sectors and estimates of historical trends in emission factors (Pacyna, 1991). The estimated cumulative inputs to unpaved urban areas, agricultural lands, and forests over the period from 1950 to

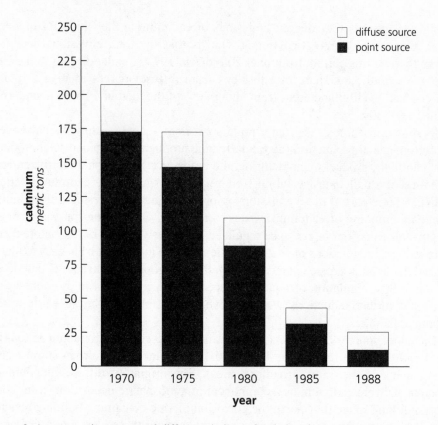

Figure 3. Aqueous point-source and diffuse emissions of cadmium in the Rhine Basin since 1970 (source: Behrendt, 1993). Reprinted with permission from IIASA.

1988 were 830 tons, nearly 4000 tons, and about 1500 tons, respectively. These inputs translate into 420 g/ha for unpaved urban areas, 400 g/ha for agricultural lands, and 220 g/ha for forest areas.

Cadmium sorption to soils is proportional to the inverse of the square root of the soil water proton concentration (Boekhold *et al.*, 1990). Based on this relationship, the velocity at which the cadmium that is deposited on the soil surface infiltrates into soils can be estimated as a function of the pH (Stigliani and Jaffé, 1993). The results are shown in Figure 4. Unlike forest and urban soils, which are more acidic, agricultural soils have been maintained typically at a pH of 6 or more by the practice of liming. As shown in Figure 4, cadmium has essentially been accumulating in the root zone of agricultural soils (top 20 cm), whereas it has been leaching at a significant rate from the more acidified soils in urban centers and forests.

A major concern of this buildup of cadmium in the root zone of agricultural soils is its effect on the cadmium content in foods. The average cadmium intake for the population in the European Community in the mid-1970s was on the order of 30 μg/day or 210 μg/week, compared to the World Health Organization's (WHO) recommended maximum intake of no more than 400 to 500 μg/week

Table 2: *Sources and trends in aqueous point-source and diffuse emissions of cadmium in the Rhine Basin (tons per year)*[1]

	1970	1975	1980	1985	1988
Point sources					
Non-ferrous metal production	81	71	43	3	>1
Iron and steel production[2]	32	31	24	10	3
Cadmium product manufacturing	18	10	3	1	>1
Phosphate fertilizer production	28	27	15	15	10
Small industry	12	6	2	>1	>1
Municipal sewage	2	1	1	1	1
Subtotal	172	146	89	30	14
Diffuse sources					
Urban runoff and leaching	23	17	12	8	6
Agricultural runoff and erosion	12	10	8	7	7
Subtotal	35	27	20	15	13
Grand total	207	174	109	45	27

[1] Some totals do not add up precisely due to rounding differences.

[2] Includes coke production.

(Hutton, 1982).[1] Stigliani and Jaffé (1993) conducted first order calculations on ingestion of cadmium for the hypothetical situation in which a person obtained his entire food supply from crops grown in the basin. They assumed that cadmium intake is directly proportional to the cadmium content in crops, and that the cadmium content in crops is directly related to the dissolved cadmium in soils (Hutton, 1982), although it is known that atmospheric deposition also directly affects the cadmium content in crops (Keller and Brunner, 1983).

According to the calculations of Stigliani and Anderberg (in press), the cadmium content in agricultural soils has increased during the period between the mid-1970s to the late 1980s from 630 g/ha to 700 g/ha. Under the assumption that at a soil pH of 6 the dissolved cadmium in soil solution, and hence uptake by crops, is proportional to the concentration of cadmium in the soil, then the calculated increase in cadmium intake by ingestion is calculated to increase by 12%, still far below the WHO's recommended maximum intake. If we assume a cadmium increase in agricultural soils of 4.7 g/ha-yr, which was observed in the late 1980s, it will take on the order of 120 years for the average intake of cadmium to reach the maximum allowable intake rate. Changes in the pH of agricultural soils will have a much more important effect on the average cadmium intake. Following our simplified first order analysis, Figure 5 shows the estimated average cadmium intake for different years if the pH of agricultural soils were to decrease from 6.

---

[1] Editor's note: Improved sampling and analysis techniques indicate that actual cadmium intake in the 1970s was significantly less than 30 µg/day.

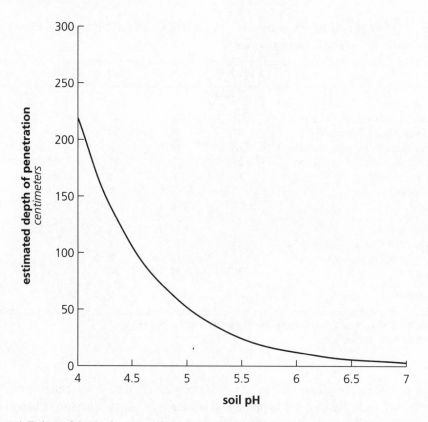

Figure 4. Estimated depth of penetration of cadmium in soils over 40 years as a function of pH.

The calculations, for a constant pH of 6, are in agreement with those done by Keller and Brunner (1983) for Swiss soils, and show that the average cadmium concentration in agricultural soils has not build up to levels that would be considered unsafe, and will not reach such levels for many decades. What these calculations have shown is that cadmium levels have built up to such levels that shifts in pH that are quite feasible (e.g., from 6 to 5.5) may pose a problem today, whereas a similar decrease in the 1950s would not have resulted in exceeding the WHO limits. These elevated levels of intake, however, would decrease over time, given that at lower values of pH the cadmium would leach out of the root zone at a higher rate, thus resulting in decreasing soil concentrations of cadmium.

Conclusions

In summary, analysis of pollution trends in the Rhine Basin since the 1950s has provided several insights. One is that industrial production is not of necessity linked to high levels of pollution, as had been the case in the 1950s and 1960s. A second conclusion is that although large reductions in overall pollution have been achieved; emissions from diffuse sources of pollution from agricultural lands and

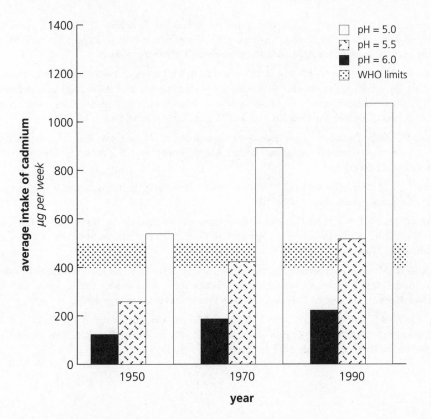

Figure 5. Average estimated cadmium intake under different conditions of increasing agricultural soil acidity (from pH 6.0 to pH 5.0). Estimates assume that the sole food intake of cadmium is from crops grown in the Rhine Basin.

urban storm runoff are responsible for an increasing share of the total emissions. Finally, despite the large gains in control of chemical pollution, we must be concerned about the potential impacts from the cumulative chemical inputs to soils in the basin over previous decades. An important function of soils is their capacity to buffer ecosystems against the uptake or mobilization of trace metals and other pollutants. The impact of long-term accumulation of cadmium in agricultural soils has been to increase the uptake of cadmium in crops. Changes in the pH of agricultural soils today will have a much more important effect on the average cadmium intake than was the case a few decades ago.

References

Alcamo, J., J. Bartnicki, K. Olendrzynski, and J. Pacyna. 1992. Computing heavy metals in Europe's atmosphere: Part 1. Model development and testing. *Atmospheric Environment 26A*, 3355–3369.

Behrendt, H. 1993. *Point and Diffuse Loads of Selected Pollutants in the River Rhine and its Main Tributaries*. Research Report No. RR-93-1, International Institute for Applied Systems Analysis, Laxenburg, Austria.

Behrendt, H. 1988. *Changes in Nonpoint Nutrient Loading into European Freshwaters: Trends and Consequences Since 1950 and Not-Impossible Changes until 2080*. Report No. WP-88-026, International Institute for Applied Systems Analysis, Laxenburg, Austria.

Boekhold, A. E., S. E. A. T. M. Van Der Zee, and F. A. M. De Haan. 1990. Prediction of cadmium accumulation in heterogeneous soils using a scaled model. In *Proceedings of Conference: Calibration and Reliability of Groundwater Modelling*, September 1990, The Hague, Netherlands. IAHS Publication No. 195, IAHS, The Hague, Netherlands, 211–220.

Hutton M. 1982. *Cadmium in the European Community: A Prospective Assessment of Sources, Human Exposure and Environmental Impact*. MARC Report No. 26, Chelsea College, University of London, London, U.K.

Keller, L., and P. H. Brunner. 1983. Waste-related cadmium cycle in Switzerland. *Ecotoxicology and Environmental Safety 7*, 141–150.

Pacyna, J. M. 1991. *Emission Factors of Atmospheric Cd, Pb, and Zn for Major Source Categories in Europe in 1950 through 1985*. Report No. 0-1512, Norwegian Institute for Air Research, Lillestrom, Norway.

Stigliani W. M., and S. Anderberg. Industrial metabolism at the regional level: The Rhine Basin. In *Industrial Metabolism: Restructuring for Sustainable Development* (R.U. Ayres and U.E. Simonis, eds.), The United Nations University Press, Tokyo, Japan, in press.

Stigliani W. M., and P. R. Jaffé. 1993. *Industrial Metabolism and River Basin Studies: A New Approach for Analysis of Chemical Pollution*. Research Report No. RR-93-6, International Institute for Applied Systems Analysis, Laxenburg, Austria.

21

Emissions and Exposure to Metals: Cadmium and Lead

Valerie Thomas and Thomas Spiro

Abstract

Cadmium and lead are examined in terms of their damage to ecosystems and human health, and in terms of their uses and environmental emissions. In both cases, the major uses of the metals have only a small known effect on ecosystems or human health. For cadmium, non-occupational human health and ecosystem damage has been found only in areas of acute localized contamination, and future risk to the general population appears to be remote. What risk there is stems more from trace cadmium contamination in phosphate fertilizers and in coal than from cadmium products. For lead, there are major detrimental health effects both on the human population and many bird populations. But while the major use is lead-acid batteries, the major known health and environmental effects are due to leaded gasoline, lead-based paint, and lead shot, all of which are minor uses of lead. This conclusion underscores the importance of including careful analysis of exposure and pollutant effects in industrial ecology, and not just emissions or use of toxic materials.

Introduction

The goal of industrial ecology is to create the basis for an industrial system that is increasingly compatible with the health of environmental systems and human beings. In an ideal industrial ecology scenario, materials would be reused rather than extracted from the earth and dissipated into the environment. The mining of these materials would essentially be stopped, and replaced by efficient recycling; inherently dissipative products, such as paints and ammunition, would be made only from materials that are environmentally safe.

Before any such industrial ecology can be planned, much less implemented, it is important to understand the environmental and human health consequences of current practices, as well as to understand what the effects of changing these practices might be.

This chapter focuses on two metals, cadmium and lead. For these two metals, we will survey the entire system of extraction, use, and emission into the environment,

297

and the resulting effects on living things. Cadmium and lead present instructive contrasts in their effects on human health and ecosystems, and in the sources and mechanisms of the most harmful exposures. However, the framework of the analysis is not specific to metals, and may be helpful for further studies in industrial ecology.

The analysis is organized around three basic questions: (1) How does the metal damage human health or ecosystems? (2) How is the metal used by industries and consumers? (3) For each case of damage to human health or ecosystems, what specific pathways are especially important?

Cadmium
Effects of Cadmium on Humans and Ecosystems

Most recent studies in industrialized countries report average cadmium intakes of 10–20 µg/day; there have been few studies in developing countries (Gunderson, 1988; Vahter *et al.*, 1991; Louekari *et al.*, 1991; Dabeka and McKenzie, 1992; Buchet *et al.*, 1983). Smokers have about twice the average exposure because tobacco concentrates cadmium and because cadmium is absorbed more efficiently when inhaled than when ingested.

Cadmium accumulates in the liver and kidneys, increasing over the course of a lifetime. Based on studies of cadmium-exposed male workers, slight kidney damage (urinary excretion of certain proteins) has been thought to occur in about 10% of people who ingest more than 200 µg/day of cadmium, over a period of 25 to 50 years (Kjellstrom, 1986). The health significance of these changes in kidney function is not understood, but it has been assumed that these changes could lead to further kidney problems, and that cadmium exposure should be kept below the level that might cause these effects. Consequently, the World Health Organization (WHO) has recommended a maximum cadmium intake of 60 to 70 µg/day (1 µg/day per kilogram of body weight) (ATSDR, 1989). However, a recent cross-sectional population study in Belgium suggests that the WHO's recommended maximum intakes may be too high (Buchet *et al.*, 1990). At higher intakes (600 µg/day) a bone disease called itai-itai can develop. It has been found only in Japan, primarily in the Jinzu area, where rice was grown in water polluted by cadmium from a nearby mining operation. It occurred mostly in women who were elderly, had a vitamin D deficiency, and had more than one child (Fergusson, 1990).

Inhaled cadmium has also been found to cause lung cancer in cadmium-exposed workers (Stayner *et al.*, 1992).

In terrestrial ecosystems, the primary documented effect of heavy metal contamination has been damage to soil organisms, which decompose organic matter and recycle nutrients. For cadmium, these effects have only been observed in the vicinity of smelters, with cadmium concentrations of about 100 ppm (a thousand times higher than typical soil concentrations), and in combination with high levels of other heavy metals such as zinc and lead (Tyler, 1972; Watson *et al.*, 1976; Giller and McGrath, 1988; Friedland *et al.*, 1986). But there may also be significant effects at lower cadmium concentrations. In laboratory studies, cadmium has been shown to be toxic to at least one soil microorganism at cadmium concentrations only about ten times those typically found in soils (Van Straalen *et al.*, 1989).

Hunters are often advised not to eat the livers of wild deer and moose because of their high cadmium content. Studies of deer in New Jersey and in Ontario have found a correlation between high cadmium levels and low pH or low buffer capacity of the soil (Stansley *et al.*, 1991; Glooschenko *et al*, 1988). However, livers and kidneys are naturally high in cadmium (about 0.1 µg/g for a typical beef liver), and the extent to which these high levels are the result of anthropogenic cadmium emissions is unknown. Moreover, although cadmium levels of deer livers are often high enough to be a health hazard to the people eating them, there is no evidence that these cadmium levels are harmful to the deer. Indeed, deer livers deemed unsafe to eat have about the same cadmium concentrations (1.5 µg/g wet weight) as typical adult human livers.

There are only a few records of cadmium toxicity in polluted aquatic ecosystems. In freshwater systems, salmonids (including trout) and cladocerans (including water fleas) are the most sensitive to cadmium, showing adverse health effects (reproductive and developmental) at cadmium concentrations of about 1 µg/l in soft water, and about 20 µg/l in hard water.[1] Typical freshwater levels are less than about 0.1 µg/l cadmium, well below the threshold for toxic effects (Sprague, 1987; Stoeppler, 1991).

Ocean cadmium levels are generally similar to those in freshwater systems (Simpson, 1981). However, ocean species are typically less sensitive to cadmium, with mollusk and crustacean larvae, the most sensitive species, affected at about 10 µg/l. Shellfish often have accumulated levels of cadmium (~ 600 µg/kg) that could be a health hazard to humans if eaten in abundance, but these levels have not been reported to be toxic to the shellfish themselves (McLeese *et al.*, 1987; McKenzie-Parnell *et al.*, 1988; Delos, 1985).

Sources and Use of Cadmium

Figure 1 is a diagram of world cadmium extraction and use. Cadmium is mainly extracted from zinc ores, where it is found as a minor constituent; refined cadmium is a byproduct of zinc refining. Over one-half of the refined cadmium is used in batteries, with the remainder used in plating, in pigments, as a stabilizer of polyvinyl chloride plastics, and in alloys. Because consumer nickel–cadmium batteries are often sealed within devices, and because there is little infrastructure for their collection or recycling, cadmium recycling is essentially limited to industrial batteries, and amounts to only about 5% of annual consumption. The rest of the cadmium ends up in landfills or waste water, or as particles transported through the atmosphere, subject to wet or dry deposition. Airborne particles are produced

[1] Cadmium is less toxic in hard water because the hard water makes gills less permeable to cadmium.

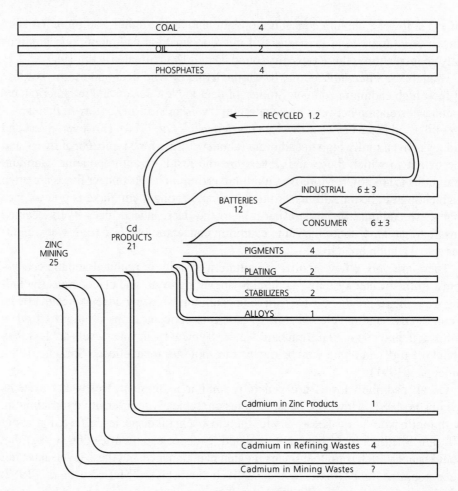

Figure 1. World extraction, use, and disposal of cadmium, 1989 (thousand tons). Data from Roskill, 1990; USBM, 1991; Jolly, 1988; Fulkerson and Goeller, 1973; WRI, 1990; Stowasser, 1988; Jolly, 1992; Nriagu and Pacyna, 1988.

by combustion of cadmium-containing products (e.g., in municipal incinerators), as well as from fossil fuel combustion. Corrosion and abrading of cadmium-containing materials, such as pigments and metal plate, leads to further release of cadmium into soils and water systems.

Cadmium is an impurity in several high-volume commodities: coal, oil, and phosphate fertilizer. Collectively, these commodities account for about one-half as much cadmium as is in the cadmium product stream. The cadmium in oil is discharged directly into the atmosphere upon combustion, while the cadmium in coal is partitioned between emissions and fly ash. Cadmium in phosphate fertilizer is a particular concern because, as discussed below, cadmium in food plants is the main exposure route for humans.

Improvements in pollution control have resulted in reduced air and water emissions over the past several decades. Stigliani, Jaffé, and Anderberg (this volume) estimate that in the Rhine River Basin air emissions have dropped by about 85% since 1965, and water emissions have dropped by 85% since 1970. Reduced air emissions have been reflected in a decrease in the rate of cadmium deposition to snow in Greenland by a factor of 2.5 since 1970 (Boutron *et al.*, 1991).

There have been a number of efforts to reduce the use of cadmium in products. Japan has phased out most uses of cadmium plating, beginning in 1969 (Roskill, 1990). In 1979 Sweden banned the use or import of cadmium pigments, plating, and stabilizers. Not only did this essentially eliminate these products from Sweden, but it also led to reductions in the use of cadmium to create such products elsewhere, so that cadmium-free products could be exported to Sweden (Kjällman, 1989). In the United States, more than a dozen states have passed Toxics Use Reduction Acts, which typically promote or require reductions in the use of cadmium and other toxics (Ryan and Schrader, 1991). U.S. manufacture of barium–cadmium stabilizers is expected to be phased out by 1997, being replaced by barium–zinc and calcium–zinc stabilizers (USBM, 1992). Although there has been a worldwide drop in the use of cadmium plating, pigments, stabilizers, and alloys between 1970 and 1989, the increase in cadmium use in batteries has resulted in an overall 45% increase in world cadmium use during this time (Roskill, 1990).

Cadmium Pathways

In this section we focus on pathways to human exposure; little is known about cadmium exposure for other species. Although many factors can affect uptake and absorption of cadmium, Table 1 shows the typical contribution from each of the pathways of human exposure, based on data from the United States and Europe. Except for cigarettes, food is by far the largest source of absorbed cadmium.

Cadmium is taken up by plants, and it is accumulated in the kidneys and livers of mammals and in shellfish.[2] While some people are at risk from eating large quantities of shellfish or deer or moose liver, three-quarters of typical U.S. cadmium intake is estimated to be from potatoes, grains, and cereals (Gunderson, 1988).

The amount of soil cadmium that gets into plants depends, in part, on the chemistry of the soil. For example, in Shipham, England, the site of zinc and cadmium mining in the 17th to 19th centuries, levels of cadmium in soil often exceeded 300 ppm, a thousand times higher than typical rural levels. Health inventories, however, showed only slight effects attributable to cadmium. In contrast, in the Jinzu area of Japan, levels of cadmium in agricultural soil reached a maximum of 70 ppm, but itai-itai disease developed from exposure to mining wastes (Tsuchiya, 1978). The difference in effects has been attributed to differences in the fraction of

---

[2] Leafy vegetables such as spinach and lettuce tend to have higher cadmium concentrations on a dry weight basis than grains such as wheat and rice; the uptake of cadmium by plants is determined by soil properties as well as by soil cadmium concentration (Fergusson, 1990).

Table 1: *Typical cadmium intake and absorption for adults*

Material	Intake (kg/day)	Amount Cd in Material (µg/kg)	Cd Intake from Material (µg/day)	Absorption Factor[1]	Cd Absorption (µg/day)
Food[2]	2	—	10–20	4–7%	0.4–1.4
Air	20	<0.03[3]	<0.6	25–50%	<0.3
Water	1.4[4]	<1	1	4–7%	<0.1
Soil	<0.0001	270[5]	<u><0.03</u>	4–7%	<u><0.002</u>
Total for nonsmokers			**10–20**		**0.4–1.8**
20 Cigarettes	0.002[6]	1000	2	25–50%	<u>0.5–1</u>
Total for pack-a-day smokers					**0.9–2.8**

[1] The percentage of ingested cadmium absorbed ranges from about 4 to 7% in the absence of modifying factors, such as calcium, iron, or protein deficiency. Absorption of cadmium in the lungs is known with less certainty but is estimated to be 25–50% of inhaled cadmium.

[2] Wet weight of food, excluding tap water.

[3] Concentrations in the United States range from 0.034 µg/m^3 to 0.001 µg/m^3.

[4] Daily consumption of tap water, including water consumed in juices, coffee, etc. The U.S. maximum allowed level is 5 µg/l of drinking water.

[5] The average U.S. topsoil concentration is about 0.27 ppm, although typical soils can range from 0.1 to 2 ppm.

[6] About 10% of the cadmium in cigarettes is inhaled during smoking.

Data from Elinder, 1986; Nordberg *et al.*, 1986; Gartrell *et al.*, 1986; Gunderson, 1988; Federal Register, 1985; Page *et al.*, 1987; Elinder *et al.*, 1983; Watanabe *et al.*, 1987; Stoeppler, 1991.

cadmium in the dissolved phase in these soils: in the Jinzu area there was a relatively low pH (~5.1), low calcium carbonate content (~0.4%), and very low content of hydrous oxides of iron and manganese. All of these factors tend to increase the fraction of soil cadmium that is in solution. In contrast, in Shipham, there was a high pH (~7.5), high calcium carbonate content (6–14%), and high hydrous oxide content of the soils, all of which act to decrease the concentration of cadmium in soil solution (Morgan, 1988).

As shown by Stigliani, Jaffé, and Anderberg (this volume), cadmium is relatively immobile in agricultural soils, which are typically maintained at a pH of 6 or higher by the addition of lime. Thus the continuing addition of cadmium to soils is expected to result in an increase in cadmium in soils and in crops. Some experimental studies have found an increase in cadmium levels in crops over the past century. The cadmium content in plants in a plot in England, as well as the cadmium content of Swedish wheat and barley, increased by about a factor of 2 between the turn of the century and the 1970s (Jones *et al.*, 1992). However, a study of U.S. agricultural soils from about 1930 to 1980 found no increase (Mortvedt, 1987). Another study found no increase in cadmium in bread cereals grown in Germany during this century (Lorenz *et al.*, 1986).

Nevertheless, it has been estimated that the cadmium content of U.S. cropland is increasing by about 1% annually, with about three-quarters of this from phosphate fertilizer and the rest from deposited air emissions.[3] As of the mid-1980s, coal and oil combustion were estimated to account for 70% of cadmium air emissions in the United States, followed by municipal waste incineration (Federal Register, 1985). In municipal waste, batteries and stabilizers in plastics account for over 60% of the cadmium (Franklin Associates, 1989).

Stigliani, Jaffé, and Anderberg (this volume) have estimated a similar buildup of cadmium in soil in the Rhine River Basin. They estimate about a 1% annual increase of cadmium in food, assuming that the pH of the soil remains constant. If this model is correct, if current cadmium ingestion is about 15 µg/day, and if cadmium input remains constant, average dietary intakes will remain well below the WHO threshold for many decades. However, as noted earlier, it has been suggested that the WHO threshold should be lowered.

Application of sewage sludge to agricultural land can result in much greater localized increases in soil cadmium concentrations. Sewage sludge applied to agricultural land typically has a somewhat lower cadmium content than phosphate fertilizers, but is applied in much larger amounts per hectare. In the United States, new regulations of the Environmental Protection Agency (EPA) have set a limit of 39 ppm for cadmium in sludges that can be applied to land year after year, and a maximum annual cadmium addition to land of 1900 g/ha (Federal Register, 1993). This would triple the average cadmium content in the top 15 cm of soil. But there is not enough sewage sludge to affect more than a small fraction of U.S. cropland; if all of the 5 million tons of U.S. sludge were spread on land, at a typical application rate of 7.4 tons/ha, this cadmium would be concentrated on a little over 1% of the U.S. cropland. Currently about one-third of U.S. sewage sludge is spread on the land (Federal Register, 1993).

Although the continuing deposition of cadmium onto agricultural land leads to the expectation of gradually increasing cadmium exposure, over the past decade measurements of average cadmium intake per day have dropped, from 30 µg/day or higher to 15 µg/day or less (Elinder, 1986). Although much of this decrease can be attributed to improvements in analytical techniques, there is some indication that there has been a real drop in cadmium intake in places where high air concentrations of cadmium have been reduced (Ducoffre *et al.*, 1992). A Belgian study found that blood cadmium levels decreased by about 14% annually between 1984 and 1988; a German study reportedly found a decrease of about 10% annually between 1979 and 1986. In one area of Japan dietary intakes dropped by about 30% between the late 1970s and late 1980s, and a decrease was also

---

[3] Phosphate fertilizer typically contains about 20 µg/g of cadmium but can contain 100 µg/g or more. The average U.S. topsoil concentration of cadmium is about 0.27 ppm. Assuming a soil density of 1.5 kg/l and a 6-in (0.15-m) plow depth, cropland has about 600 g/ha of cadmium. The average cadmium addition to U.S. cropland is estimated to be 5 g/ha annually; thus the annual increment of soil cadmium is about 1% of the total (Delos, 1985).

found in the cadmium content of rice (Watanabe *et al.*, 1992). The findings of decreased cadmium exposure in Belgium, Germany, and Japan correlate with the decreasing cadmium emissions, but they are contrary to the expectation that cadmium intake will increase in proportion to the total amount of cadmium in the soil. The relationship of cadmium exposure to soil cadmium deserves further study.

Implications for the Industrial Ecology of Cadmium

What lessons can we draw for the industrial ecology of cadmium? First, the extent of human health damage from cadmium is not well established, although a study from Belgium indicates that 10% of the population may have slight kidney damage attributable to cadmium. There is little knowledge of ecosystem damage, and it might be greater than we now realize.

Most of the concern about long-term cadmium exposure has been that a buildup of cadmium in soil might lead to dangerous exposures in the future. However, there is great uncertainty about how much cadmium exposure is due to anthropogenic emissions, and about whether exposure is increasing or decreasing. If cadmium exposure is increasing, it is probably due almost entirely to cadmium from coal and phosphate fertilizers. While there may be significant human or ecosystem exposures due to cadmium in products or in mining or manufacturing wastes, such exposures have yet to be identified and understood. Clarification of these issues is needed before we can understand which cadmium products, if any, might be leading to increased exposure, and what steps, if any, should be taken to reduce cadmium exposure.

Lead
Effects of Lead on Humans and Ecosystems

In the United States, lead poisoning is considered to be one of the most common and preventable pediatric health problems. As of 1984, 17% of all children, and 30% of children in the inner cities, had blood lead levels greater than 15 µg/dl (U.S. Department of Health and Human Services, 1988). (Children are more at risk than adults because they absorb lead more efficiently and because they are exposed to more lead per unit of body mass.) Over the past 20 years, the criteria for "elevated" blood lead levels in the United States have dropped from 40 µg/dl to 10 µg/dl (Centers for Disease Control, 1991). At 10 µg/dl and less, effects on the intelligence, growth, and hearing of children have been found.

In the animal kingdom, the main victims of lead poisoning are waterfowl, eagles, and other birds, who die from the ingestion of lead shot or lead fishing sinkers. A study of 1429 deaths of bald eagles in the United States from 1963 to 1984 found that 6% had died of lead poisoning (National Wildlife Health Laboratory, 1985; Kaiser *et al.*, 1980). As of the early 1980s, 2–3% of the U.S. fall population of waterfowl were dying each year from lead poisoning; similar effects have been found worldwide. In popular hunting areas, lead poisoning can be much more prevalent (U.S. Fish and Wildlife Service, 1985; Pain, 1992).[4]

Lead is also one of the most frequent causes of poisoning of domestic animals. Causes of death include grazing near lead smelters as well as consumption of lead-based paint, lead batteries, etc. (Oskarsson *et al.*, 1992; Ewers and Schlipkoter, 1991).

Concentrations of lead in soil above 1000 ppm, such as are found near lead smelters (100 times greater than rural background levels), have been found to decrease soil microorganism activity, resulting in slower decay of organic matter. However, at levels below 1000 ppm, such as result from even fairly heavy deposition of lead from leaded gasoline combustion, effects on ecosystems have not been found (Friedland *et al.*, 1986; Friedland and Johnson, 1985; Doelman, 1978). And, as for cadmium, incidents of harm to aquatic organisms are not well documented (Wong *et al.*, 1978).

Sources and Uses of Lead

Figure 2 is a diagram of world lead use, showing a somewhat different pattern than for cadmium. Lead consumption is about a hundred times larger, and about 40% is recycled, mostly from lead-acid batteries, in contrast to about 5% for cadmium. As with cadmium, the main use is batteries; other uses include ammunition, pigments, rolled and extruded products, solder, gasoline additive, and cable sheathing. Much of the lead ends up in landfills, but there is a substantial soil-, water-, and airborne component.

The recycling rate for lead-acid batteries in the United States (about 80%) is even higher than the worldwide average (about 65%) due to lead-acid battery recycling laws in many states; as of 1991, 30 U.S. states had such laws (Woodbury, 1992).

Between 1960 and 1990 in Western countries, lead consumption decreased in all product categories except batteries and ammunition. In the United States, a number of lead products have effectively been phased out, including leaded gasoline, lead solder for food cans and plumbing systems, and lead-based interior paints. However, because lead use in batteries has increased by more than a factor of 2 worldwide, overall lead use has increased by about 40% in this interval (International Lead and Zinc Study Group, 1992).

Although leaded gasoline has essentially been banned in the United States, it is still used in most countries, as shown in Table 2 and Figure 3. Lead in gasoline accounts for only about 2% of the total annual consumption of lead, but most of it is emitted directly to the air. Table 2 shows that, as of 1990, only Japan, the United States, Canada, Guatemala, and Brazil had phased out leaded gasoline. However,

---

[4] The effects and extent of lead poisoning have been determined through the examination of gizzards of hundreds of thousands of birds of many different species in numerous areas by many investigators (Sanderson and Bellrose, 1986).

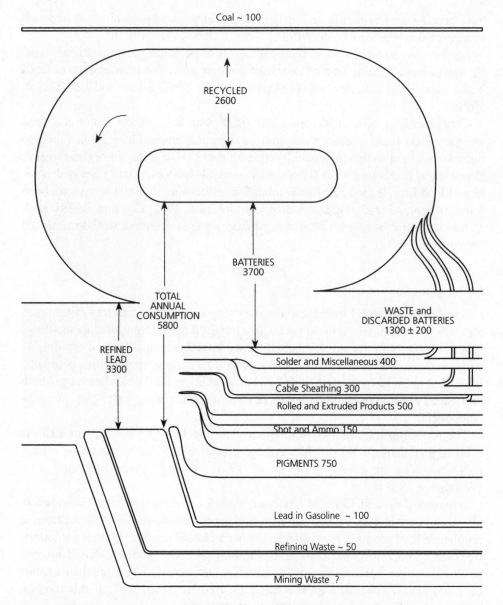

Figure 2. World extraction, use, and disposal of lead, 1990 (thousand tons). Data from International Lead and Zinc Study Group, 1992; Woodbury, 1992; Valkovic, 1983; and Table 2.

many countries have reduced the concentration of lead in leaded gasoline since 1990, including most European nations, Mexico, Argentina, South Africa, Iran, Israel, Taiwan, and Thailand (Octel, 1992). In addition, catalytic converters are now required on all new cars in the European Economic Community and in Mexico, and these converters are damaged by lead, so the use of unleaded gasoline can be expected to increase as new cars replace the old. Currently, the former Soviet Union appears to be the largest consumer of leaded gasoline in the world.

Almost no unleaded gasoline is available in many parts of Africa, Asia, and South America. As shown in Figures 3a and 3b, while the United States and Canada account for close to half of world gasoline consumption, they account for only a tiny fraction of the worldwide emissions of lead from gasoline. In Table 2, countries are listed in order of their contribution to the regional emissions of lead from gasoline.

Lead Pathways

Birds

The reason that so many birds die of lead poisoning is that waterfowl eat lead shot, perhaps mistaking it for seeds or small mollusks. Although many species of birds are poisoned by lead, waterfowl are particularly vulnerable because their populations and waterfowl hunting are highly concentrated geographically. Lead shot is ground up in the gizzard and absorbed in the blood; death can be caused by ingestion of one or two pellets (U.S. Fish and Wildlife Service, 1985).

Lead from shot can be passed up the food chain when birds containing the shot are eaten by other animals. The available evidence indicates that waterfowl are the primary source of lead in the bald eagle's diet. Aside from lead embedded or ingested in live birds (sampling studies typically find that 10–30% of U.S. geese and ducks have shot embedded in their bodies), eagles and other scavengers can get a concentrated dose of lead by eating the substantial number of birds that are killed by hunters but not retrieved (U.S. Fish and Wildlife Service, 1985).

As of 1991, nontoxic steel shot is required for waterfowl hunting in the United States, and use of lead shot has been banned in a number of areas since 1976. Lead shot is entirely banned for hunting in the Netherlands as of February 1, 1993, and its use is banned for waterfowl hunting in Norway. Lead shot is also banned at some sites in Australia, Canada, and Denmark (Pain, 1992). But some hunters are reluctant to use steel shot, despite numerous studies showing that it is a technically acceptable alternative. For example, in 1985 it was estimated that U.S. hunter compliance with existing steel shot regulations was less than 50% (U.S. Fish and Wildlife Service, 1985). The sale of lead fishing sinkers has also been banned in Britain and the United States because of their potential for poisoning birds (EDF, 1992).

Humans

While many factors can affect the lead absorption rates, especially for children, Table 3 shows the typical contribution to lead intake and absorption from each of the pathways of human exposure. It shows that for adults, the main source, as for cadmium, is food. Unlike cadmium, however, lead is not taken up significantly by plants: concentrations in food plants are about a thousand times less than the concentrations in rural soils. Typically, lead in food comes from soil particles adhering to food, as well as from tableware, lead-soldered food cans, lead crystal, lead wine seals, etc.

Table 2: *World use of leaded gasoline, 1990*

	Consumption of Motor Gasoline (× 1000 barrels/day)	Lead Content (g/l)	Total Lead (tons/yr)	% Leaded If Not 100%
North America				
Mexico	440	0.2	5000	90
United States	7200	0.026 (leaded)	1400	2
		0.003 (unleaded)		
Canada	560	0.003	97	~0
Total	7800		7000	
Central and South America				
Venezuela	170	0.85	8400	
Columbia	110	0.8	5100	
Argentina	91	0.4	2100	
Peru	30	0.84	1500	
Chile	30	0.8	1000	
Puerto Rico	50	0.13	380	
Virgin Islands	4	1.12	260	
Brazil	300	0.003	50	~ 0
Guatemala	7	0.003	1	0 (?)
Other	150	0.8	7000	
Total	1400		26,000	
Western Europe				
Italy	330	0.3	5700	95
France	470	0.25	4600	85
Spain	190	0.4	4400	99
United Kingdom	570	0.15	3500	66
West Germany	640	0.15	1700	30
Sweden	98	0.15	430	50
Switzerland	87	0.15	380	50
Netherlands	81	0.15	350	50
Other	534	0.15	700	
Total	3000	0.15	22,000	
Eastern Europe and USSR				
USSR	1800	~ 0.2	20,000	95
Romania	48	0.3–0.6	1000	
Poland	64	0.4	1500	
East Germany	90	0.27	1400	
Hungary	37	0.4	860	
Czechoslovakia	40	0.15	340	99
Total	2100		25,000	
Middle East				
Saudi Arabia	160	0.4	3700	
Iran	140	0.3	2400	
Iraq	75	0.4	1700	

Table 2: *(contd.)*

	Consumption of Motor Gasoline (× 1000 barrels/day)	Lead Content (g/l)	Total Lead (tons/yr)	% Leaded If Not 100%
Middle East *(contd.)*				
Kuwait	20	0.53	620	
UAE	23	0.4	530	
Syria	26	0.24	360	
Israel	36	0.15	310	
Other	80	0.4	1900	
Total	540		12,000	
Africa				
Nigeria	110	0.66	4200	
South Africa	110	0.3	2000	
Algeria	45	0.6	1700	
Libya	34	0.8	1600	
Egypt	51	0.4	1200	
Other	110	0.4	2600	
Total	460		13,000	
Far East and Oceania				
China	460	0.01–0.48	300–6000	46
Australia	300	0.15–0.84	2000	70
India	83	0.56	2700	
Thailand	64	0.4	1500	
New Zealand	43	0.45	770	70
Malaysia	65	0.15	570	
Japan	760	0.003	130	~ 0
Taiwan	84	0.026	130	
Sri Lanka	4	0.2	50	
Singapore	10	0.15	49	56
Hong Kong	6	0.15	32	62
Other	330	0.4	7700	
Total	2200		25,000	
World Total	16,412		130,000	

Data on gasoline consumption from U.S. Department of Energy, 1991, and on lead content from Octel, 1992, except as follows: USSR, Russian Standard, 1987; Guatemala, CONCAWE, 1992 (unconfirmed). (Note: 1 barrel = 42 gallons.)

Toxics and the Environment

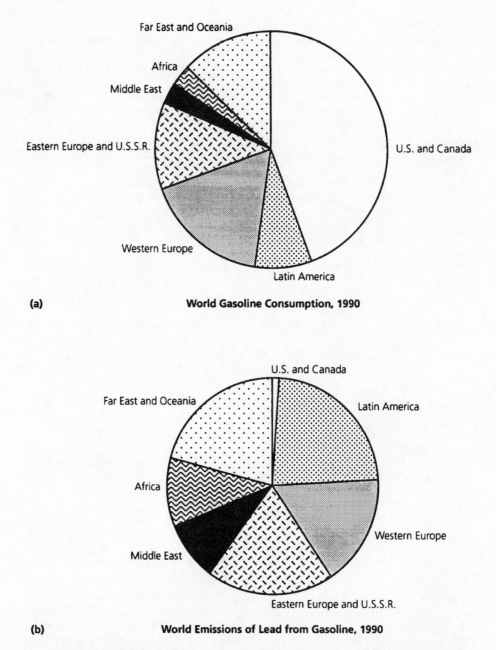

Figure 3. (a) World gasoline consumption, 1990. (b) World emissions of lead from gasoline, 1990.

For children, the main source of elevated intakes typically is soil or dust, which they ingest from dirt on their hands, toys, and other objects. Sources of high lead levels in urban and household dusts include lead-based paint and residues from leaded gasoline. Lead in drinking water comes from lead pipes, lead solders, or brass fixtures in the plumbing system.

310

Table 3: *Typical lead intake and absorption for adults and children*

Material	Intake (kg/day)	Amount Pb in Material (µg/kg)	Pb Intake from Material (µg/day)	Absorption[1]	Pb Absorption (µg/day)
Adults					
Air	20	0.1	2	15–30%	<0.6
Soil/dust	<0.0001	20,000–200,000[2]	<20	7–15%	<3
Water	1.4	5–15	7–21	7–15%	0.5–3
Food	2	<0.2–2500	30–50	7–15%	3–5
Total			**~50**		**4–11**
Children, 2–4 Years					
Air	6	0.1	0.5	25–45%	<0.2
Soil/dust	0.0001[3]	20,000–200,000	2–20	30–40%	0.6–6
Water	0.7	5–15	3.5–10	30–40%	1–4
Food	1[4]	<0.2–2500	10–30	30–40%	3–12
Total			**30–60**		**5–22**

[1] The percentage of ingested lead that is absorbed has been determined through studies of lead in diet and feces and through animal studies. Absorption of lead in the lungs has been determined through studies with radioactive lead and studies of the typical size distributions of lead-bearing particles in urban air (EPA, 1989b).

[2] Soil and household dust levels range from 5 ppm in some rural soils to 3000 ppm in urban household dusts; higher levels can be found in the vicinity of urban point sources (EPA, 1989b).

[3] Soil ingestion has been measured by a mass balance measurement of fecal excretion of poorly absorbed soil minerals (aluminum, silicon, and titanium). It can range up to 5g/day.

[4] Wet weight of food for a two-year-old, excluding all tap water.

Data from EPA 1989a, 1989b, 1990; Gunderson, 1988; Dabeka and McKenzie, 1992; Vahter *et al.*, 1991; Barnes, 1990.

Absorption of about 10 µg of lead per day, as tabulated above for the typical child, results in a blood lead level of about 5 µg/dl. In order to have a blood lead level of 15 µg/dl, a child typically must absorb about 40 µg of lead per day (EPA, 1989b).[5] In the presence of lead-based paint, lead levels in indoor dust can reach 2000 ppm. At this level, ingestion of 0.1 g/day of dust would result in absorption of as much as 100 µg of lead daily, well above the threshold for health effects.

Although use of interior lead-based paint has been banned in the United States, it remains in place in older housing stock and continues to be ingested by children, even when painted over with the newer unleaded paint. Acute lead poisoning can occur when paint is stripped from walls or furniture during renovation.

[5] Blood lead levels do not increase linearly with increasing lead absorption, although for the range of values in Table 3, a linear approximation is adequate.

Prior to the phase-out of lead in gasoline in the United States in the 1970s, gasoline accounted for 85% of the atmospheric emissions of lead, and typical urban air lead levels were about an order of magnitude higher than they are today (EPA, 1991, 1990). This resulted not only in greater inhalation of lead, but also in a higher concentration of lead in urban dust. Average blood lead levels for the adult U.S. population fell from about 18 µg/dl in 1976 to about 7 µg/dl in 1990 (EPA, 1983). Figure 4 shows the close correlation between the decline in use of leaded gasoline and the decline in blood lead levels between 1976 and 1980. In countries that still use leaded gasoline, it is likely to be the main source of air emissions of lead.

In regions near smelters and mines, soil lead levels can reach 20,000 ppm or more. However, studies have shown that uptake of lead from mining wastes is lower than from smelter sites or urban locations. In Butte, Montana, a mining district since the 1800s, soil lead levels are often above 2000 ppm, but the children have an average blood lead level of only 3.5 µg/dl. Similar results have been found in other communities located near mining waste sites (Bornschein *et al.*, 1990). At smelter sites, in contrast, average children's blood lead levels are greater than 10 µg/dl (Bornschein *et al.*, 1992). These differences in bioavailability have been attributed to the fact that lead at mining sites typically consist of sulfides, which can be encapsulated by secondary minerals. In contrast, lead from smelters shows greater chemical diversity, and is more easily absorbed when ingested (Davis *et al.*, 1992).

Bernstein (1991) has written an instructive report on the trials of a mobile home community in Aspen, Colorado, which is located near an old silver mine. The ground contains high levels of lead, over 20,000 ppm in some places, from the mine tailings. This ground has been designated a Superfund site by the EPA, which proposes to dig up the soil and dump it in a nearby boulder field. Apart from the cost and inconvenience of this solution, there is a great deal of concern among the residents that digging up the soil will stir up lead-bearing dust, which may be far more dangerous than the soil is. The blood lead levels of the residents have been found to be substantially lower than the national average. The Aspen residents marshaled spirited opposition to the EPA plan, which has now been shelved, pending further study. This conflict is likely to multiply in many locales as elevated soil lead levels are found, with no satisfactory abatement procedure. It is important to define the magnitude of the hazard more carefully, through well-conceived studies of bioavailability.

Implications for the Industrial Ecology of Lead

Known human and ecosystem exposure to lead is primarily due to a small number of specific products: leaded gasoline, lead shot, and lead-based paint. Exposure from lead-acid batteries, the largest use of lead, is thought to be much smaller. Because of the well-documented adverse health and environmental effects of these leaded products, reduction in their use could have significant environmental and health benefits.

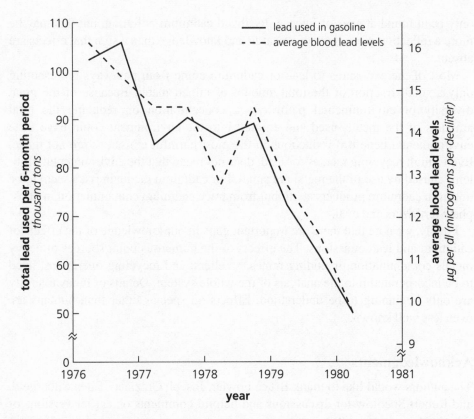

Figure 4. Parallel decreases in blood lead values and amounts of lead used in gasoline during 1976–80 (from U.S. Department of Health and Human Services, 1988).

Comparison and Conclusion

It is not straightforward to compare the extent to which human health has been affected by cadmium and by lead. Lead has been most thoroughly studied in the United States, where, due to historical use of lead-based paint and leaded gasoline, typical lead exposures may be greater than in most countries. In the United States, 17% of children were estimated to have blood lead levels greater than 15 μg/dl, the level at which the U.S. Centers for Disease Control recommend intervention and at which detrimental effects on children's intelligence and development have been found. The ecosystem effects of lead are also significant, due especially to the ingestion of lead shot and fishing weights by birds.

The health consequences of cadmium exposure in the general population have been more thoroughly studied in Belgium, where, due to the concentration of cadmium industries, typical exposures may be greater than in most countries. In Belgium, it was found that about 10% of the general population has slight kidney damage that could be attributed to cadmium, although, unlike lead, the health consequences of this damage are not clear. Ecosystem damage from cadmium has only been found at sites of intense, localized cadmium pollution, but this may be more a reflection of our lack of research and knowledge than a sign that effects are absent.

Most of the exposures to lead or cadmium come from pathways representing only a small fraction of the total amount of mined metal. Because of the great diversity of environmental pathways, a concentration on reducing the total amounts of the metals used and emitted into the environment could have little environmental benefit if reductions in the most harmful emissions are not made. Even a high recycling rate, as for lead, does not mean that the environmental problems caused by use of the metal are small. For cadmium the main risks seem to be not from cadmium products at all, but from trace cadmium contamination in phosphate fertilizers and coal.

Finally, we note that there are important gaps in our knowledge of the effects of cadmium and lead emissions. The effects of the numerous point sources of heavy metals contamination, including mines, smelters, and recycling operations, need to be incorporated into the analysis of the whole system. Details of bioavailability are only beginning to be understood. Effects on species other than humans are even less well known.

Acknowledgments

The authors would like to thank Bruce Fowler, Joseph Graziano, Ellen Silbergeld, and Robert Socolow for discussions and helpful comments on earlier versions of this chapter.

References

ATSDR (Agency for Toxic Substances and Disease Registry). March 1989. Toxicological Profile for Cadmium. ATSDR/TP-88/08, PB89-194476.

Barnes, R. M. 1990. Childhood soil ingestion: How much dirt do kids eat? *Analytical Chemistry 62(19)*, 1023–1033.

Bernstein, J. 1991. Report from Aspen. *The New Yorker*, 121–136.

Bornschein, R., *et al.* 1990. *Midvale Community Lead Study: Final Report*. University of Cincinnati, Cincinnati, Ohio.

Bornschein, R., *et al.* 1992. *The Butte-Silver Bow County Environmental Health Lead Study, Final Report*. Butte-Silver Bow Department of Health, and the Department of Environmental Health, University of Cincinnati, Cincinnati, Ohio.

Boutron, C. F., *et al.* 1991. Decrease in anthropogenic lead, cadmium, and zinc in Greenland snows since the late 1960s. *Nature 353*, 153–156.

Buchet, J. P., *et al.* 1983. Oral daily intake of cadmium, lead, manganese, copper, chromium, mercury, calcium, zinc and arsenic in Belgium: A duplicate meal study. *Food and Chemical Toxicology 21(1)*, 19–24.

Buchet, J. P., *et al.* 1990. Renal effects of cadmium body burden of the general population. *The Lancet 336*, 699–702.

Centers for Disease Control. 1991. Preventing Lead Poisoning in Young Children. U.S. Department of Health and Human Services.

CONCAWE (The Oil Companies' European Organization for Environmental and Health Protection). 1992. Motor Vehicle Emission Regulations and Fuel Specifications—1992 Update. Report No. 2/92, Brussels, Belgium.

Dabeka, R. W., and A. D. McKenzie. 1992. Total diet study of lead and cadmium in food composites: Preliminary investigations. *Journal of the Association of Official Analytical Chemists International 75(3)*, 386–394.

Davis, A., M. V. Ruby, and P. D. Bergstrom. 1992. Bioavailability of arsenic and lead in soils from Butte, Montana, Mining District. *Environmental Science and Technology 26(3)*, 461–468.

Delos, C. 1985. *Cadmium Contamination of the Environment*. U.S. Environmental Protection Agency, PB85-221679, EPA-440/4-85-023.

Doelman, P. 1978. Lead and terrestrial microbiota. In *The Biogeochemistry of Lead in the Environment* (J.O. Nriagu, ed.), Elsevier/North Holland Biomedical Press, Amsterdam, 345.

Ducoffre, G., *et al.* 1992. Decrease in blood cadmium levels over time in Belgium. *Archives of Environmental Health 47(5)*, 354–456.

EDF (Environmental Defense Fund). 1992. Letter from Bruce Manheim *et al.* to EPA Administrator William K. Reilly.

Elinder, C. G. 1986. Cadmium: Uses, occurrence, and intake. In *Cadmium and Health: A Toxicological and Epidemiological Appraisal*, Volume I, CRC Press, Boca Raton.

Elinder, C. G., *et al.* 1983. Cadmium exposure from smoking cigarettes: Variations with time and country where purchased. *Environmental Research 32*, 220–227.

EPA (U.S. Environmental Protection Agency). 1983. Air Quality Criteria for Lead. Volume III, EPA-600/8-83-028A, U.S. EPA, 11–26.

EPA. 1989a. *Exposure Factors Handbook*, EPA 600/8-89/043.

EPA. 1989b. Review of the National Ambient Air Quality Standards for Lead: Exposure Analysis Methodology and Validation. EPA-450/2-89-011, Office of Air Quality Planning and Standards Staff Report, U.S. Environmental Protection Agency, A–2, A–6, A–18.

EPA. 1990. *National Air Quality and Emissions Trends Report, 1988*. EPA 450/4-90-002, U.S. EPA, Washington, D.C., 12–13, 20–22.

EPA. 1991. National Air Pollutant Emission Estimates 1940–1990. EPA-450/4-91-026, U.S. EPA.

Ewers, U., and H.-W. Schlipkoter. 1991. Lead. In *Metals and Their Compounds in the Environment* (E. Merian, ed.), VCH, New York.

Federal Register. 1985. Assessment of Cadmium Under Section 122 of the Clean Air Act. 50FR200, October 16, 1985, 42,000–42,004.

Federal Register. 1993. Standards for the Use or Disposal of Sewage Sludge; Final Rules. 58FR32, February 19, 1993, 9392.

Fergusson, J. E. 1990. *The Heavy Elements: Chemistry, Environmental Impact and Health Effects*. Pergamon Press, New York, 549.

Franklin Associates. 1989. Characterization of Products Containing Lead and Cadmium in Municipal Solid Waste in the United States, 1970 to 2000. Prairie Village, Kansas.

Friedland, A. J., and A. H. Johnson. 1985. Lead distribution and fluxes in a high-elevation forest in Northern Vermont. *Journal of Environmental Quality 14(3)*, 332–336.

Friedland, A. J., *et al.* 1986. Coniferous litter decomposition on Camels Hump, Vermont: A review.

Canadian Journal of Botany 64, 1349–1354.

Fulkerson, W., and H. E. Goeller (eds.). 1973. *Cadmium: The Dissipated Element*. Oak Ridge National Laboratory, ORNL-NSF-EP-21, Oak Ridge, Tennessee.

Gartrell, M., *et al*. 1986. Pesticides, selected elements, and other chemicals in adult total diet samples, October 1980–March 1982. *Journal of the Association of Official Analytical Chemists International 69(1)*, 146–161.

Giller, K. E., and S. P. McGrath. 1988. Pollution by toxic metals on agricultural soils. *Nature 335*, 676.

Glooschenko, V., C. Downes, R. Frank, H. E. Braun, E. M. Addison, and J. Hickie. 1988. Cadmium levels in Ontario moose and deer in relation to soil sensitivity to acid precipitation. *The Science of the Total Environment 71*, 173-186.

Gunderson, E. L. 1988. FDA total diet study, April 1982–April 1984, dietary intakes of pesticides, selected elements, and other chemicals. *Journal of the Association of Official Analytical Chemists International 71(6)*, 1200–1209.

International Lead and Zinc Study Group. 1992. Principal Uses of Lead and Zinc, 1960–1990. International Lead and Zinc Study Group, London. Based on data from France, Germany, Italy, Japan, U.K., and the United States.

Jolly, J. 1988. *Zinc 1988, Minerals Yearbook*. U.S. Bureau of Mines.

Jolly, J. 1992. Zinc. In *Annual Report 1990*. U.S. Bureau of Mines.

Jones, K. C., *et al*. 1992. Evidence of an increase in the cadmium content of herbage since the 1860s. *Environmental Science and Technology 26(4)*, 834–836.

Kaiser, T. E., *et al*. 1980. Organochlorine pesticides, PCBs, and PBB residues and necroscopy data for bald eagles from 29 states—1975–1977. *Pesticide Monitoring Journal 13(4)*, 145–149.

Kjällman, A. 1989. The effects and experiences of the Swedish ban on cadmium—Actions concerning the collection of spent nickel–cadmium batteries. In *Cadmium 89*, The Cadmium Council, London, 36–37.

Kjellstrom, T. 1986. Critical organs, critical concentrations, and whole body dose–response relationships. In *Cadmium and Health: A Toxicological and Epidemiological Appraisal*, Vol. II, CRC Press, Boca Raton, Florida.

Lorenz, H., *et al*. 1986. Content of cadmium in cereals of the past compared with the present. *Zeitschrift für Lebensmittel-Untersuchung und-Forschung 183*, 402–405.

Louekari, K., *et al*. 1991. Estimated dietary intake of lead and cadmium and their concentration in blood. *The Science of the Total Environment 105*, 87–99.

McKenzie-Parnell, J. M., *et al*. 1988. Unusually high intake and fecal output of cadmium, and fecal output of other trace elements in New Zealand adults consuming dredge oysters. *Environmental Research 46*, 1–14.

McLeese, Sprague, and Ray. 1987. Effects of cadmium on marine biota. In *Cadmium in the Aquatic Environment* (J.O. Nriagu and J.B. Sprague, eds.) John Wiley and Sons, New York.

Morgan, H. (ed.). 1988. The Shipham report. *The Science of the Total Environment 75(1)*, 1, 44.

Mortvedt, J. J. 1987. Cadmium levels in soils and plants from some long-term soil fertility experiments in the United States of America. *Journal of Environmental Quality 16(2)*, 137–142.

National Wildlife Health Laboratory. 1985. Bald Eagle Mortality from Lead Poisoning and Other Causes 1963–1984. Madison, Wisconsin.

Nordberg, G. F., *et al*. 1986. Kinetics and metabolism. In *Cadmium and Health: A Toxicological and Epidemiological Appraisal*, Volume I, CRC Press, Boca Raton.

Nriagu, J. O., and J. M. Pacyna. 1988. Quantitative assessment of worldwide contamination of air, water, and soils by trace metals. *Nature 333*, 134–139.

Octel. 1992. *Worldwide Survey of Motor Gasoline Quality*.

Oskarsson, A., *et al.* 1992. Lead poisoning in cattle—Transfer of lead to milk. *The Science of the Total Environment 111*, 83–94.

Page, A. L., *et al.* 1987. Cadmium levels in soils and crops in the United States. In *Lead, Mercury, Cadmium and Arsenic in the Environment* (T.C. Hutchinson and K.M. Meema, eds.), SCOPE 31, John Wiley and Sons, New York, 123.

Pain, D. J. (ed.). 1992. Lead poisoning in waterfowl. In *Proceedings, International Waterfowl and Wetlands Research Bureau Workshop, Brussels, Belgium, 1991*. International Waterfowl and Wetlands Research Bureau Special Publication 16, Slimbridge, U.K.

Roskill. 1990. *The Economics of Cadmium*. Seventh edition, Roskill Information Services, London, 209.

Russian Standard. 1987. Motor Gasoline. Russian Standard GOST 2084-77, Technical Conditions.

Ryan, W., and R. Schrader. 1991. *An Ounce of Prevention: Rating States' Toxic Use Reduction Laws*. Center for Policy Alternatives, Washington, D.C.

Sanderson, C. G., and F. C. Bellrose. 1986. A review of the problem of lead poisoning in waterfowl. *Illinois Natural History Survey*, Special Publication 4.

Simpson, W. R. 1981. A critical review of cadmium in the marine environment. *Progress in Oceanography 10*, 1–70.

Sprague, J. B. 1987. Effects of cadmium on freshwater fish. In *Cadmium in the Aquatic Environment* (J.O. Nriagu and J.B. Sprague, eds.), John Wiley and Sons, New York, 139–169.

Stansley, W., D. E. Roscoe, and R. E. Hazen. 1991. Cadmium contamination of deer livers in New Jersey: Human health risk assessment. *The Science of the Total Environment 107*, 71–82.

Stayner, L., *et al.* 1992. A dose–response analysis and quantitative assessment of lung cancer risk and occupational cadmium exposure. *AEP 2(3)*, 177–194.

Stoeppler, M. 1991. Cadmium. In *Metals and Their Compounds in the Environment* (E. Merian, ed.), VCH, New York.

Stowasser, W. F. 1988. Phosphate rock. In *Minerals Yearbook—1988*. U.S. Bureau of Mines.

Tsuchiya, K. (ed.). 1978. *Cadmium Studies in Japan: A Review*. Elsevier/North Holland Biomedical Press, New York, 25.

Tyler, G. 1972. Heavy metals pollute nature, may reduce productivity. *Ambio 1(2)*, 52–59.

U.S. Department of Energy. 1991. *International Energy Annual 1991*. Energy Information Administration, DOE/EIA-012(91), U.S. Government Printing Office, Washington, D.C.

U.S. Department of Health and Human Services. 1988. The Nature and Extent of Lead Poisoning in Children in the United States: A Report to Congress. U.S. Department of Health and Human Services report PB-100184.

U.S. Fish and Wildlife Service. 1985. Draft Supplemental Environmental Impact Statement on the Use of Lead Shot for Hunting Migratory Birds in the United States. U.S. Department of the Interior, DES 85-53 I 1.98 B76/draft/supp., III–35, III–65.

USBM (U.S. Bureau of Mines). 1991. Cadmium in 1990. *Mineral Industry Surveys*.

USBM. 1992. *Cadmium in the 4th Quarter of 1992*. Mineral Industry Surveys, U.S. Bureau of Mines.

Vahter, M., *et al.* 1991. Methods for integrated exposure monitoring of lead and cadmium. *Environmental Research 56*, 78–89.

317

Valkovic, V. 1983. *Trace Elements in Coal*. Volume I, CRC Press, Boca Raton, Florida.

van Straalen, N. M., *et al*. 1989. Population consequences of cadmium toxicity in soil microarthropods. *Ecotoxicology and Environmental Safety 17*, 190–204.

Watanabe, T., *et al*. 1987. Cadmium and lead contents of cigarettes produced in various areas of the world. *The Science of the Total Environment 66*, 29–37.

Watanabe, T., H. Nakasuka, and H. Satoh. 1992. Reduced dietary cadmium intake in past 12 years in a rural area in Japan. *The Science of the Total Environment 119*, 43–40.

Watson, A. P., *et al*. 1976. *Impact of a Lead Mining–Smelting Complex on the Forest Floor Litter Arthropod Fauna in the New Lead Belt Region of Southeast Missouri*. Oak Ridge National Laboratory, Environmental Sciences Division Publication No. 881, ORNL/NSF/EATC-30, Oak Ridge, Tennessee.

Wong, P. S., *et al*. 1978. Lead and the aquatic biota. In *The Biogeochemistry of Lead in the Environment* (J.O. Nriagu, ed.), Elsevier/North Holland Biomedical Press, Amsterdam, 345.

Woodbury, W. D. 1992. Lead. In *Annual Report 1990*, U.S. Bureau of Mines.

WRI (World Resources Institute). 1990. *World Resources 1990–91*. Oxford University Press, New York.

22

Nuclear Power: An Industrial Ecology That Failed?

Frans Berkhout

Abstract

Nuclear power systems have been specifically designed to control the emissions of nuclear material, and so would seem to have conformed to the industrial ecology paradigm. Nevertheless, the nuclear industry has stagnated in many parts of the world, public opposition is widespread, and prospects for a revival are not great. One lesson from this example is that control of materials is not enough, for a technology that appears to be unacceptable to society. A further lesson is that the social, political and psychological factors must be taken into account in determining the acceptability of a technology. And, finally, the nuclear industry is an example of a technology in which recycling is deemed to pose a greater risk than disposal. The overall lesson is that the grand ideas of industrial ecology must be applied with caution.

Industrial ecology, while not yet a clearly defined field of research and action, can already be said to build on certain basic tenets: that industry-related environmental risks can be scientifically assessed; that such assessments can inform regulation and technical innovation towards cleaner production; and that cleaner production is achieved through greater control over waste streams and enhanced recycling of materials. To a great extent, all of these features are already represented in the nuclear industry. This short chapter discusses how far the experience of the nuclear industry can shed some light on the ambitious project of industrial ecology.

The Nuclear Fuel Cycle

Nuclear fuel cycles (NFC) are large-scale industrial systems with two principal products: electricity and warheads for nuclear weapons. Only a small proportion of the aggregate radioactivity of the NFC—uranium, plutonium, and associated radioactive waste products—is discharged directly into the environment.[1]

---

[1] Large volumes of mill tailings with low specific activities are generated in the mining of uranium. Making general estimates for health effects to the general public and workers associated with these wastes is difficult since much depends on specific conditions and management techniques employed at the mine. One estimate estimates the normalized exposure to radiation associated with mining wastes as constituting about one-tenth of the total exposure related to effluents from the nuclear fuel cycle (UNSCEAR, 1988).

Expressed in terms of the total activity (or toxicity) within the system these typically add up to small fractions of 1%.[2] A huge effort is expended in achieving control of materials. Indeed, nuclear technology is to a large extent a "control" technology, organized to contain nuclear materials and minimize dissipation. In this sense it serves as a metaphor for good practice in the industrial handling of toxic materials. This is not to say that the housekeeping at nuclear facilities has always been beyond criticism, or to understate the catastrophic consequences of a major accident. But the basic principle embodied in nuclear technology is that all materials should, more or less, be absolutely controlled (see Berkhout, 1991).

There are two main reasons for these intensive controls. First, ionizing radiation is toxic to humans. While, in principle, a thorough mixing into the ambient environment (for instance, the sea) of waste radioactivity produced by nuclear fuel cycles would not significantly increase background levels of radiation, such mixing is impractical. A management strategy known as "containment and concentration" has been adopted instead for radioactive wastes. This approach aims at complete engineered control of radioactive waste streams, thereby preventing doses to humans that are above a low maximum.

Second, nuclear materials control is important because plutonium and highly-enriched uranium (HEU) are weapons materials. Only a few states have acquired nuclear weapons. Many have forsworn the acquisition of nuclear weapons and submitted their nuclear activities to international inspection to verify this commitment. Stability within the world political order depends on international confidence that non-nuclear weapon states are not diverting civil nuclear materials to weapons use; hence the need for verifiable materials controls.

Despite all this effort to control materials and to internalize environmental costs, pursued more effectively than has any other industry, the nuclear industry is in a poor state in most countries. Its growth has stagnated in many parts of the world, public opposition is widespread, and the prospects for a revival seem poor, even at a time when paths toward reductions of greenhouse gas emissions are being sought.

For these reasons, we take the nuclear fuel cycle to be an example of an ecological approach to an industry (an industrial ecology, however partial and flawed) which failed. A systematic attempt to reduce environmental impacts did not apparently reduce the controversy evoked by those impacts, and did not produce a sustainable industry in many parts of the world.

---

[2] Even small fractions of 1% represent a large amount of activity. The fuel cores of the world's commercial reactors discharge a total of some 10^{12} Curies (Ci) of radioactivity each year. Much of this decays rapidly, although even after 10 years of storage this fuel will still contain some 10^9 Ci. One curie is a quantity of radioactive material in which 3.7×10^{10} decays occur per second (Nero, 1979). The relationship between curies of activity released to the environment and final health detriment is complex since it depends on the exposure pathways, the dosimetry, and the modeling of risks of health effects. Radiation detriment will be proportional to the dose of radiation absorbed by body tissue. An important reference is NRC (1990).

Vulnerabilities Generated by Nuclear Fuel Cycles

Large-scale technological systems like nuclear power have deep-seated effects, both physical and cultural. In describing the vulnerabilities created by such technologies, we must use a suitably wide definition, including not just direct economic or health effects, but also wider political, institutional, and cultural stresses generated in societies. In doing so, we are also seeking to move away from a narrowly risk-based definition of vulnerability, and to embrace one which can represent more fully the many aspects of the relationship between technology and society.

Human Health Effects

Humans, and especially children, are sensitive to ionizing radiation. Small increases in radiation doses delivered over long periods of time are believed to cause increases in cancer rates, and empirical evidence for this is growing. Radiation also affects germ cells so that genetic damage may be passed on from one generation to the next.

Only doses to humans are controlled today, on the assumption that a risk acceptable to human populations will not cause unacceptable risks for other biota. National regulations are almost all based on the recommendations made by a single international body of scientists—the International Commission on Radiological Protection (ICRP). The ICRP makes recommendations for the maximum acceptable doses to workers who handle radioactive materials and to the general public. Dose limits for workers are generally put at ten times the limit for the general public, with the justification that the public receives its dose involuntarily.

Few people receive the maximum permitted dose from civil nuclear activities.[3] Populations at greatest risk are those living near fuel reprocessing plants and nuclear weapon complex facilities. Even in these "critical groups," doses above the permitted level are today rare in most countries (Russia may have been an exception). Maximum doses from solid waste repositories are expected to be limited even further.[4] The accident at Chernobyl, which released a substantial part of the inventory of activity in the reactor, caused higher exposures and increased rates of cancer over a wide area.[5] On average, the greatest human exposure to humanmade radiation is from medical uses of radiation (about 50% of the exposure from natural sources).

---

[3] "Acceptable" maximum annual doses of humanmade radiations to the general public have been defined as those associated with a risk of death in that year of 10^{-6}. The dose was set at 1 milliSievert (mSv) per year in 1990 by the ICRP. Average per capita doses from natural background sources of radiation is estimated at 2.4 mSv/yr. A Sievert is a measure of the "effective dose equivalent" which takes into account both the energy deposited in tissue (Gy) and the relative biological effectiveness (RBE) of the ionizing radiation. One Sievert is 100 rem.

[4] To between 0.1 and 0.3 mSv/yr.

[5] The per capita dose equivalent commitment in southeastern Europe associated with the Chernobyl accident is estimated at 0.38 mSv in the first year (UNSCEAR, 1988).

Nuclear Weapons Proliferation

The second great vulnerability connected with nuclear fuel cycles is the risk of the illegitimate use for weapons of nuclear materials.[6] It is generally assumed that international security would be harmed by further proliferations, although this view is not universal.[7] States' national security would also be threatened by the theft of nuclear materials by sub-national groups or individuals. Controls on access to sensitive nuclear materials, enforced internally by states, are therefore another prerequisite of a nuclear fuel cycle.

Centralization and Secrecy

Many commentators have argued that because nuclear fuel cycle technologies are massive and usually highly concentrated, state and private sector institutions connected with them have been distorted, to the detriment of other social goals. Specifically they argue that alternative energy technologies have not received the same political and commercial support as nuclear power, and that the huge investments made (human and capital) in nuclear energy could not easily be transferred to other activities.

An adjunct of this political and economic concentration, as well as of the sensitive nature of the materials and technologies being exploited, was a pervasive secrecy surrounding nuclear activities. As public concern over nuclear safety mounted in many countries through the 1970s, the industry was forced to become more open in its affairs and to submit itself to intense public scrutiny. Significantly, while the nuclear industry in most countries is as open as any other commercial sector, the charge of secrecy still sticks.

Materials Flows in the Nuclear Fuel Cycle

Some simple estimates of materials flows in civil nuclear fuel cycles around the world can be made. In the early 1990s world annual demand for natural uranium stood at about 55,000 metric tonnes. This is fabricated into both natural uranium and enriched uranium fuel—some 11,500 tonnes in total. Once irradiated, this fuel contains about 74 tonnes of plutonium and about 80 tonnes of fissile uranium (uranium-235) embedded in about 11,000 tonnes of uranium-238. Of this, no more than about one-fifth (10 tonnes of plutonium and 11 tonnes of U-235) is likely to be separated from fission products in reprocessing plants. Only a proportion of this—perhaps half—is likely to be recycled as fuel in commercial reactors (see

---

[6] The Nuclear Non-Proliferation Treaty (1968) defines the five states: the United States, Russia, the United Kingdom, France, and China, which may legitimately have nuclear weapons.

[7] The dominant view is represented in the U.N. Security Council statement of 31 January 1992 that: "The proliferation of all weapons of mass destruction constitutes a threat to international peace and security" (U.N. Security Council, 1992). The classic statement of the dissenting view is Waltz (1981).

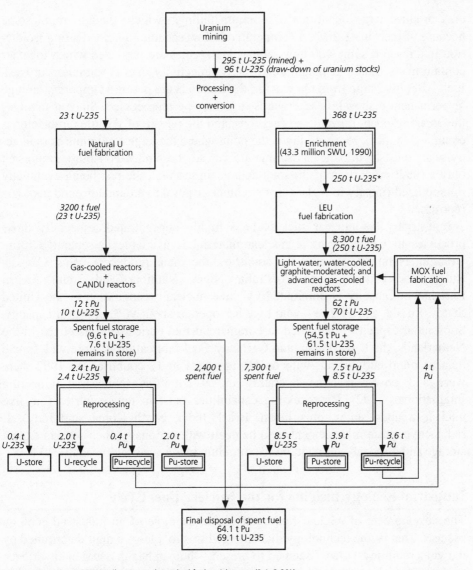

Figure 1. Schematic picture of 1991 flows of fissile material (U-235 and Pu) through world civilian fuel cycles. Reproduced with permission, from the *Annual Review of Energy and the Environment* volume 18, © 1993, by Annual Reviews Inc.

Figure 1 for a schematic picture of fissile materials flows through the world's civil fuel cycles) (Berkhout and Feiveson, 1993).

The nuclear fuel cycle consists of six steps: uranium mining, enrichment (omitted in some cases), fuel fabrication, irradiation in a reactor, fuel storage or reprocessing, and disposal of radioactive wastes. The mining and enrichment of uranium ores produce large quantities of uranium tailings, which, though troublesome because of their mass, have a comparatively low specific toxicity. During irradiation in a reactor, some 400 new radioactive species are produced which together contain many orders of magnitude more radioactivity than the uranium in fresh fuel. After discharge from the reactor, the fuel rods are put into temporary storage or permanent disposal, or afterwards they may be reprocessed. Since it involves the breaking-open of irradiated fuel rods, and the release of volatile radioisotopes contained in them, reprocessing is the point where the largest amounts of radioactivity are released into the environment. Even so, these released isotopes represent only a small proportion of the total activity in the fuel, the rest being eventually immobilized (mostly in a glass) for eventual disposal to an underground geologic repository.

Industrially the nuclear fuel cycle is highly concentrated, especially those processes in which sensitive nuclear material is available in separated form. Uranium is mined in about 15 countries, the main producers being Canada, Russia, the United States, Australia, Niger, Namibia, and South Africa. Enrichment capacity is dominated by three nuclear weapon states—the United States, Russia, and France—who together operate about 90% of world capacity. Substantial capacities dedicated to commercial fuel enrichment also exist in the Netherlands, the U.K., Japan, and Germany. Fuel fabrication is more widespread than uranium enrichment, with operating plants in 17 countries. In 1992 there were 433 power reactors operating in 29 countries (Nuclear Engineering International, 1992). Reprocessing capabilities exist in ten countries (the five nuclear weapon states, plus Japan, Israel, India, North Korea, and Italy). If radioactive waste management is to be dealt with nationally, then we can expect storage and disposal sites containing plutonium in upwards of 40 countries.[8]

Industrial Ecology Insights for the Nuclear Fuel Cycle

The development of nuclear fuel cycles is an example of an industrial ecology process. That is, the technology that emerged was to a large extent determined by the vulnerabilities which it seemed to present—human health and genetic effects, the risk of nuclear war, and later the institutional threats posed by centralization and state secrecy. The industry was contained and concentrated, and the best minds were put to the task of assessing safety and environmental consequences. Reactors became physically massive as multiple safety devices were clamped onto them, while very little radioactivity was permitted to escape routinely from the fuel cycle. Stringent standards were imposed, first from within the industry and later by national and international regulators. Material recycling was a basic goal from the beginning, and a global industrial infrastructure for the purpose was gradually put in place. A professional corps of experts and analysts developed

---

[8] Research reactors are either operating or under construction in 55 countries (IAEA, 1989).

around the industry, some as regulators, others as watchdogs and critics. The industry became freer with systematic information about its environmental performance. In short, it made many of the changes which are now thought of as symptomatic of "eco-industrial" restructuring.

There is of course a flaw in this argument. For while the nuclear industry clearly was reformed, the basic technology and structure of material flows was not greatly altered. The industry everywhere decided to try to adapt existing technologies and concepts, rather than to pursue a deep reappraisal of process and practice. It is stuck with the problem of controlling radioactivity ever more tightly.

But can materials control ever go far enough? Those critics of nuclear power who propose phase-outs argue that no amount of reform will make the industry acceptable in an environmentally conscious, democratic society. Risks of catastrophic accidents can never be eliminated and are unacceptable. Nor can secrecy and centralization of control ever be fully diffused. If these arguments are accepted, then the only remaining question is how existing nuclear capacity can be replaced in a "sustainable" way.

Opponents to such argument assert that nuclear materials cycles, properly managed, do not threaten global sustainability, in a way that climate change might. Furthermore, they argue that the relative risks associated with a simple once-through Light Water Reactor (LWR) fuel cycle are much lower than many similar industrial risks. It would, for instance, be regarded as completely unacceptable for 20% of preschool children in the United States to have their health impaired by humanmade radiation in the environment, yet this is a conservative estimate of how many children are affected by lead poisoning.[9]

Nuclear Insights for Industrial Ecology

It now seems clear that simply focusing on the control of materials cycles is not enough to build a sustainable industry. An industrial ecology perspective cannot ignore several other factors that determine the evolution of industrial practice, and that are at the root, in many countries, of the failure to convince enough people that nuclear power is benign and necessary. An industrial ecology perspective must address the unique connection of the nuclear fuel cycle with nuclear weapons production—evoking the nuclear holocaust—that makes this materials cycle quite different from all others.

However, the basic message—that the list of vulnerabilities (or threats to social control, or threats to sustainability) assessed must include social, political, and psychological factors—is surely applicable to all cases. Moreover, achieving consensus on what are acceptable risks, or what is a sustainable form of industrial activity, will never be easy. No final answer can ever be reached on these questions because knowledge develops and tastes change. Implicit in the idea of sustainability is the

---

[9] This is the proportion of children estimated to have blood lead concentrations of greater than 15 µg/dl (U.S. Department of Health and Human Services, 1988).

notion that a state of long-run equilibrium can be achieved in human relationships with the environment. The example of the nuclear industry appears to show that such an equilibrium will never be found, and that thinking of environmental effects as only physical and chemical phenomena will never do justice to the essence of our relation to nature. Systems for ranking environmental risks are already attempting to take account of this lesson (see Graedel, Horkeby, and Norberg-Bohm, this volume).

Lastly, the nuclear case raises questions about one of the basic assumptions of "eco-restructuring": the value of recycling. Recycling has been part of the vision of nuclear power since the outset, although it has traditionally focused only on fissile materials (that is, U-235 which forms about 0.7% of natural uranium, and Pu-239 which forms about 1% of typical irradiated fuel). But there is now a consensus that the fuel cycle that minimizes vulnerability (that is, minimizes health detriment and the risks of proliferation) is a fuel cycle without recycling, since less radioactivity is dispersed into the environment, occupational doses are reduced, and separated fissile materials do not arise in the fuel cycle.

Conclusion

The nuclear industry proves to be an ambiguous test case for an industrial ecology perspective. This is partly because of its many special characteristics: the fear of radiation; the link with nuclear weapons; and the uniqueness of its fuel cycle. Nevertheless, many components of an industrial ecology analysis and materials cycle management have been applied to the nuclear fuel cycle, with mixed results. The application of such a perspective in other industrial systems or materials cycles should be accompanied by a suitable degree of caution. The alternatives to what already exists often have unexpected and undesirable consequences.

Acknowledgments

This chapter is based on discussions held in the Working Group on Exotic Intrusions at the Global Change Institute on Industrial Ecology and Global Change, July 1992. I am particularly grateful to Robin Cantor and Steve Fetter.

References

Berkhout, F. 1991. *Radioactive Waste: Politics and Technology*. Routledge, London.

Berkhout, F., and H. Feiveson. 1993. Securing nuclear materials in a changing world. *Annual Review of Energy and Environment 18*, 631–635.

IAEA (International Atomic Energy Agency). 1989. *Directory of Nuclear Research Reactors 1989*. IAEA, Vienna, Austria.

Nero, A. V. 1979. *A Guidebook to Nuclear Reactors*. University of California Press, 35–37.

NRC (U.S. National Research Council). 1990. *Health Effects of Exposure to Low Levels of Ionizing Radiation: BEIR V*. National Academy Press, Washington, D.C.

Nuclear Engineering International. 1992. *World Nuclear Industry Handbook 1993*. Reed Business Publishing Group, London, 10.

U.N. Security Council. 1992. United Nations Security Council Press Release. SC/536, January 31, 1992.

U.S. Department of Health and Human Services. 1988. *The Nature and Extent of Lead Poisoning in the United States: A Report to Congress*. Agency for Toxic Substances Disease Registry, U.S. Department of Health and Human Services, PB 89-100184.

UNSCEAR (U.N. Scientific Committee on the Effects of Atomic Radiation). 1988. *Sources, Effects and Risks of Ionizing Radiation: 1988 Report*. United Nations, New York.

Waltz, K. 1981. *The Spread of Nuclear Weapons: More May Be Better*. Adelphi Paper No. 171, International Institute of Strategic Studies, London.

PART 4

INDUSTRIAL ECOLOGY IN FIRMS

23

Introduction

The Editors

Product and process design choices used to be a private concern, of primary interest within rather than outside the firm. Design decisions reflected the firm's priorities, such as low production cost, high quality, and easy manufacturability. In recent years environmental problems have resulted in public policy interventions, changed consumer preferences, and new constraints on industry behavior. Yet in many cases, environmental information is inadequate. Thus engineers rarely understand the overall environmental impacts of their design choices; government policy-makers have inadequate information for making environmental policy; and few consumers know what they buy.

Industry is recognizing the need for new approaches, in order to: (1) include environmental criteria in product and process design, (2) improve cooperation between firms and government regulators in developing more precisely targeted, cost-effective policies, and (3) anticipate future environmental constraints, liabilities, and opportunities.

New approaches will build upon the steadily improving understanding of pollutant flows and human exposures. The objective is developing fair, efficient, and stable strategies that can win public acceptance.

Part 4 contains six chapters that give insights into the state of play in firms today, as they search for new ways to take advantage of the insights of industrial ecology. To set a context for these chapters, we review below some of the evolving methods in support of environmentally sensitive decision-making.

Materials Balance Accounting

Materials used by industrial societies undergo numerous transformations in the time between their extraction from the earth as raw materials and their deposition back to the environment as wastes. After a material is extracted, it must be refined or processed into a chemical form that renders it suitable as a feedstock for the manufacture of products. The feedstock may be used up during the manufacturing process (e.g., industrial solvents and catalysts), or it may be embodied as an integral component of the final product. Some products, for example, gasoline and pesticides, are released to the environment in their totality during normal use.

Other products, for example, batteries and plastic cups, are still in consumers' hands, usually as solid wastes, when consumers have no further use for them. An important class of products falls between these two limits: the material is only partially released during normal (dissipative) use. Examples in this category include automobile tires and brake linings, and most surface coatings such as paints and roofing materials. "Product Life-Cycle Management to Replace Waste Management," by Braungart, proposes institutional innovations—waste "supermarkets" and "parking lots"—to reorganize economic activity in a way that reduces the final environmental impacts of these materials flows.

Process Design Tools

Firms need to be able to compare the environmental hazards of different design choices. Designers are constantly faced with choosing among materials and processes, as well as among the end-use characteristics of their products, such as the degree of energy efficiency and recyclability. The integration of hazard assessment into product life-cycle design would enable designers to compare design alternatives.

The accent on design is important since 80–90% of total life-cycle costs are committed by the final design stage (Fabrycky and Blanchard, 1991). Improved design tools would also help firms identify where innovation would provide the greatest environmental benefits. The conceptual development of life-cycle design tools has begun (see Allenby, 1991; SETAC, 1991; EPA, 1993). This research suggests a need for both generic and industry-specific tools. Implementation will require extensive collaboration between design tool developers and industrial engineers. Tuning the design tools to a specific company and product line will generally be accomplished at the company level. "Industrial Ecology in the Manufacturing of Consumer Products," by France and Thomas, illustrates how life-cycle ideas are being applied to redesign familiar products such as automobiles and cameras.

Management Strategies

Industrial organizations throughout the world have sought in recent years to implement Total Quality Management (TQM), a management theory that emphasizes building bridges among actors within organizations. It redefines the firm's activity as the production of end-use services rather than simple physical products, and thus focuses upon customer satisfaction as a key objective. It also seeks to improve both products and their manufacturing processes, which requires that design and process engineers communicate. "Design for Environment: A Management Perspective," by Paton, discusses managerial issues in the implementation of an environmental version of TQM.

Hazard Assessments

Firms need societal guidance on how to prioritize environmental factors. Two levels of hazard assessment play crucial roles in the implementation of industrial ecology: assessments of individual hazards and comparative assessments across hazards. Assessment of individual environmental dangers focuses on pathways of chemicals and energy through the environment, culminating in exposure pathways for humans and ecosystems.

Comparative hazard assessment combines individual hazard assessments with perceptions of risk and relative value. This requires that impacts occurring over different spatial and temporal scales, and about which we have different degrees of scientific uncertainty, be somehow ranked against each other.

Recently, several frameworks for comparative environmental hazard assessment have been developed. These are reviewed in "Prioritizing Impacts in Industrial Ecology," by Graedel, Horkeby, and Norberg-Bohm. Although further development and experimentation is needed, several characteristics emerge that are critical to the success of these tools in both public and private decision-making. They include transparency, explicit treatment of uncertainty, and ease of use.

Organizational Analysis

Changing the structure of industrial production means changing big organizations. Large enterprises must internalize external mandates such as environmental protection. How this can be accomplished will be very important in developing strategies for implementing industrial ecology. "Finding and Implementing Projects that Reduce Waste," by Nelson, emphasizes the importance of organizational factors, including employee motivation, in bringing about change within firms.

Standard economic theory suggests that a firm must trade off improvements in environmental performance and economic performance. However, a closer look shows that in most large enterprises, the functions of ownership, management, production, and sales are performed by different people, with different values, objectives, and expectations. This will frequently lead to "principal–agent" problems. Specifically, the preferences and incentive structures for managers may not be aligned with those of shareholders, resulting in inefficiencies and decisions that are not strictly profit-maximizing. "Free-Lunch Economics for Industrial Ecologists," by Panayotou and Zinnes, explores the economics of this phenomenon.

References

Allenby, B. 1991. Design for environment: A tool whose time has come. *Semiconductor Safety Association Journal*, 5–9.

EPA (U.S. Environmental Protection Agency). 1993. *Life-Cycle Assessment: Inventory Guidelines and Principles*. EPA/600/R-92/245, prepared by Battelle Columbus Labs, NTIS, Springfield, Virginia.

Fabrycky, W. J., and B. S. Blanchard. 1991. *Life-Cycle Cost and Economic Analysis*. Prentice-Hall, Englewood Cliffs, New Jersey.

SETAC (Society of Environmental Toxicology and Chemistry). 1991. *A Technical Framework for Life-Cycle Assessments*. Workshop report, SETAC Foundation, Pensacola, Florida.

24

Product Life-Cycle Management to Replace Waste Management

Michael Braungart

Abstract

As "eco-restructuring" commences, new elements in our waste management infrastructure should be created. These include a redefinition of product types to acknowledge their life-cycle environmental impacts; reallocation of responsibilities between producers and consumers for these products; product redesign for environmental compatibility; source-separation sites or "waste supermarkets," and accessible repositories or "waste parking lots."

Where We Need to Go

Industry has traditionally focused on production rather than waste management. Over time this has led to the creation of chemicals and products for which no environmentally sound method of disposal exists. Large-scale production has led in turn to significant waste disposal problems. In order to shift from a primitive, low-efficiency type I industrial ecology (see Graedel, this volume) to something more sustainable, a new infrastructure for waste management is required.

Marketplace norms today define products on a spectrum from "consumable" to "durable," depending on their useful lifetimes. However, to give prominence to environmental factors, it is useful to classify products according to their life-cycle, cradle-to-grave impacts. The crucial distinction is between *consumable products* and *service products*.

Consumable products, such as washing powder or food, are purchased to be consumed, i.e., converted by chemical reaction into energy and byproducts. They are normally put out into the natural environment after only one use. Service products, by contrast, are not consumed; rather they provide some service over and over again. Automobiles, television sets, and washing machines are examples of products providing the services of transportation, entertainment, and cleaning. Thinking prescriptively, an "eco-restructured" economy should differentiate its treatment of consumable products and service products. All consumer products should be biodegradable or abiotically degradable, non-bioaccumulative, non-carcinogenic, non-teratogenic, non-mutagenic, and (in used concentration) non-toxic to human beings. By contrast, service products could be less tightly constrained in terms of constituent materials, but more tightly constrained in terms of disposal. A consumer requiring a service product would lease it from its producer or put a refundable deposit on its purchase: consumers would not own service products. When a service product has served its function and needs to be renewed, the consumer would return it to the producer. The producer, not the consumer, would be responsible for disassembly and recycling.

"Waste supermarkets" would provide centralized locations for disassembly and recycling. These could be expanded versions of today's "buy-only" shopping markets: when shopping for new goods, the consumer would also do a "de-shopping" by returning used service products—including packaging as well as the automobiles, television sets, and washing machines just mentioned. A waste supermarket would not be a dump site but rather a source separation warehouse.

By giving products back to producers in a closed loop, aided by a deposit-return system of financing, incentive could be provided to maximize product disassembly and recycling and minimize expensive toxic waste.

Managing the Transition

Recycling markets already exist for certain service products, such as glass and paper. And some consumable products such as food already are biodegradable, so that their dissipative use is not alarming. However, many products, both service and consumable, are menaces after they are used: ultimately they will be judged "unmarketable." But they are part of current society and must be dealt with.

"Unmarketable" products—those that cannot now be consumed or used in an environmentally sound way—are of at least three kinds: (1) Service goods for which no appropriate recycling technology currently exists. Present recycling for most materials is often "down-cycling," because the materials and the products are of a lower quality after each recycling process (e.g., park benches or sound-proof barriers made out of recycled plastics). (2) Long-lived service goods, such as lead pipe, that outlast their producers in useful service, and thus need government involvement in their safe disposal because no firm is now left to be accountable. (3) Consumable goods with toxic constituents, such as nickel–cadmium batteries, that society does not want to dispose of in a dissipative manner.

In an eco-restructured economy, unmarketable products would not be manufactured. In a period of transition, however, institutions would be needed to store such products safely and retrievably. One can imagine a "waste parking lot." The waste parking lot would eliminate the current practice of dumping waste in irretrievable locations, where it is subject to leaching and high recovery costs. Waste parking lots would have to be part of a larger restructuring program that would greatly reduce waste volumes; otherwise, storage would become unmanageable.

Following the parking lot analogy, the government would build and maintain the structures, and the owners of waste would pay for renting the space within them. The lots would be organized for homogeneity, so that identical wastes from various products could be stored together. The owners of waste would remain responsible for the safety, stability, and maintenance of the stored items.

Some advantages of the waste parking lot concept include enforcement of the "polluter pays" principle, clear realignment of responsibilities, rental costs as an economic incentive to develop new environmental technologies, and simplified reclamation of useful waste constituents in the future.

Summary

Key elements of the future waste management infrastructure involve industry, government, and the consumer. Within the firm, consumable products should be designed for safe dissipative use, and service products should be designed for easy disassembly and recycling. Governments should re-align responsibilities, encouraging leasing and other mechanisms, to ensure that producers have incentives to design products for lower life-cycle environmental impacts. Governments should also encourage both waste supermarkets (to promote and simplify source separation and recycling) and waste parking lots (to store problematic wastes retrievably). These are transition steps until the economy is able to produce sustainable products.

25

Industrial Ecology in the Manufacturing of Consumer Products

Wayne France and Valerie Thomas

Abstract

Some consumer products are now being designed specifically to reduce life-cycle environmental impacts. Progress is being made in closing the materials flow loops in industrial economies; examples include firms' consumer product initiatives for single-use cameras, beverage containers, and motor vehicles. Innovative research arrangements, such as cooperative industrial partnerships, are important in helping competing firms to enhance environmentally conscious manufacturing.

Introduction

The purpose of this chapter is to illustrate the increased understanding and application of industrial ecology principles to the design, development, and manufacturing of component parts, product systems, and industrial megasystems. Some specific examples have been selected to highlight the progress that is being made toward industrial ecology in the manufacturing and use of consumer products.

Implementing Industrial Ecology

To successfully develop and implement closed-loop industrial ecology systems, this approach must be incorporated into one's thinking. For example, during product and process design, validation, and improvement, engineers traditionally make disciplined decisions based on such factors as design for manufacturability and design for assembly. Now, with the support of industry's management, design for environment (including design for disassembly, design for separability, and design for recyclability) has become another important checkpoint.

Besides the engineer's responsibility for environmentally responsible products and processes, a critical contribution to the success of any consumer-oriented program is acceptance and support by the general public. It is essential for the public to understand these concepts and become involved in making the process work, particularly as political and regulatory considerations come into play.

To give a simple example, families have traditionally practiced handing down clothes, and in this way have participated in several life-cycle elements. The flow diagram in Figure 1 illustrates this concept and serves as the basis for similar charts to follow. In this case, the flow is from raw materials to manufacturing to children's clothing as a consumer product. Once used, options include "remanufacturing" the clothes for a second child (the hand-me-down cycle), reusing the clothes in another market serviced by rummage sales and the Salvation Army, or entering a secondary-use system such as rags. From that point, options for rags include entering a tertiary use system wherein rags are used for manufacturing high-grade paper; another option is disposal by landfill or by incineration, which can provide energy. Thus, in addition to illustrating some closed-loop cycling, the example demonstrates how a coupled infrastructure can use an output from one sector as an input to another.

Consumer Product Initiatives

Single-Use Cameras

At first glance, the thought of Kodak's single-use camera being environmentally benign seems contradictory. But the single-use camera has been designed for reuse and recycling, as shown in Figure 2, and over 85% of each camera can be reused or recycled. The camera is sealed so that the entire camera must be sent to the photofinisher to have the film developed. Kodak pays 5¢ per camera, plus shipping, for the photofinisher to send the cameras back to its Rochester, New York facilities. As of 1994, Kodak reported that a little over 50% of these cameras were being returned. The main core of the camera, and the flash unit of the models with a flash, are typically reused directly; a small mark is placed on the inside of the camera core to note the number of times it has been used. Camera cores will be used a maximum of six times. Parts that do not pass inspection are ground up and fed into the raw material stream for molding into new cameras. The lens assembly is not reused by Kodak, but is sent for reuse to toy manufacturers; the cardboard case is incinerated in a waste-to-energy incinerator at the Kodak Plant. The polycarbonate outer shell of the Fun Saver Weekend 35, a waterproof model, is made from virgin polycarbonate, and on return, this shell is sent for recycling elsewhere. Through a separate, non-reimbursed program, the photofinisher may also recycle the polyethylene film spool and steel film magazine. (Eastman Kodak 1992a; 1992b; 1994; Moscowitz 1993; Van DeMoere, 1992).

Single-use cameras have been a financial success for Kodak and, according to Kodak's environmental spokesman James Blamphin, "The recycling program has saved us a great deal of money" (Moscowitz, 1993).

Beverage Containers

The material used extensively for plastic soft drink bottles in the U.S. is polyethylene terephthalate (PET). The life cycle of this consumer product is shown in the

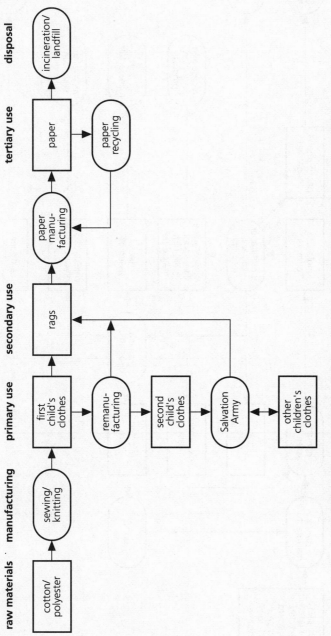

Figure 1. Clothing ecology: family hand-me downs. Rectangles indicate products; ovals indicate processes.

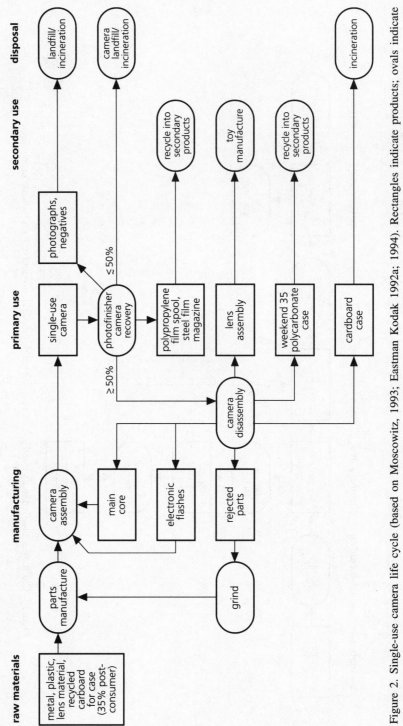

Figure 2. Single-use camera life cycle (based on Moscowitz, 1993; Eastman Kodak 1992a; 1994). Rectangles indicate products; ovals indicate processes.

flow diagram of Figure 3 (Frosch and Gallopoulos, 1992). As of 1992, the nation-wide recycling rate of these bottles is over 40%; about 24% of all PET products are recycled (American Plastics Council, 1993). The recycling of PET can involve grinding and extruding the material to make PET fibers for carpets, or for insulation for sleeping bags, pillows, and ski jackets. As of 1991, about 35% of all synthetic carpeting sold in the United States was made partly from old soft-drink bottles (Holmes, 1991). Alternatively, PET can be chemically treated and used to make polyurethane foam insulation or polyester resin bathtubs, or it can be repolymerized into PET. Thus, this is a good example of the output from one product system providing the input to other product systems. Plastics recycling is still in its infancy, but economics may eventually turn in its favor (Kirkman and Kline, 1991; Stilwell *et al.*, 1991; Cairncross, 1992).

Motor Vehicles

The motor vehicle is one of the most highly recycled consumer products. An infrastructure has been developed, as illustrated in Figure 4, which handles about 90% by weight of the 9 to 10 million vehicles removed from service each year in the United States. About 75% by weight of the vehicle scrap that is processed is recycled (primarily ferrous and nonferrous metals) or reused. Examples of reuse include parts needed by auto body shops, such as body panels, doors, and fenders, or by second-hand shops, such as wheel covers, tires, radiators, and radios. Critical components of this infrastructure are some 2000 automotive scrap recyclers, manufacturers of durable goods such as washers and dryers that use the recycled materials, and the recipient iron and steel plants that use this scrap as raw material for foundry operations (Frosch and Gallopoulos, 1992).

The remaining 25% by weight is vehicle shredder-residue called "fluff," which is typically composed of plastics (34%), fluids and lubricants (17%), rubber (12%), glass (16%), and other materials (21%) as described elsewhere (A.D. Little, 1992). The disposal of fluff in landfills represents less than 2% of the total volume of waste material sent to municipal landfills each year. As indicated by the faint flow line from "vehicle scrap" to "pyrolysis" (that is, combustion in the absence of oxygen), techniques are being developed to convert fluff to products such as oil, gas, and fillers for thermoplastics (Automotive Engineering, 1992).

The recycling of motor vehicle components and materials includes a variety of integrated systems for secondary use. For example, in the United States about 80% of lead-acid batteries are recycled (Thomas and Spiro, this volume); the use of "core" deposits (typically $5) helps ensure the return of used batteries. For platinum on catalytic converters, the recycling rate is now nearing 40% (Gabler, 1991). Once at the recycling facility, the technology exists to recover 90% of the platinum.

Recently there has also been an increase in the rate of recycling of tires and bumpers. The dashed lines in Figure 4 for these systems indicate that such systems are under development. Tires can be recycled, or incinerated to produce energy

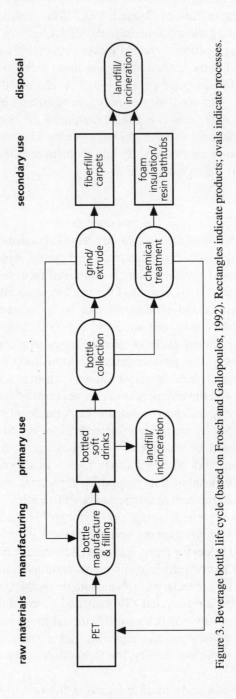

Figure 3. Beverage bottle life cycle (based on Frosch and Gallopoulos, 1992). Rectangles indicate products; ovals indicate processes.

(OTA, 1989; *Daily Japan Automotive News*, 1992). For polypropylene bumpers, recycling into plastic parts is being developed (FHI, 1992).

Another agent of recycling is the automobile dealer. As seen in Figure 4, dealers now have the equipment to recycle, reuse, and replace such automotive fluids as the Freon air-conditioning refrigerant and the ethylene glycol engine coolant. Such procedures minimize the release of these chemicals to the environment during repairs.

Cooperative Industrial Partnerships

Research on environmentally conscious manufacturing is being stimulated by a new ethos of cooperation. The pooling of technical and financial resources by industry, government, academia, and other interested partners is resulting in a synergism that is accelerating the development and implementation of technologies essential to industrial ecology. The approach has been facilitated by the National Cooperative Research Act of 1984, which permits and encourages precompetitive cooperative research without many of the restrictive antitrust regulations. In response to such legislation, a number of cooperative industrial initiatives dealing with manufacturing and the environment have been formed. Good examples include the Environmentally Conscious Manufacturing program of the National Center for Manufacturing Sciences (NCMS) and the Vehicle Recycling Partnership. The benefits of such collaboration include leveraging resources to reduce costs, generating information rapidly and without duplication, and sharing findings to facilitate implementation of newly developed technologies.

NCMS is a not-for-profit collaborative research, development, and technology transfer consortium with a membership of more that 160 corporations committed to making U.S. manufacturing globally competitive. In 1991, a program was initiated to reduce the impact of manufacturing on the environment (France, 1992). This program on Environmentally Conscious Manufacturing has been designed to focus, initially, on some immediate, pressing issues. Some of its first projects deal with solvents used in manufacturing and their substitutes, the handling and disposal of metal-working fluids, and reduced lead use in manufacturing. In addition, the participants have initiated a strategic project on life-cycle design for environmental compatibility. They intend to assess current practices and develop the knowledge needed to implement the concepts and principles of industrial ecology. By bringing together all the various interested participants, a multidisciplinary network has been established with critical linkages. The result is a concrete effort to identify critical issues, to develop pollution prevention technologies, and then to ensure that such technologies are implemented.[1]

---

[1] In addition to general discussion about this consortium during the 1992 Global Change Institute on Industrial Ecology and Global Change, several participants cited the potential benefits to NCMS of establishing an environmental science advisory committee to provide advice and recommendations about its strategic research agenda. This committee has now been formed.

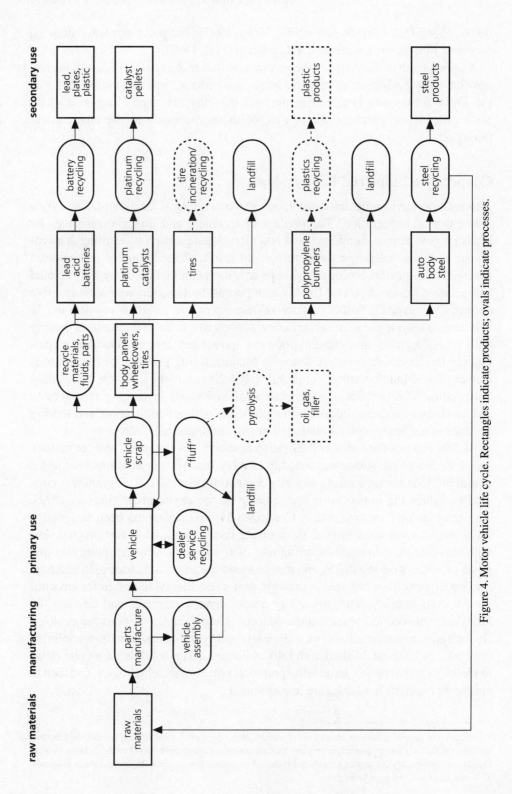

Figure 4. Motor vehicle life cycle. Rectangles indicate products; ovals indicate processes.

The Vehicle Recycling Partnership (USCAR, 1992) was formed in November 1991, with a mission to identify and pursue joint research and development efforts pertaining to recycling, reuse, and disposal of motor vehicles and their components. Increased use of recyclable and recycled materials in motor vehicle design also will be promoted. The goals of this partnership are as follows:

- Reduce the total environmental impact of vehicle disposal
- Increase the efficiency of the disassembly of components and materials to enhance vehicle recyclability
- Develop material selection and design guidelines
- Promote socially responsible and economically achievable solutions to vehicle disposal.

This partnership will examine the megasystem associated with motor-vehicle recycling (see Figure 4) and its many potential feedstreams for secondary markets. Possible cooperative projects with manufacturers of other consumer products, such as washers, dryers, and other durable goods are also envisioned.

Conclusion

Progress is being made in planning, designing, and implementing industrial ecology systems where the goal is to close the loop to achieve near-zero waste. In the production of consumer products, engineers are applying a systems approach as they optimize individual processes and integrate subsystems with megasystems. The recognition that participation in this process makes good business sense is a motivating factor. Consortium partnerships are providing a unique and innovative approach to solving critical environmental problems and will expedite removing manufacturing from the pollution equation.

Acknowledgments

The review comments from William Schlesinger and Bette Hileman are very much appreciated. The authors have benefited from discussions with James Blamphin of Eastman Kodak Co. and Joe Simmer of GM Research and Development Center.

References

A. D. Little, Inc. 1992. Recycling state-of-the-art for scrapped automobiles. Report to the American Iron and Steel Institute, Reference 67110, 100 pp.

American Plastics Council. 1993. *Post Consumer Plastics Recycling Rates*, June 11.

Automotive Engineering. 1992. Recycling the automobile. *Automotive Engineering*, 41–57.

Cairncross, F. 1992. How Europe's companies reposition to recycle. *Harvard Business Review*, 34–45.

Daily Japan Automotive News. 1992. Leading tire manufacturers using old tires as fuel. May 19.

Eastman Kodak, 1992a. Kodak FunSaver cameras: A recycling success story. *Quality: News for Kodak People (Oct.)*

Eastman Kodak, 1992b. Film container recycling. *EnviroNotes.*

Eastman Kodak, 1994. Kodak recycles single use cameras (press release, Feb.).

FHI (Fuji Heavy Industries). 1992. FHI (Fuji Heavy Industries) to recycle bumpers to produce interior parts. *Daily Japan Automotive News* (April 20).

France, W. D. 1992. Green manufacturing and the role for industrial partnerships. In *Proceedings of the First International Congress on Environmentally Conscious Design and Manufacturing*, The Management Roundtable, Boston, Massachusetts, 4–5 May 1992.

Frosch, R. A., and N. E. Gallopoulos. 1989. Strategies for manufacturing. *Scientific American 261(3)*, 144–152.

Frosch, R. A., and N. E. Gallopoulos. 1992. Towards an industrial ecology. In *The Treatment and Handling of Wastes* (A.D. Bradshaw, R. Southwood, and F. Warner, eds.), Chapman and Hall for the Royal Society, London, 269–292.

Gabler, R. C., Jr. 1991. *A Platinum-Group Metals Consumption and Recycling Flow Model*. U.S. Bureau of Mines, IC 9303, Washington, D.C.

Holmes, H. 1991. Recycling plastics. *Garbage*, 32–39.

Kirkman, A., and C. H. Kline. 1991. Recycling plastics today. *Chemtech* (October), 606–614.

Moskowitz, M. 1993. Recycling: Kodak's program really clicks. *Plastics World*, 13.

OTA (U.S. Congress Office of Technology Assessment). 1989. Facing America's Trash: What Next for Municipal Solid Waste? OTA-O-424, U.S. Congress, Office of Technology Assessment, Washington, D.C.

Stilwell, E. J., R. C. Canty, P. W. Kopf, and A. M. Montrone. 1991. *Packaging for the Environment: A Partnership for Progress*, American Management Association, New York, 262 p.

USCAR. 1992. Media information: Vehicle recycling partnership, Dearborn, Michigan.

Van DeMoere, A. 1992. Design for recyclability: Kodak's single-use cameras. In *Proceedings of the First International Congress on Environmentally Conscious Design and Manufacturing*, 4–5 May 1992. The Management Roundtable, Boston, Massachusetts.

26

Design for Environment: A Management Perspective

Bruce Paton

Abstract

Management must consider a number of factors when seeking to internalize environmental values within design and production processes. Success depends on an integrated management approach that includes a clear vision of what is to be accomplished, a workable business plan, effective business processes, and an understanding of the financial impact of reuse and recycling.

Introduction

A quiet revolution is sweeping corporate environmental management. After decades of managing environmental emissions from their plants, corporations are turning their attention to the environmental effects of their products. This new emphasis is being driven by a combination of market and government forces. More than ever before, customer demands and competitor initiatives are introducing environmental issues into product design decisions. At the same time, governments around the world are imposing broad new requirements that address the environmental effects of products and of the processes used to make and support them.

These requirements are raising the stakes for design decisions and posing unprecedented challenges for manufacturing companies. They are forcing manufacturers to reexamine their approaches to a wide range of concerns, including product and packaging design, material selection, production processes, and energy consumption, to name just a few. In an era of brutal international competition, companies must reconcile these new requirements with other imperatives, particularly pressures to compete on cost and time-to-market.

Design for environment is a discipline that provides strategies and techniques for designing and producing environmentally responsible products, which can compete in the international marketplace. To be effective, design for environment requires an integrated management approach. This framework must include, at a minimum, a clear vision of what needs to be accomplished, a coherent business plan, effectively managed business processes, and a thorough understanding of the financial impact of reuse and recycling. This chapter presents a management-oriented overview of design for environment programs, based on these four elements.

Vision

A well-implemented design for environment program should improve a company's environmental record and its business performance. Achieving this balance requires a clear vision that addresses both the business and environmental impacts of product decisions.

The vision should specify at least six environmental characteristics. Products should:

• Minimize environmental impact
• Be safe for their intended uses
• Optimize consumption of energy and materials
• Meet or exceed all applicable legal requirements
• Be reusable and/or recyclable
• Ultimately be disposed of in an environmentally safe and responsible manner.

The vision should also specify the desired business effects of decisions regarding the environmental characteristics of the product. Environmentally responsible products should provide a measurable source of competitive advantage and business success, by:

• Contributing to revenues profits and growth
• Minimizing delays in market introduction
• Avoiding mistakes that harm sales
• Eliminating barriers that prevent world-wide acceptance.

This balanced vision can help all parts of the business focus on succeeding in the marketplace, by meeting the market's environmental requirements. Achieving this vision will certainly not be easy, but the absence of a clear vision almost guarantees a future of repeated and ineffective reaction.

Business Planning

Businesses in markets affected by environmental requirements should address these needs in their business plans. The business plan can create a framework that provides clear marching orders for all business functions based on a realistic assessment of current and future business conditions. It clarifies the importance of the company's environmental practices to its business success. This clarity helps define the environmental features that will be necessary for product families to be successful over the length of the planning period. The plan should address at least four topics: motivations, financial and strategic importance, strategies and goals, and timetables for implementation.

The plan should specify the motivations for action. The magnitude of the changes that may be needed to meet environmental requirements calls for realistic appraisal of which problems need to be addressed, and when they need to be solved. Companies will typically be motivated by some combination of company values, shareholder expectations, legal requirements, customer expectations, and competitive pressures. A careful inventory of motivations can be particularly useful in persuading different functions (e.g., manufacturing, marketing, R&D) to contribute to the solutions.

The business plan should assess the importance of environmental requirements in relation to other factors influencing the strategy. Severe cost pressures might constrain design for environment efforts for some products. Conversely, the opportunity to gain advantage against a key competitor could lead to faster adoption of certain elements of environmental design.

Assessing the financial impact of an environmental requirement can motivate design for environment efforts that might otherwise be stifled by other priorities. For example, several computer companies have assessed the potential financial risk from proposed product return requirements and concluded that design for disassembly is a much more urgent priority than market data would indicate.

Approaches to design for environment will not and probably should not be implemented all at once in most businesses. The business plan should address the timing for implementation of specific design for environment elements. The plan will be most useful if it identifies the obstacles to success, specific responses to those obstacles, and quantifiable goals to be achieved by specific dates.

Business Process Management

Business process management offers many opportunities to influence the environmental impact of products and the processes used to create and support them. Six categories of business processes offer distinct opportunities: product design, manufacturing process design, materials management, supplier management, order fulfillment, and service and support.

Product Design

Product design is the process that harnesses the company's creative energies to satisfy customer and market requirements. At least six elements of product design can help to assure that products are environmentally sound: product definition, product design reviews, platform design, design for reduced material usage, design for reuse, design for recyclability. These approaches allow the product development team to fulfill the environmental requirements specified in the business plan.

Environmental requirements must be incorporated in stable product definitions. A product definition is a statement of the features that a specific product should have when development is completed. It serves as the basic specification that the product development team relies on throughout the design process. This step is crucial to clarify where environmental requirements fit in the hierarchy of product requirements that designers must address. If environmental requirements are not factored into the basic definition, designers will have great difficulty finding time to work on them and will not know how to make appropriate tradeoffs with other design objectives.

Product design decisions must be subject to rigorous design reviews at all phases of development to assure that customer requirements have been met and that they have been implemented in a way that does not impose unreasonable delays or costs on the final product. Design reviews are a valuable safety net to assure that strategies and approaches are consistent across the functions involved in producing a product.

Design of the basic concept (or platform) will determine many of the environmental characteristics of the initial product and subsequent modifications. For example, manufacturers of some personal computers and minicomputers have begun designing products so that they can be upgraded to future products by exchanging circuit boards. In addition to the business motivations for this approach to platform design, those businesses have created the potential for dramatically reducing solid waste from ultimate disposal of the product.

Reduction of material usage offers the greatest potential for minimizing environmental impact, particularly for high-volume products. Reduction of materials often translates directly into cost savings while offering environmental improvements. A design with less material requires smaller initial purchases, less shipping expense, less energy to build and transport, and a smaller burden when products are ultimately returned.

Design for reuse (particularly for items requiring frequent replacement, such as batteries and cartridges) can reduce material costs, encourage repeat purchases from the original manufacturer, provide relief from future requirements for recycled content, and reduce recycling costs. Effective design for reuse requires a thorough analysis of the product and the system for recovering it from customers. Particularly for high-volume products, the environmental and financial stakes can be very high.

Cost-effective design for recycling requires at least two types of care. First, materials that are likely to be recycled (such as plastic parts) need to be marked for efficient sorting. For example, protocols for marking plastic parts have already been suggested in the plastics design community. These identify the basic resins used and any material added to change the appearance or engineering properties of the plastic.

Second, products must be designed for cost-effective disassembly. This typically involves classic elements of design for manufacturability, such as reducing the number and variety of fasteners. Design for disassembly may also require some innovation to overcome some effects of design for manufacturability. For example, products assembled with snap-fit fixtures will be easier to assemble, but harder to disassemble than comparable products fastened with screws or bolts. One relatively new solution uses snap-fit fasteners that can be broken apart when disassembly is required.

Manufacturing Process Design

In many industries the rate of change in manufacturing process technologies is extremely rapid. Often the most effective strategy for improving environmental performance is to focus on new processes. (Recent experience with removing chlorofluorocarbons from existing processes illustrates the high cost of retro-fitting.) Two strategies are important in new process design: selection of new process technology, and new equipment selection.

Careful analysis of new process technologies prior to purchase can eliminate many environmental problems. Often multiple approaches are available, each with different tradeoffs among cost, effectiveness, and environmental impact. Selection of the technology defines the broad parameters affecting environmental performance.

Chemical use management is an important component of this new technology selection. Careful scrutiny of any chemicals used in new processes is critical to prevent losing ground or creating new problems. Often this can involve major efforts to find appropriate substitutes.

For many technologies, equipment is available from a wide selection of vendors. These vendors typically compete on a variety of factors, including cost and environmental performance. Proprietary systems are often available to minimize environmental impact. Incorporating environmental effects in the analysis of competing vendors can be a very effective process for assessing tradeoffs between environmental objectives and other business goals.

Materials Management

Materials management is one of the most important levers for influencing product recyclability and waste emissions. Three strategies are particularly important: material selection, hazardous materials elimination, and chemical use evaluation.

To avoid adverse environmental and cost effects, all products should be designed with materials that permit reuse or recycling. Currently, circuit boards for many products can be reused in the service and repair channels. However, plastic enclosures and many other parts may be difficult or expensive to recycle. Selection of appropriate materials can help create cost-effective (and sometimes profitable) relationships with recyclers and material vendors, that permit businesses to sell recovered materials.

Many products can be built using materials with some recycled content, particularly in enclosures. This approach can provide environmental benefits and promote sales. (Government procurement standards in many areas seem to be headed toward requirements for specified percentages of recycled materials to be included in each product.) Design specifications in many current products inhibit the use of materials with recycled content. Product lines may gain significant savings by changing specifications to permit inclusion of recycled content, when feasible and appropriate.

Hazardous materials, such as cadmium or mercury, frequently inhibit sales in particular countries or to particular customers (especially government agencies or defense contractors). Products should be designed, to the extent possible, without using materials that may restrict sales. This is particularly critical for products that rely on worldwide markets to create demand for high volumes.

One relatively new approach incorporates a chemical use evaluation as a basic part of new product design reviews. This approach identifies potential chemical problems early enough in the design process to permit reengineering, if necessary, to improve the environmental impact of the product.

Supplier Management

Suppliers can play a critical role in the creation of environmentally sound products. Manufacturers should consider at least three supplier management strategies: technology partnerships, reuse or recycling relationships, and supplier evaluation.

Meeting environmental requirements may require development of new materials to avoid problems created by existing materials. For example, many plastics vendors have acknowledged that their long-term survival depends on their ability to create materials that can be recycled conveniently and in a cost-effective manner. These vendors are actively developing plastic resins with engineering properties to meet new requirements, while eliminating ingredients that inhibit cost-effective recycling. Many are also establishing business relationships conducive to cost-effective return and recycling of the resins.

Cost-effective recycling requires coordination between manufacturers and their suppliers. Manufacturers may be able to improve the economics of reuse or recycling by developing explicit agreements with their suppliers. These may specify the degree of segregation required, the lot sizes required by the supplier, and possibly agreements to accept the recovered materials in future material purchases.

Conformance with many environmental requirements will require careful scrutiny of suppliers' environmental practices. For example, a recent U.S. regulation requires manufacturers to label products "made with" ozone depleting chemicals. A product is considered to be "made with" those chemicals if it, or any of its component or subassemblies, comes into contact with any of these compounds at any time. Manufacturers are forced to understand their suppliers' practices to evaluate whether labeling requirements will apply.

The manufacturing firm will need to be clear and consistent in its scrutiny of suppliers. It should have a clear policy stating the purposes, methods, and consequences of its supplier evaluation. Manufacturers may have different motivations for evaluating suppliers. These may include:

- Assuring that supply lines are not interrupted by fires, explosions or regulatory shutdowns
- Reducing liabilities for suppliers' misdeeds
- Certifying that suppliers comply with all applicable laws
- Choosing suppliers with the best environmental practices.

The manufacturer's motivations will determine the type and the depth of supplier evaluation required. Emerging methods including survey questionnaires, audits, and third party certifications.

Order Fulfilment

Order fulfilment processes are the means by which businesses convert orders for existing products into completed sales and deliveries. They are not part of the physical product, but their design has a major effect on any product's environmental impact. Three processes are particularly important to consider in design for environment systems: packaging, forecasting and inventory management, and distribution.

Packaging is the most visible design for environment issue in most parts of the world today. A detailed examination of this topic is beyond the scope of this chapter. However, several design strategies described above are beginning to show impressive results. Firms in the electronics industry have gained significant savings and prevented waste by experiments in packaging design that reduce material use, improve material selection, design for reuse, and design for recycling.

Inventory management has become an important element of manufacturing strategy in recent years. Companies have worked very hard to improve their forecasting systems to minimize the volume of unsold products. However, inventory management remains a huge source of unnecessary costs to many manufacturing firms. At the same time it is a very large, hidden source of environmental burdens. Unneeded products that are scrapped or transported for reprocessing are unlikely to appear in any of the existing environmental standards, but will become very visible as product recycling requirements become more prevalent. By bringing increased visibility to the high costs of poorly managed inventory, design for environment can make a very significant contribution to overall business success.

Energy management has always been a factor in the economics of distribution. However, until recently little attention has been paid to the resource usage implications of companies' distribution systems. As the focus on life-cycle analysis increases, this area will become increasingly important. This new scrutiny will raise difficult questions about where products are made and how they are distributed.

Relatively little has appeared in the design for environment literature to help firms address the environmental aspects of distribution. However, many of the existing approaches to distribution management can be adapted to address additional constraints such as energy efficiency.

Service and Support

Businesses' environmental strategies often overlook service and support processes. However, these processes are taking on increasing importance financially and gaining increasing visibility in environmental requirements. Two

355

elements of service and support have become particularly important: material sourcing for maintenance and repair, and collection/recycle/disposal systems.

Material sourcing for maintenance and repair can provide an important outlet for reused materials. The shortening of product life cycles has simultaneously created a problem and an opportunity for manufacturing firms. Products such as printers or personal computers typically function long after the manufacturer has stopped making them. As a result, supplying replacement parts can be expensive for the manufacturer. Discarded products may be an excellent source for reusable components or circuit boards to refurbish. (This is similar to the long-standing practice of going to junk dealers to get parts for old cars.) The large volume of products to be serviced, and the high profitability of servicing them, can make maintenance and repair services a ready outlet for a large volume of parts from discarded products.

Both legal and market requirements are forcing many businesses to collect and recycle or dispose of packages and products. The economics of these systems depends heavily on product design, as described above. However, some firms are realizing significant savings by finding *ad hoc* markets for materials recovered from products that were not designed for reuse or recycling. For example, Digital Equipment Corp. has publicized its success in recycling plastic computer housings into roofing shingles for resale. Revenues from sale of the shingles help offset the costs of recycling. Each business will need to evaluate the economics and the logistical needs for effectively complying with product return or recycle requirements. Legal and business constraints may force manufacturers to supplement their in-house capabilities with third party vendor arrangements.

Since product return/recycle requirements are likely to become much more widespread, current business and product plans should include arrangements for addressing these needs. In addition, requirements for recovering costs should be addressed for each product line.

Financial Impact of Design for Environment

Corporate managers have been conditioned to regard environmental management purely in terms of its costs. This view portrays additional requirements as a drag on corporate profits. Conversely, pollution prevention activities are sometimes viewed as opportunities to reduce costs, by eliminating the waste of costly raw materials.

Environmental requirements that apply to products can have more complex financial impacts. The design for environment practices described above have the potential to affect profits, by influencing both costs and revenues. The net effect on profitability depends substantially on the approaches to design for environment taken by a firm and by its competitors.

On the cost side, reuse or recycling can be very expensive for products that have not been designed with these processes in mind. Collection and disassembly can be costly, and the resulting materials may have little or no recoverable value.

Conversely, products that are appropriately designed may be inexpensive to disassemble and may yield parts or materials with high recoverable value.

On the revenue side, products that are not designed with environmental requirements in mind can cause disappointing sales for a variety of reasons. For example, failure to address specific requirements may prevent sales in a particular country or may preclude consideration for specific bidding opportunities. Conversely, products designed to meet environmental requirements may be able to increase revenues by assuring worldwide acceptance or by taking market share away from competitors who are less able to respond to changing requirements.

The potential for design for environment choices to affect both revenues and costs independently makes the net effect on profitability quite uncertain. The actions of governments, customers, and competitors will affect both revenues and costs in ways that the individual firm cannot control. This uncertainty underscores the critical importance of planning and implementing a systematic approach to the design of products and the processes used to build, ship, support, and recover them. Design for environment can provide the framework for systematically identifying the choices and tradeoffs to be made in balancing environmental requirements with other business priorities.

Acknowledgments

This chapter draws on my experience at Hewlett-Packard and on the work of a design for environment task force sponsored by the American Electronics Association.

Suggested Readings

Allenby, B. R., and A. Fullerton. 1991. Design for environment—A new strategy for environmental management. *Pollution Prevention Review*, Winter 1991–92.

Berkman, B. N. 1991. European electronics goes green. *Electronic Business*, August 19, 1991.

Berube, M. R. 1992. *Integrating Environment into Business Management: A Study of Supplier Relationships in the Computer Industry*, Ph.D. thesis, Massachusetts Institute of Technology, Cambridge, Massachusetts.

Borsboom, T. 1991. The environment's influence on design. *Design Management Journal*, Fall 1991.

Brady, W. J. 1991. Environmentally conscious products—An IBM case study. In *Design for Environmentally Responsible Products*, Conference Proceedings, Washington, D.C.

International Chamber of Commerce. 1990. *The Business Charter for Sustainable Development, Principles for Environmental Management*, Publication 210/356A, Paris, France.

Nussbaum, B., and J. Templeman. 1990. Built to last, until it's time to take it apart. *Business Week*, September 17, 1990.

Office of Technology Assessment, Congress of the United States. 1992. *Green Products by Design, Choices for a Cleaner Environment*, U.S. Government Printing Office, Washington, D.C.

27

Prioritizing Impacts in Industrial Ecology

Thomas Graedel, Inge Horkeby, and Victoria Norberg-Bohm

Abstract

The implementation of industrial ecology often involves choices among differing materials or technologies, each of which embodies a set of potential impacts on raw materials supplies, energy use in manufacture and in service, and air, water, and soil quality. Reasoned choices among product and process options can only be made if impacts are prioritized, a task that has received insufficient attention to date and that must be done by cooperative efforts among a number of interested parties. Several prioritization efforts are reviewed and compared with recommendations for optimal approaches to prioritization in industrial ecology.

Introduction

All industrial activities have some effect on the external environment. Energy and raw materials are consumed, materials are transformed, byproducts are generated, and some degree of waste is produced. Many of the impacts of industrial activity on the environment can be eliminated by thoughtful product and process design and execution. At some point, however, the straightforward actions have all been taken, and choices between impacts present themselves. Product designers, for example, may want to consider whether a particular metal, a plastic, or a composite would have the lowest environmental impacts. It is here that prioritizing impacts becomes vital.

Such prioritization cannot be accomplished by industry alone. Rather, industrial efforts must ultimately be related to larger societal efforts concerning risk comparison and prioritization. The participation of the community of environmental scientists is needed to define and evaluate the different types of risks posed by different environmental hazards. Broader public participation is needed to weigh the relative importance of different types of impacts (human health, ecosystem, and economic), which occur over different time scales, and about which there are different degrees of scientific uncertainty.

While it has been much more common for environmental scientists and others to study the impacts of individual environmental hazards than to attempt to rank them in priority order, several systematic efforts to prioritize have been made in the past few years. In this chapter, we summarize these approaches and discuss their advantages and disadvantages. The variety of approaches demonstrates the use of prioritization in different settings: within a country, for cross-country comparisons, and within industry. For illustration, we will show some of the rank ordering of environmental problems that emerges from these efforts. The purpose of this chapter is not to present or defend specific ordering of impacts, however, but to illustrate the process. We conclude by discussing several of the characteristics of methods for prioritization that are necessary if these systems are to be utilized in industrial ecology.

The Challenge of Relative Risk Assessment and Management

An important consideration that is often overlooked but that plays a role in the prioritization of impacts is that a single impact generally has many sources and a single source generally has many impacts. As a consequence, the industry–environment interaction embodies a summation of impacts of different magnitudes and different spatial and temporal scales. The result is that a choice between two designs may involve choosing between impacts of different types, as in choosing between having an influence on a single impact of high importance or on several impacts of lower importance. A recent example involved the substitution in the electronics industry of organic solvents for some of the chlorofluorocarbon cleaning operations; this change substituted modest impacts on photochemical smog and air toxics for a substantial impact on stratospheric ozone.

A qualitative approach to this situation has been presented by Graedel (this volume) in a series of matrix displays, the axes of the matrices being sources (specific activities or industries) and critical properties (specific impacts). The assessments at each point in the matrix provide a sense of a specific impact of a specific source, and the diagrams also permit one to derive a sense of the overall source structure leading to a specific impact and the overall impact structure of a particular source. While useful as general guidance, such an approach does not provide the specificity of advice often needed by industrial design and development engineers.

Three Approaches to Prioritizing Environmental Problems
Setting National Priorities: Unfinished Business and Reducing Risk

One of the first systematic efforts to assess and compare environmental hazards was undertaken by the U.S. Environmental Protection Agency (EPA) in 1986 and 1987. The goal of this study was to give the EPA a scientific basis for tackling the environmental problems that are the most serious and thus to make the best use of its limited resources. The resulting report, *Unfinished Business* (1987), provided an assessment and ranking of environmental problems along four dimensions: human cancer risk, human noncancer health risk, ecological risk, and welfare risk. In 1990, the Science Advisory Board of the EPA reviewed the original study and produced a follow-on report, *Reducing Risks*, to overcome several of the shortcomings perceived in the original effort. One charge to this group was to develop evaluations on only two dimensions, merging cancer and noncancer health risks, and ecological and welfare risks. The follow-on study also addressed the issue of consistent and inclusive problem definition and the evaluation of data used in the rankings.

Five ranking parameters were used by the Ecological and Welfare Risks subgroup:

• Spatial scale of the impact (large scales being of more concern than small)
• Importance of the ecosystem that is exposed
• Severity of the hazard
• Intensity of exposure, including the number of people potentially exposed and the degree of exposure of this group
• Temporal dimension, including the time scales of effects and ecological recovery.

In contrast, the Health Subgroup in the follow-on study felt that the methods and data for ranking health risks were inadequate at this point. They thus limited themselves to identifying which of the high-risk health rankings of the first study were fully supported by the available data.

Table 1 shows the rankings from the *Reducing Risk* study. Perhaps the most important result is that the EPA has begun to realize the long-term importance of relative risk assessment in environmental management. Key findings of the study include:

Table 1: *Prioritization of risks from the EPA study* Reducing Risks

Risks to the Natural Ecology and Human Welfare

Relatively high risk:
Habitat alteration and destruction; species extinction and overall loss of biological diversity; stratospheric ozone depletion; global climate change

Relatively medium risk:
Herbicides/pesticides; toxics, nutrients, biochemical oxygen demand, and turbidity in surface waters; acid deposition; airborne toxics

Relatively low risk:
Oil spills; groundwater pollution; radionuclides; acid runoff to surface waters; thermal pollution

Risks to Human Health

High risk:
Ambient air pollutants; worker exposure to chemicals in industry and agriculture; pollution indoors; pollutants in drinking water

Potential risk:
Pesticide residues on food; toxic chemicals in consumer products

Reprinted with permission from the Science Advisory Board, U.S. EPA.

- Environmental problems given high rankings for one type of impact did not uniformly receive high rankings in others, leading to the conclusion that the relative severity of an environmental problem depends on the type of adverse effect with which one is concerned.
- There was a disparity between those problems ranked as having the highest risk and the related resource allocations at the EPA. However, there was a positive correlation between the EPA's priorities and the public perceptions of risk. These results point to the involvement not only of environmental scientists but also of the public and policy-makers in relative risk assessment.
- The information on environmental risks is often inadequate for the task of comparative hazard assessment, particularly in the areas of the extent of human and ecological exposures to pollutants and exposure–response relationships.

In addition to the EPA reports, summaries of the studies can be found in Morgenstern and Sessions (1988) and Roberts (1990).

International Comparisons of Environmental Hazards

The second approach that we discuss is a method (Norberg-Bohm *et al.*, 1992) designed to be used by international institutions, nongovernmental organizations, and governments that are involved in setting national environmental agendas or developing environmental programs that require international coordination. It was developed and evaluated through its application to four country case studies: India, Kenya, the Netherlands, and the United States.

The method begins with a simple linear model of hazard causation. For each stage in this model, key indicators are defined and used to characterize environmental concerns (see Figure 1). The indicators are designed to reflect both the causes and the consequences of environmental problems, and also to pose realistic demands on available data. For a given environmental problem in a given country, each of these indicators is scored on a scale of 1 to 9. The research team also defined 28 environmental hazards, a list designed to be understandable and meaningful to policy-makers and other nonspecialists, comprehensive (including virtually all of the potentially significant environmental problems faced by any nation), and at a level of resolution that is neither too general nor too complex for utility.

By analyzing indicator data in various ways, this method provides one source of input for hazard prioritization. While it gives a systematic way of ordering key data about a range of environmental problems, it does not specify a manner for interpreting or aggregating these data. Rather, in recognizing the central role of values in creating hazard prioritizations it can be used to illuminate the implications of different preferences. The use of the method can be illustrated with two examples. Figure 2 compares those environmental problems that are thought to pose the greatest current hazards with those problems that are expected to have the greatest future consequences. Problems that fall in the upper right-hand corner of the graph are those that are among the worst current problems and are also expected to become significantly more severe in the future. For the United States,

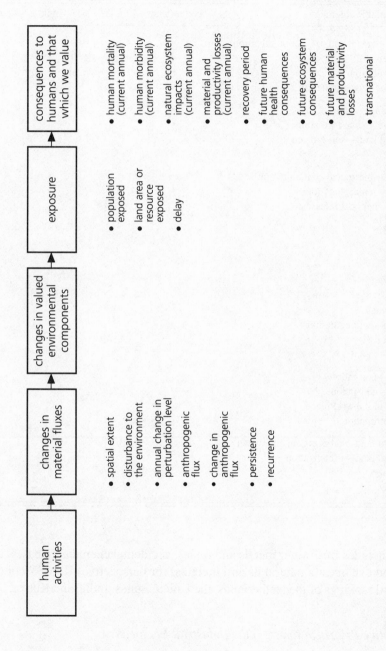

Figure 1. A causal taxonomy for comparative hazard assessment of environmental problems (from Norberg-Bohm *et al.*, 1992).

Table 2: *A comparison for the Netherlands, the United States, India, and Kenya of hazards ranked as most severe in terms of total consequences. Dots indicate a high-ranking hazard (from Norberg-Bohm et al., 1992).*

Environmental Problem	Netherlands	U.S.	India	Kenya
Freshwater—metal and toxic contamination	•	•	•	
Ocean water	•	•	•	
Toxic air pollution	•	•	•	
Stratospheric ozone depletion	•	•		
Climate change	•	•		
Ground-level ozone formation	•			•
Agricultural land—soil erosion			•	•
Forests			•	•
Freshwater—biological contamination			•	
Freshwater—eutrophication	•			
Accidental chemical releases		•		
Wildlife				•
Fish				•
Floods			•	
Droughts				•
Cyclones		•		
Pest epidemics				•
Freshwater—sedimentation				
Acidification				
Indoor air—radon				
Indoor air—non-radon				
Radiation—non-radon				
Chemicals in the workplace				
Food contamination				
Agricultural land—salinization, alk., water				
Agricultural land—urbanization				
Groundwater				
Earthquakes				

this list includes freshwater metals and toxics, accidental chemical releases, ocean water, and cyclones (a natural hazard included for perspective). Table 2 compares the hazard rankings of the Netherlands, the United States, India, and Kenya.

Quantitative Prioritization from the Industrial Perspective

Assessing the overall environmental impact of a product as complex as are most modern products is a daunting task. Even as presumably simple a question as the relative desirability of paper, foam, or china coffee cups took one research group

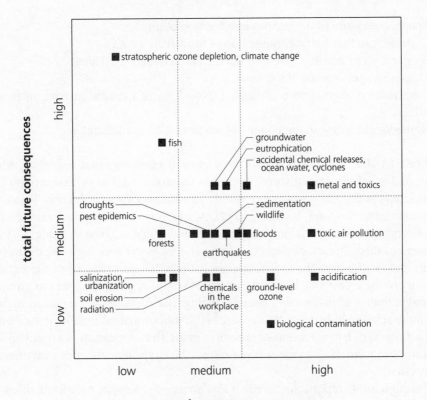

Figure 2. A comparison of U.S. hazards: "total current consequences" vs. "total future conse-quences." "Total current consequences" is an aggregation of the scores for human mortality, human morbidity, natural ecosystem impacts, and material productivity losses. "Total future consequences" is an aggregation of future human health consequences, future ecosystem consequences, and future material and productivity losses (from Norberg-Bohm *et al.*, 1992).

123 pages to discuss, and ended with the conclusion that the answer depends on details of product manufacture and use (*Economist*, 1992). To address product design choices in a formal and relatively efficient manner, Volvo has developed an analytic tool called the Environment Priority Strategies for Product Design (EPS) system that allows product designers to select product components that minimize environmental impact (Ryding et al., 1993).

Analytically, the EPS system is quite straightforward, though detailed. An envi-ronmental index is assigned to each type of material used in automobile manufac-ture. Different components of the index account for the environmental impact of this material during product manufacture, use, and disposal. The three components are summed to obtain the overall index for a material in "environmental load units" (ELUs) per kg of material used. The units may vary. For example, the index for a paint used on the car's exterior would be expressed in ELU/m^2.

When calculating the components of the environmental index, the following factors are included:

- Scope: a measure of the general environmental impact
- Distribution: the size or composition of the affected area
- Frequency or intensity: extent of the impact in the affected area
- Durability: persistence of the impact
- Contribution: significance of impact from 1 kg of material in relation to total effect
- Remediability: cost to remediate impact from 1 kg of material.

The ELU factors are calculated by a team of environmental scientists, ecologists, and materials specialists for every raw material and energy source (with their associated pollutant emissions) used in automobile production (Swedish Environmental Research Institute, 1991). A selection of the results for raw materials, energy use, and emissions into air, soil, and water is given in Table 3. A few features of this table are of particular interest. One is the very high values for platinum and rhodium in the raw materials listings. These result from the extreme scarcity of these two metals. Similarly, the use of oil is given a higher environmental index than coal in the raw materials listing, not because the emissions impacts are more severe but because the resource is so much smaller. Finally, in the "emissions to water" listing, the assumption is made that the metals are emitted in a mobilizable form. If an inert form is emitted, the environmental index may need to be revised.

Throughout its report, the Swedish Environmental Research Institute discusses the need for more research to complete the list of indices. It also points out that continuing updates are required as environmental impacts are reevaluated. In many cases, uncertainty in the magnitude of impacts leads to a range of up to a power of ten in the recommended ELU factors. ELU factors having a range rather than a single recommended value should be regarded as variable parameters in the overall calculation, and sensitivity studies should be performed to assess whether or not the uncertainties are important to the final choice of materials and processes.

Once the environmental indices are agreed on, they are multiplied by the materials uses and process parameters (in the appropriate proportions) to obtain environmental load units for processes and finished products. An important feature of this system is that it performs a life cycle analysis, assessing a product's environmental impact through all its stages, including materials extraction, construction and manufacturing, use, and disposal.

As an example of the use of the EPS system, consider the problem of choosing the more environmentally responsible material to use in fabricating the front end of an automobile. Two options are available: galvanized steel and polymer composite. They are assumed to be of comparable durability, though differing durabilities could potentially be incorporated into EPS. The steel design is heavier but more rugged.

Based on the amount of each material required, the environmental load values at each stage of the product life cycle are calculated. Table 4 illustrates the total life cycle ELUs for the two front ends. There are several features of interest. One

Table 3: *A selection of environmental indices (ELU/kg)*[1]

Raw Materials:		Emissions-Water:	
Co	1.2×10^4	Suspended matter	1×10^7
Cr	22.1	BOD[4]	1×10^4
Fe	0.38	COD[5]	1×10^5
Mn	21	TOC[6]	1×10^5
Mo	4.2×10^3	Oil	1×10^5
Ni	700	Phenol	1
Pb	260	Phosphorus	2
Pt	4.2×10^7	Nitrogen	10
Rh	4.2×10^7	Al	1
Sn	4.2×10^3	As	0.01
V	42	Cd	10
Oil	0.17	Cr	0.5
Coal	0.1	Cu	5×10^{-3}
		Fe	1×10^{-7}
Emissions-Air		Hg	10
CO_2	0.04	Mn	1×10^{-7}
CO	0.04	Ni	1×10^{-3}
NO_x	250	Pb	0.01
N_2O	0.6		
SO_x	6.0	Energy	
VOC[2]	10	Oil	0.33
PAH[3]	600	Coal	0.28
Aldehydes	20		
F	1×10^7		
Hg	10		

[1] Swedish Environmental Research Institute (1991).
[2] VOC = volatile organic carbon.
[3] PAH = polycyclic aromatic hydrocarbons.
[4] BOD = biochemical oxygen demand.
[5] COD = chemical oxygen demand.
[6] TOC = total organic carbon.

is that the steel front end has a larger materials impact during manufacturing, but is so completely reusable that its overall ELU on a materials basis is slightly lower than that of the composite. However, it is more than twice as heavy as the composite unit, and that factor results in much higher environmental loads during product use, due to greater fuel consumption. The overall result is not intuitively obvious: the composite front end is the better choice in terms of environmental impacts during manufacture; the steel unit is the better choice in terms of recyclability; and the composite unit is the better choice overall because of lower product use impacts. Attempting to make the decision on the basis of an analysis of only part of the product life cycle would result in an incompletely guided and potentially incorrect decision.

The EPS system is currently being refined and tested by several organizations within the Volvo Car Corporation. Other companies in Sweden and elsewhere have

Table 4. *Calculation of environmental load values*

Materials and Processes	Production			Product Use			Waste						Total ELU
							Incineration			Reuse			
	ELU/kg	kg	ELU	ELU/kg	kg	ELU	ELU/kg	kg	ELU	ELU/kg	kg	ELU	
Plastic composite													
GMT	0.58	4.0	2.32				−0.21	3.7	−0.78	−0.58	0.3	−0.17	1.37
Compression molding	0.03	4.0	0.12										0.12
Petrol				0.82	29.6	24.27							24.27
Total sum			2.44			24.27			−0.78			−0.17	25.76
Galvanized steel													
Galvanized steel	0.98	9.0	8.82							−0.92	9.0	−8.28	0.54
Spot welding	0.004	48	0.19										0.19
Painting (m²)	0.03	0.6	0.02										0.02
Steel stamping	0.06	9.0	0.54										0.54
Petrol				0.82	48.0	39.36			0				39.36
Total sum			9.57			39.36						−8.28	40.65

Adapted from Ryding et al., 1993.

expressed great interest in developing EPS systems that are specific to their products and manufacturing procedures. The EPS's greatest strength is its flexibility; raw materials, processes, and energy uses can be added easily. If a manufacturing process becomes more efficient, all products that utilize the process will automatically possess upgraded ELUs.

Implementing Impact Prioritization in Industrial Ecology

Although none of the techniques summarized above is an ideal solution to the need for producing priority rankings of environmental impacts, each provides useful guidance toward improved solutions. While each system uses somewhat different parameter and problem definitions and different degrees of quantification, many of the same key parameters appear in each of the efforts: temporal and spatial scale, exposure, and severity of health, ecosystem, and economic impacts. Such convergence on what data are needed for comparative risk assessment is in itself a quite useful step. One of the key results of these early efforts is to identify the types of additional research and information needed on individual environmental hazards in order to engage in comparative hazard assessment.

Without making a judgment on the appropriateness of specific rankings, the methodology exemplified by the Volvo/Swedish Environmental Research Institute/Federation of Swedish Industries collaboration provides the best model now available of the type of system needed for implementing impact prioritization in industry. Its approach, based on individual chemicals and materials, and the quantitative results that it provides make it straightforward to choose among alternative materials and processes, provided that the uncertainties in the ELU factors are recognized. However, it is important not to let the quantitative results imply a greater degree of certainty than exists or to obscure the essential value choices involved in the system. The rankings themselves, destined always to be in a state of flux, must be designed so that they can be readily modified in light of new information or new perspectives.

In the long run, industry would like a system for comparative hazard assessment that can be integrated into process design and product life cycle design tools. Ideally, such a system would give hard and fast rules about tradeoffs and provide some type of "environmental index" for designers to use for comparison of different options. Given current trends in globalization, industry would also prefer an international prioritization to a national one. It is fair to conclude from this review that such systems are not currently within our grasp, due to a combination of lack of information, uncertainty, differing value systems, and different environmental conditions in different parts of the world. Having said this, it is important to emphasize that current efforts not only are important experiments in method, but also provide usable results. From a societal viewpoint, we are clearly able to distinguish high-risk problems from those that pose lower risks. From an industry viewpoint, firms can readily identify strategies for the implementation of environmentally responsible processes or products.

The following characteristics emerge as important for risk prioritization systems:

- Transparency and changeability: The values that are used for key parameters should be clear to system users, and the users should be able to revise them. This latter provision allows for the incorporation of new information.
- Sensitivity analysis: Even with improved knowledge, we can expect to be faced with continuing uncertainties regarding environmental hazards. Thus, any method for risk assessment should provide the ability to perform sensitivity analyses.
- Legitimacy and participation: Experience demonstrates that comparative risk assessment carried out single-handedly by any single group will not stand the test of universal acceptability. We may be a long way off from societal agreement about any given scheme of risk prioritization. What is clear is that movement in that direction will require the involvement of a broad range of interested parties: industrial product and process design engineers, environmental scientists, and experts in society, economics, and government. These groups, in combination and partnership, constitute the industrial ecology community.

References

Economist (The). 1992. Washed up. Aug. 1, London, U.K.

Environmental Protection Agency. 1987. *Unfinished Business: A Comparative Assessment of Environmental Problems*. Overview Report and Appendices I–IV, Office of Policy Analysis and Office of Policy, Planning and Evaluation, Washington, D.C.

Environmental Protection Agency. 1990. *Reducing Risk: Setting Priorities and Strategies for Environmental Protection*. Report SAB-EC-90-021 and 021A, Science Advisory Committee, Washington, D.C.

Morgenstern, R., and S. Sessions. 1988. Weighing environmental risks, EPA's unfinished business. *Environment 30 (6)*, 14–17 and 34–39.

Norberg-Bohm, V., W. C. Clark, B. Bakshi, J. Berkenkamp, S. A. Bishko, M. D. Koehler, J. A. Marrs, C. P. Nielsen, and A. Sagar. 1992. International comparisons of environmental hazards: Development and evaluation of a method for linking environmental data with the strategic debate on management priorities. In *Risk Assessment for Global Environmental Change* (R. Kasperson, and J. Kasperson, eds.), United Nations Press, New York.

Roberts, L. 1990. Counting on science at the EPA. *Science 249*, 616–618.

Ryding, S., B. Steen, A. Wenblad, and R. Karlson. The *EPS System—A Life Cycle Assessment Concept for Cleaner Technology and Product Development Strategies, and Design for the Environment*. Paper presented at EPA Workshop on Identifying a Framework for Human Health and Environmental Risk Ranking, Washington, D.C., June 30–July 1, 1993.

Swedish Environmental Research Institute. 1991. *Environmental Indices for Swedish Industry*. Swedish Environmental Research Institute, Gothenburg, Sweden.

28

Finding and Implementing Projects that Reduce Waste

Kenneth Nelson

Abstract

Success stories are important to share, especially when they contradict the assumptions of simple economic theory. Annual waste reduction contests among employees of the Louisiana Division of The Dow Chemical Company have been held for the past decade. This contest has continued to find significant, highly cost-effective energy- and materials-saving projects each year, implying that even well-managed firms do not automatically optimize their use of resources. The additional efficiencies squeezed out of the firm's plants suggest that great potential exists to improve the efficiency of the industrial sector, if appropriate leadership is provided and if organizational and other internal barriers can be overcome.

The Louisiana Division Program

Keeping employees interested in saving energy and reducing waste is a constant challenge. Ideally, all employees should be personally committed to a philosophy of continuous improvement. Improving operations should be seen as part of their normal jobs.

The Louisiana Division of The Dow Chemical Company began an energy conservation program in 1981. It took the form of an annual contest for employees to improve energy efficiency. In 1983, the scope of the contest was expanded to include yield improvement, and in 1987, Dow's Waste Reduction Always Pays (WRAP) program was added. The contest has been enormously successful in motivating employees to find ways of saving energy and reducing waste. In the late 1970s and early 1980s many companies jumped on the energy conservation bandwagon and started energy conservation programs. But as energy prices dropped, most of these programs were phased out. The program at Dow's Louisiana Division is an exception. Not only did it not disappear, it is far stronger now than it was ten years ago.

The Louisiana Division of The Dow Chemical Company is located in Plaquemine, Louisiana, ten miles from Baton Rouge, the state capital. The division has about 2500 employees and more than 20 plants making products, such as ethylene, propylene, chlorine, caustic, polyethylenes, glycol ethers, and

371

chlorinated solvents. It supplies internal steam and power needs via cogeneration, primarily using gas turbine combined cycle units.

In 1981, when fuel gas prices were rising and projected to rise even further, Dow's Louisiana Division, like most companies, set up an energy conservation program. It took the form of an annual "Energy Contest," and a newly formed energy evaluation committee took on the task of administering it. The contest was aimed at capital projects, and the requirements were simple:

- Projects had to cost less than $200,000. This simplified and expedited the authorization process.
- Projects had to have a return on investment (ROI) greater than 100%. While this might seem high now, most companies had the same requirement in the early 1980s. It was unclear how long fuel prices would remain high, and the philosophy was to get our money back in a year or less.
- Routine maintenance projects did not qualify.

The contest was aimed at engineers in the production plants and began without much fanfare. It was announced in late 1981, appropriate forms were distributed, and a deadline was set. Since projects were to be funded in 1982, the contest was called the 1982 Energy Contest.

A total of 39 projects were submitted. Members of the energy evaluation committee met with the plants and reviewed each entry in detail. When the reviews were completed, only 27 projects were approved. Half of the claimed energy savings (as submitted in the original entries) did not exist. Still, the final results were impressive: 27 winners, a capital investment of $1.7 million, and a 173% return on investment.

Once these 27 projects were found, many believed there could be no others with such high returns, but the contest was allowed to continue for another year. The 1983 Energy Contest was run in a similar manner and produced even more spectacular results: 32 winners, capital investment of $2.2 million, and a 340% return.

Some changes were then made. First, the general manager removed the $200,000 limit. "If there's a million-dollar project out there with a 100% return on investment, we ought to be looking for it," he said. Second, he expanded the scope of the contest to include *yield savings*—making the same amount of product using fewer raw materials. Today, we would call this *waste reduction*. Audits of projects after they are in operation were also begun. The 100% return on investment cutoff was maintained. Again the 1984 results were impressive: 38 winners, $4.0 million capital investment, and a 208% return.

A historical summary of the contest is presented in Table 1. Since 1984, the return on investment hurdle has been lowered, but the contest format has remained essentially the same. Three rule modifications were:

- Both projects requiring capital funds and projects that can be financed out of current-year budgets are now included.
- Projects must save at least $10,000 per year. This helps focus attention on large savings.

Table 1: Summary of winning projects with capital investment less than $2 million (ROI is return on investment)

	1982	1983	1984	1985	1986	1987	1988	1989	1990	1991	1992	1993
Winning Projects	27	32	38	59	60	90	94	64	115	108	109	140
Average ROI	173%	340%	208%	124%	106%	97%	182%	470%	122%	309%	305%	298%
ROI Cut-Off	100%	100%	100%	50%	40%	30%	30%	30%	30%	30%	50%	50%

- Projects that reduce maintenance costs are now included. Routine maintenance projects, however, still do not qualify. A project must eliminate or substantially reduce a maintenance problem.

At the request of the Louisiana Division Environmental Services department, Dow's WRAP program become a part of the 1988 contest. It was renamed the Energy/WRAP Contest. WRAP projects do not require a return on investment, but they must reduce waste. Adding emission control devices does not qualify. A historical summary of just the WRAP projects is presented in Table 2.

In practice, virtually all cost-saving projects have qualified as long as they have resulted in *real* savings greater than $10,000 per year. Participation in the Energy/WRAP Contest has continued to grow, and the last four years have been outstanding, with over 100 winning projects each year.

Gradually, until 1992, the return on investment (ROI) cut-off was lowered. This does not mean that a lower quality of project has been allowed. Rather, this has been a response to falling costs of fuel gas and a belief that they may rise again within the project life. For example, an energy saving project with a 30% return on investment in 1987 would have a 100% return on investment if evaluated at 1983 fuel prices.

The year 1988 was unusual in that most plants ran at capacity. Every extra pound of product meant extra profit. The contest was therefore deliberately downplayed that year so that plants could focus their efforts on production reliability. Although the contest was given virtually no publicity, participation remained excellent. Because many of the projects involved incremental capacity increases (high profit pounds), the average return on investment for the 1989 Energy/WRAP Contest was the highest ever achieved.

Credit was given for yield savings and waste reduction in earlier contests, but formally including the WRAP program in 1988 gave the contest an added boost. Although WRAP projects do not require a return on investment, all were above the 30% cutoff except for one project, which had a return on investment of 25%. In Table 2, note the high average returns of WRAP projects. Even though many projects are received with low or negative returns on investment, the weighted average return is usually over 100%. Waste reduction can really pay.

Table 2: *Summary of WRAP projects with capital investment less than $2 million (ROI is return on investment)*

	1988	1989	1990	1991	1992	1993
WRAP Projects with ROIs above Cut-Off*	23	22	37	35	35	39
WRAP Projects with ROIs below Cut-Off	1	3	16	18	19	28
Total Winning WRAP Projects	24	25	53	53	54	67
Average ROI of WRAP Projects	200%	106%	108%	97%	111%	82%

*These projects are also included in Table 1 and Table 3.

Table 3 and Figure 1 summarize the cost and annual savings from virtually all contest projects.[1] One of the curious results of requiring a $10,000 savings rather than a minimum cost has been the number of infinite return on investment projects—those requiring no capital or expense funds. The term infinite return on investment, of course, is an oxymoron since no return on investment is possible without investment. But the phrase has caught on, and having such projects is a distinction. It has also allowed the recognition of work by computer programmers, often the unsung heroes in a production plant. In the past three years, 97 infinite return on investment projects have been implemented. Cumulatively they save more than $16.6 million per year.

Audits occur after a project has been implemented and running for a period of time. The purpose of audits is to compare each project's actual cost and performance with the values used to justify the project in the original contest application. Every project is audited.

Fluctuating fuel gas costs or product prices are not important in audits. The main interest is in actual project costs, the pounds or BTUs saved, and anything new that was learned. Through 1993, 575 projects were audited. The average return on investment was 204% vs. 202% predicted after review by the energy evaluation committee. On average, projects perform a little better than expected. Total audited savings for the 575 projects are over $110 million per year.

There have been very few technology breakthroughs. Nearly all projects are applications of old principles. The types of projects range from simple changes in the way plants are controlled to major equipment revisions. Many of the energy-saving and waste-reduction ideas have been published (Nelson, 1990a, 1990b). Some general suggestions follow:

- It pays to work with the supplier of a raw material to reduce the impurity levels. Impurities often are converted to waste products.
- A distributing reactor feeds better. Adding a properly designed distributor at the inlet of a reactor vessel improves the uniformity of flow through the reactor, improving conversion and increasing yield.
- Providing a separate, smaller reactor for recycle streams allows optimization of temperature and pressure to maximize conversion of the recycle streams to desired products.
- Recycling seal flushes and purges back into the process reduces waste. This can usually be done with little difficulty.
- Preheating the feed to a distillation column reduces energy requirements and improves the efficiency and quality of distillation.

---

[1] Although the capital investment per project is no longer limited, projects costing more than $2 million are not included in Tables 1, 2, and 3. There have been 14 such projects over the past 12 years. Including them would inappropriately skew the data. These projects are included in Figure 1.

Table 3: *Summary of winning contest projects (project costs of less than $2 million) (in thousands of dollars per year)*

	1982	1983	1984	1985	1986	1987	1988	1989	1990	1991	1992	1993
Cost	1700	2200	4000	7100	7100	10,600	9300	7500	13,100	8600	6400	9100
Savings												
Fuel gas	2970	7650	6903	7533	7136	5530	4171	3050	5113	2109	5167	4586
Yield & capacity	83	−63	1506	2498	789	3747	13,368	32,735	8656	17,909	11,645	20,311
Maintenance	10	45	−59	187	357	2206	583	1121	1675	2358	2947	2756
Miscellaneous	0	0	0	0	0	19	−98	154	2130	5270	518	788
Total	3063	7632	8350	10,218	8291	11,502	18,024	37,060	17,575	27,647	20,277	28,440

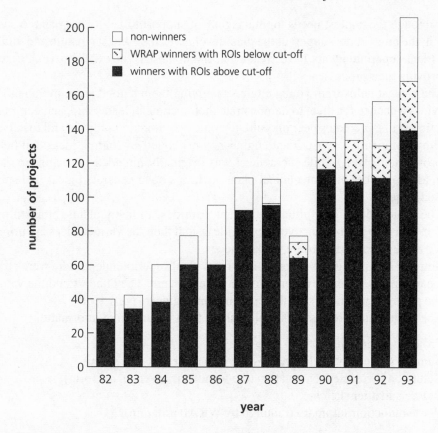

Figure 1. Twelve-year history of Louisiana Division Contests. ROI is return on investment.

Why Does the Program Work?

The success of the Louisiana Division program depends on management support, the composition of the energy evaluation committee, and grass-roots support. The program receives continuing support from management because it does not conflict with other objectives or organizational responsibilities. When the program began in 1981, it did not require a new department, redeploying people, or a multi-million-dollar budget. Rather than setting up separate functions, the goal was to use people in their existing positions and to make project development and evaluation an integral part of their jobs.

Every program needs a focal point, and the energy evaluation committee is the focal point for the contest. The committee includes representatives from process engineering, economic evaluation, and various production areas. Having a representative from economic evaluation is extremely important. The committee needs a person who knows the procedures for getting projects authorized, and who works with managers in allocating the division's capital resources. We want to work *within* the system, taking advantage of the existing procedures and framework.

377

Although the contest needs top management support to get it going and to sustain it, the grass-roots support at the plant level gives the contest strength and vitality. Plant superintendents have found that the contest is an excellent vehicle for improving their processes.

The contest helps train young engineers, giving them a good sense of values. It provides a chance for them to demonstrate their technical, leadership, and communication skills. It creates peer pressure to come up with good projects and fosters a healthy, competitive spirit. Plant engineers work together, sharing ideas that help improve their plant. While the contest was originally intended for engineers, an increasing number of nontechnical plant operators have generated ideas and submitted projects.

The contest does not conflict with plant priorities. In many plants, brainstorming sessions are used to uncover project ideas, and then the various ideas are prioritized and evaluated according to plant needs.

Some of our strongest support comes from plant superintendents who were contest participants when they were production engineers. They understand the value of the program and make sure their production engineers participate.

The contest entry forms are simple, asking for a minimum of information:

- Project description
- Summary of utility, yield, and cost savings (unit costs are given)
- Return on investment calculation (the formula is given on the form)
- Before and after sketches
- Waste reduction summary (required by WRAP program).

Only one copy need be submitted. Minor corrections and updates are made on the original. The contest receives compliments for its minimal paperwork.

Project reviews have become an important part of the Energy/WRAP Contest. Nobody wants to install a project that does not work. Our purpose, then, is to ensure, as best we can, that projects will perform as expected. Typically, four or five members of the energy evaluation committee attend a review.

Throughout the review process, the focus is on projects, not people. When there are errors, our job is to correct them, not to attack the people who made them. Having experienced people on the energy evaluation committee is important. They understand the importance of helping people explain and develop their ideas. The tone during the reviews must be supportive. If a project is "shot down," we want the submitters to feel thankful that we stopped them from wasting their time installing a project that was uneconomical or would not work. This sense of working together to a common end must prevail.

In the early days of the contest, a few plant superintendents did not like the idea of "outsiders" coming into their plant and evaluating their projects. Maintaining the integrity of reviews, requiring people to justify their projects, and treating everyone equally helped plant superintendents to appreciate that the objective was to help them, not to be an obstacle. As one plant superintendent put it: "You give us a hard time, but then, you give everybody a hard time."

While the Energy/WRAP Contest has become an important tool for uncovering and prioritizing projects, neither the contest nor the energy evaluation committee controls any capital. That function rightly remains in the hands of the managers and is administered through the economic evaluation/capital planning department. The contest avoids conflict by being integrated into the budgeting process and into decisions about capital needs and project timing. Many plants view the contest as a good way to get capital for projects because funds are normally made available for contest winners.

After seeing the spectacular results of the contest, many people ask what cash awards we give the winners. There are none—at least none that are directly related to the contest. Instead, a formal awards ceremony is held in which winners are presented with engraved plaques.

There are several reasons we have avoided any type of cash award:

- It is impossible to be fair. Who thought up the idea? Who worked on it? Who implemented it? How is the size of the award determined? Is a $10,000 project with a 1000% return on investment better than a $1 million project with a 50% return on investment? And what if the project does not succeed? If we wait until a project is installed and operating, the motivational benefits are lost, and personnel changes may occur that further obscure who should receive cash awards.
- Cash awards for contest projects can be demotivating. People with outstanding projects will feel they should have received more, and those who were unable to submit projects because of their particular work situation, such as assignment to a team building a new plant, may feel cheated.
- In a plant situation, rewarding individuals with cash will quickly inhibit communications and stifle teamwork. Employees will be reluctant to share information and ideas, fearing that someone else will get the credit and the cash.
- Within any company, there are established procedures for giving monetary rewards for good performance. Normally, each individual's supervisor has the primary input on the size of that reward, whether it is in the form of an annual salary increase or a bonus. If a separate system is established, such as cash awards for contest winners, this system literally competes with the boss for an employee's loyalty, resulting in conflicting project priorities. The effect may be subtle, but the net outcome will be loss of support from supervisors.
- Cash awards imply that finding significant improvements is not part of the regular job. This is exactly the opposite of the message we want to give. We want employees to feel that continuous improvement is a normal and important part of their jobs, not something separate. This may be the most important reason of all.

People who submit good projects are appropriately rewarded, but within the existing system. An individual's supervisor is in the best position to evaluate his or her contribution to improved plant operations (through the Energy/WRAP Contest or in other ways) and to put it into the context of total job performance.

Having a committee rather than a special office administer the contest avoids creating new levels of hierarchy or bureaucracy. The committee works with existing departments and follows existing procedures whenever possible. The annual Energy/WRAP Contest has become an integral part of those procedures—a part of our culture—adding to grass-roots support. It has not become a threat to anyone's job security.

Part of the energy evaluation committee's strategy is to see that credit for projects goes to the people and plants that thought them up and implemented them. Neither the contest nor the committee takes credit for savings.

The educational and training benefits of the Energy/WRAP Contest are subtle, yet substantial. In addition to learning the technical design aspects of a project, many engineers are able to follow their project from the idea stage through preliminary evaluation, formal evaluation, authorization, engineering, construction, and startup. They learn how to get through the "system." Along the way, they gain valuable experience in dealing with various departments, disciplines, vendors, other engineers, supervisors, and operators. This develops technical ability, people skills, and self-confidence.

Documenting why projects fail to live up to their expectations helps avoid making the same mistake twice (and keeps the energy evaluation committee humble). Audits were begun back in 1983 at the request of the division's general manager, who may have been skeptical about the high returns on investments. Some resistance from the plants was anticipated, but it never materialized. Instead, the auditing of project performance was strongly supported by our plant superintendents. Furthermore, being able to document project performance has given the Energy/WRAP Contest an enormous amount of credibility and has helped sustain its vitality.

No numerical goals, such as number of participants, number of projects, dollars spent, or dollars saved have ever been set. Instead, it was recognized early that the important thing to work on was the mechanism—the process—by which projects are conceived, designed, and implemented. We needed to provide an organizational structure that functioned smoothly to make things happen. Once an effective mechanism is in place, specific objectives and goals are unnecessary.

The contest has often been cited for the professional way in which it is run. Part of that professionalism includes a lack of gimmicks. Slogans, posters, decals, etc., to publicize saving energy and reducing waste have been avoided. There is nothing intrinsically wrong with such items, but they do little, if anything, to generate additional projects.

We have found three highly effective means of stimulating good project ideas:

- A complete listing of all Energy/WRAP Contest projects, both winners and nonwinners (we do not call them "losers") is published every year. Project descriptions are designed to give an understanding of the principles involved without including many details. This summary is read by almost everyone involved with the contest. They want to know what others are doing, and often find ideas applicable to their own plant—to submit in the next contest.

380

- A *Waste Elimination Idea Book* is published, organized by subject—such as pumps, heat exchangers, distillation, and compressors. Project ideas are taken from past contests and from work done at other Dow locations. The book is regularly updated.
- A two-day Continuous Improvement Workshop is held every year or two, aimed at giving people practical ideas for improving plant processes. It includes analytical, creative, and pragmatic approaches to finding cost-saving projects as well as specific information about various unit operations. One session, "How Winners Think," is based on interviews with our top 12 project generators, who describe how they came up with so many good projects year after year. Workshop participants are challenged to apply the same techniques in their own plants and to generate at least one new project. The group reconvenes after two weeks to discuss what each participant experienced when trying to apply the principles learned in the workshop. Response to the workshop has been excellent.

Conclusions

Hundreds of excellent projects have been implemented, and on the average they perform better than expected. The program is supported by the division general manager, managers, plant superintendents, plant engineers, and operators. It has become part of our culture. Finding improvements is accepted as an important part of the job.

We work in an atmosphere where coming up with projects is fun and challenging. There is a minimum of red tape. Teamwork and cooperation within and among plants flourishes, and we continually build momentum toward bigger and better projects with higher returns on investment. The number of projects with high returns on investment submitted each year is an ongoing source of amazement.

People often ask why such a program is not implemented more widely. This is an interesting and challenging question. There appear to be three main obstacles. The first is fear of losing control over funds used for financing projects. This fear is ungrounded. We have deliberately avoided controlling or distributing money. That function remains in the hands of our managers. They are not forced to spend money on contest projects.

The second obstacle is related to priority setting. Some managers and plant superintendents may feel that low-priority projects will be submitted and approved. This conflict is easily avoided by holding plant meetings where employees work together generating and prioritizing projects. In many cases, plant superintendents review contest projects before they are submitted.

The third obstacle has something to do with the abundance of high-return projects. Perhaps admitting that such a multitude of untapped projects exists is seen as an indictment of existing management and management policies. It shouldn't be, and fortunately has never been a problem in Dow's Louisiana Division. Belief that good projects exist may not ensure that they will be found, but disbelief and inaction certainly ensures that they will not.

One final recommendation. A bottom-up program like the Louisiana division contest needs a champion—someone who is committed to its purpose and works to make it prevail.

Addendum

It is important to recognize that *all* waste reduction projects are not cost effective and that it takes time to develop waste reduction technology. Such technology is highly desirable because it *permanently* reduces waste.

In the past, regulatory emphasis and mandates have been directed at waste *treatment* rather than waste *reduction*, because treatment technology is known and conversion of undesirable products is virtually complete.

Treatment devices, however, are usually very expensive, they are *not* cost effective for the firm (they raise product costs), and they incur on-going operating expenses. Waste *reduction* options should be thoroughly exhausted before treatment technology is turned to. Recognize, however, that treatment is sometimes the best option.

It is hoped that regulatory requirements for further treatment devices will be limited to those areas where there is a *clear* and *urgent* need. This will enable our stretched and limited resources to be more effectively directed at reducing waste.

References

Nelson, K. E. 1990a. Reduce waste, increase profits. *Chemtech 20(8)*, 476–482.

Nelson, K. E. 1990b. Use these ideas to cut waste. *Hydrocarbon Processing 69(3)*, 93–98.

29

Free-Lunch Economics for Industrial Ecologists

Theodore Panayotou and Clifford Zinnes

Abstract

The industrial ecology debate is examined in terms of economic games between firms and their employees. On the one hand, investments for better industrial ecology are subject to "incentive-compatibility" problems that could stymie their implementation; on the other hand, a great deal of improvement may be achievable by taking advantage of existing win–win situations ("free lunches"), i.e., changes that improve both economic growth and the environment. Moreover, properly assessing the economics of industrial ecology requires a careful treatment of dynamic issues. For example, the total cost of pollution abatement becomes dependent on the historical sequence of regulation.

Descriptively, industrial ecology is the study of the interactions among industries and between industries and their environment; it seeks to understand what industry is doing to itself and to the environment in which it operates. Prescriptively, some authors have stated, industrial ecology seeks to "optimize" the total materials cycle from virgin material, to finished material, to component, to product, to waste byproduct, and to ultimate disposal. Factors to be "optimized" include resources, energy, and capital. Prescriptive industrial ecology advocates a deliberate restructuring of industrial activity to achieve this optimization. The nature of this optimization, however, is not as simple as it first might seem. Does "optimization" imply minimizing resource use, minimizing environmental degradation, or ensuring that economic subsystems are environmentally sustainable? And what about a firm's profits? Are they also to be optimized or are they to be maximized? Each of these objectives leads to a different pattern of input use and production (see chapters by Graedel and Socolow in this volume).

Whether to describe or to prescribe, one must have an appropriate theory of the firm. In the theoretical, neoclassical world of competitive markets, the individual firm chooses its installed capacity and input and output mix so as to maximize profits (appropriately defined). The firm faces parametric input and output prices; its physical plant is constrained by a set of existing technology alternatives; and its future options are governed by a production function. This formulation has two implications: (1) no stone that will increase the firm's profit is left unturned, and (2) no actions or projects are undertaken that lower profits.

Neoclassical theory confirms that such profit maximizing may not, however, be in the public interest. Firms will overexploit costless inputs—resources such as air and water that are not privately owned, or at least are not given value by being traded in the marketplace. This is the well-known "tragedy of the commons" and corresponds to the economic concept of pollution as a negative "externality." If the society wants firms to generate less waste, treat waste further, and dispose of waste more safely, the society must make it profitable for firms to do so. For society at large, the solution is to create schemes that force the firm to internalize externalities to the greatest extent possible at the least cost. Among the available policy instruments are quantity rationing (permits and quotas), regulations (standards), and pricing (fiscal incentives via taxes and subsidies). Discussion of such policy tools is widely available (e.g., Pearce and Turner, 1990) and will not be pursued here.

A second problem—and one missed by the neoclassical theory of the firm—is that the firm does not act as a homogeneous decision-making unit. Extensions of the theory of the firm to describe conflicting objectives within the firm take the form of constrained optimization games among three optimizing agents: the individual firm (or its shareholders), the firm's employees, and the society. The economics of industrial organization extends the neoclassical theory of the firm and gives particularly relevant insights into the behavior of the firm when confronted with the challenge of implementing industrial ecology. A number of these extensions compose the subject of this article.

The Incentive–Compatibility Problem

Nelson (this volume) documents the existence of highly profitable, efficiency-enhancing, and pollution-reducing projects within well-managed firms that only surface with environmental regulations. That such projects remained previously undetected and untapped raises questions about the validity of the neoclassical theory of the firm, or at least calls for an explanation. From an industrial ecology perspective, the presence of "free lunches" suggests that the costs of industrial restructuring needed to reduce the environmental impacts of industry may not be as high as conventionally thought. It may be that, up to a point, what is good for ecology may be good for industry and that environmental regulations play the role of a reconciliation catalyst (outside coordinator).

Even more paradoxical is that regulations do not simply produce a rush of environmental projects submitted in the first year that the regulations are promulgated and none thereafter, but instead a steady—and even growing—stream of project proposals year after year. There could be several explanations for this paradox. First, it takes time for the extent of regulations to become known and their implications fully appreciated within the firm. Second, regulations are rarely introduced in full force overnight; usually they are gradually phased in over a period of time to allow for smooth adjustment. Third, there is learning-by-doing in identifying and formulating projects that are both profitable and in line with the regulations.

Fourth, environmental regulations tend to become stricter over time or new regulations are introduced to control previously unregulated pollutants. For all these reasons, "free lunches" are not a one-time affair, but a steady stream of profitable, yet previously untapped, opportunities.

It is here that the game between the firm's owners (e.g., the shareholders) and the employees comes into play. The firm's owners may be thought of as "principals" and the employees and managers as "agents," each motivated, typically, by agent-specific incentives. When the incentives of principals and agents are not convergent (what is called "incentive incompatibility"), inefficient or suboptimal investment decisions may result, including those in the environmental arena. Based on such a game perspective, the sections below explore a number of explanations for the paradox of free lunches within well-managed, competitive firms and their policy implications. The next section examines the simplest case: the acceptance process for a single project in which both agent and principal possess "full information." The following section shows how the presence of uncertainty further drives a wedge between agent and principal. A formal treatment of these issues can be found in Panayotou and Zinnes (1993).

Full Information

Consider a company with a manager and shareholders. Shareholders, who own the firm, offer the manager an employment contract with a base salary w_0 for running the firm's nondiscretionary activities on a daily basis. In addition, the shareholders offer an incentive bonus, B, to the manager for identifying, on a discretionary basis, potentially profitable projects and seeking permission to carry them out. We focus on a possible investment project with favorable environmental consequences.

We assume in this section that the shareholders can directly view the effort applied by the manager in implementing the project, that shareholders know the manager's utility function (see immediately below), and that the manager knows the firm's profit function. In short, we assume there is full information.

We first focus on the manager's view of the project, expressed in terms of the manager's *utility*, or satisfaction. A manager chooses to pursue a project rather than not to pursue it if her utility increases. By modeling the decision in terms of utility (instead of money) we first are able to account for the empirically observed fact that the consumption of each additional unit of a good brings smaller additional increases in satisfaction (decreasing marginal utility) and, second, allow for considerations, such as the loss of "free" time and the increase in hassle associated with extra effort.

A plot of her utility vs. her income is curved as shown in Figure 1. Let us further assume that the manager can attach a specific loss of utility, L, independent of income, to any discretionary project she considers undertaking. This disutility captures her view of the effort required to overcome resistance to change within the organization for projects in general (of which more shall be said below) as well as her view of the special effort required for the specific project. Figure 1 shows the

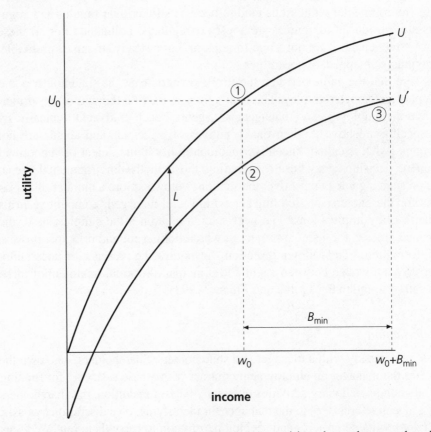

Figure 1. The manager's utility functions from income (comprising a base salary, w_o, plus a bonus) in a world where she does not undertake a project (upper line), U, and in a world where she does (lower line), U'. The second curve is displaced by the constant amount, L, which is her income-independent (by assumption) disutility for undertaking this project. If, with no project undertaken, she is at point 1, then she will be at point 2 by undertaking the project with no bonus (and she will not propose it). She will agree to undertake any project resulting in her utility increasing beyond U_o, i.e., her getting a bonus of at least B_{min} so as to land her somewhere along U' to the right of point 3.

manager's view of undertaking the project in terms of a pair of functions of utility versus income, the one, U', for undertaking the project a fixed distance, L, below the one, U, for not undertaking it. The minimum bonus, B_{min}, she would require to regard the project as worth undertaking is the amount that maintains her utility at its preproject level, U_o; both B_{min} and U_o are also shown in Figure 1.

From the point of view of the firm, we assume the project offers a certain return, R, that takes into account all costs other than the bonus to the manager, B, and a profit, $R–B$. Further, we assume that the profitability of the other activities of the firm is unaffected by whether or not such a project is undertaken. The firm, therefore, will regard the project as worth doing if the profit is greater than zero.

If the minimum bonus required by the manager is so large that the profit to the firm is negative, the project will not be undertaken. On the other hand, if the minimum bonus required by the manager is smaller than the project's return, then the project is profitable and the manager will submit it for consideration, it will be implemented, and the manager will receive the bonus. Note that it would never pay for the firm to offer a bonus greater than B_{min}.

It is clear from Figure 1 that the minimum bonus the manager requires to undertake a project gets smaller if there are changes in her world that reduce the disutility (L) that she attaches to undertaking the project. Consider three possible sources of change that could reduce that disutility: the introduction of regulation, an increase in her environmental awareness or the awareness of those working with her, and a change in the preferences of the firm's consumers toward products with better ecological properties.

Regulation

It is a characteristic of human nature that discretionary changes encounter more resistance than nondiscretionary ones. For example, if your boss asks you to wear a tie, you might resist less if you knew she was simply carrying out a new dress code of the company. Similarly, changes in a company to comply with a new environmental regulation will encounter less resistance than changes that seem to come from a manager out of the blue. In light of the discussion above, it is clear that a government environmental regulation, in reducing a manager's discretion (and therefore, indirectly, L) could both raise the profits of an already profitable firm (and/or increase the bonuses of managers) and lead previously unprofitable projects to become profitable. Assuming the new projects undertaken due to the regulation indeed improve the environment as the regulation intended, then a "free lunch" has been created; both the economy and the environment benefit.

Environmental Awareness

An increase in general environmental awareness could act in the same way regulations do: those who would normally impede change would be less likely to do so when they perceive change to be for a good cause. An increase in the environmental awareness of the manager herself (say, through education) could reduce her disutility, L, of proposing and implementing a project with environmental benefits. Either way, some projects cross the boundary from unprofitable to profitable.

Consumer Sovereignty

Customer demand for environmental improvements in the production process and in postconsumption pollution management, as well as competitor initiatives in response to these demands, reduce resistance to discretionary changes within the firm. Paton (this volume) describes how, at Hewlett-Packard and other firms, environmental issues are being introduced into product design through the articulation of a new corporate objective called design for environment. Like new regulations and heightened environmental awareness, such an evolution in corporate objectives is very likely to reduce the disutility the manager associates with a project bringing environmental benefits.

When the Project Outcome Is Uncertain

In the section above, we saw that it was possible for a "profitable" environmental project not to be carried out even when the outcome of the project was certain. This result, however, should not have been surprising. This situation is simply a variant of the case where an input of production is just too expensive to use as, for example, would be cement from a neighboring country due to transport costs. In the present case, the input is the manager's effort (and associated disutility of work) required to bring about the implementation of the project. In borderline cases, reducing the disutility of work (like reducing transport costs) would make the input affordable and the project profitable enough to undertake.

In reality, neither firms nor managers have full information about the future state of the world, so all projects have uncertain returns. As we shall see, how this risk is shared will determine whether potentially profitable environmental projects are implemented. In this section we discuss another class of principal–agent problems, where a risk-averse manager is required to assume some of the risk of a project's uncertain outcome. We find that, in such cases, a project is less likely to be undertaken than if all the risk is assumed by the firm.

We change two features of the example above. First, instead of an outcome with a return, R, to the firm with certainty, there are now two possible outcomes, successful projects with returns, R_s, and project failures with returns, R_f. For simplicity, we assume the two outcomes are equally likely and, to keep the connection to the previous example as close as possible, we assume that the expected return for this example, which is just the average of the two possible returns for the project, is the same as the certain return in the previous example: $R = (R_s + R_f)/2$. Second, to isolate the issue of risk, we assume that the manager experiences *no* disutility from the effort required for this second project, i.e., that $L = 0$.

This time, however, we assume that the manager bears some of the risk of the uncertain outcome of the project: if the project fails, her bonus will be less than if the project succeeds. To be specific, we assume the firm offers the manager a bonus contract such that she receives a (positive) bonus if the project succeeds and is *penalized* by a (positive) amount if it fails. Since no "effort" is involved in this project, the amounts of the bonus and penalty are set so that the expected bonus contract payment made by the firm to the manager equals zero. Since either project outcome is equally likely, a zero expected bonus contract payment means that the *size* of the bonus should equal that of the penalty; we denote this amount by B. As in the original problem, denote the manager's income by w_o; with $L = 0$, however, the relevant utility curve for the risky project now corresponds to the curve, U, of Figure 1 so that w_o yields a utility level of U_o as before.

For the project with the uncertain return, since the manager shares in the risk, we must now take into account (for the first time) the fact that the manager is risk averse. By risk averse we mean that the manager would refuse all bets whose expected payoff is zero. For example, the manager would find an even-money, 50:50 bet—say, where she stood to win or lose \$10,000—as offering *negative* expected utility since the increase in utility from an additional $10,000 is not as large as the decrease in utility from the loss of $10,000. That this is true is a direct consequence of the curvature of the manager's utility function, which in turn reflects diminishing marginal utility (as explained above). This can be easily illustrated using the U utility curve in Figure 2. Start at point 1 (where income is w_o yielding utility U_o) and compare the increase in utility, $U_s - U_o$, from receiving a bonus, B, with the decrease in utility, $U_o - U_f$, from being penalized by an equal amount, B, and you will see that the increase in utility is smaller than the decrease. Since the two cases are equally likely, this implies a net loss of expected utility. Clearly, the manager would never submit the project under such a bonus contract.

The amount of money the manager would require to participate in a particular bet with a zero expected return is called the *risk premium*. How big must the risk premium be in the present case in order to convince the manager to submit the risky project? It must be enough to make the manager indifferent between doing "nothing" and carrying out the present, risky project with the zero expected bonus: it must yield an expected *utility*—not an expected bonus—at least equal to the level of utility where she ignores the project, that is, U_o. Noting that the expected utility from the risky project, U_e, is equal to the utility from receiving w_1 in Figure 2 with certainty, the manager's risk premium for the risky project must be $P = w_o - w_1$ in the present case. Thus, by receiving an amount P regardless of the project's outcome *in addition to* the bonus contract described above, the manager would find the risky project minimally acceptable to carry out.

So would this risky project be implemented? If the manager were to have submitted the risky project (without receiving the risk premium), the firm's shareholders would have been happy to undertake it since the project would have had the same, positive, expected, net-of-bonus return, $R - 0$, as in the example with effort. (Note that this is due to our assumption that the firm is not risk averse, but risk neutral). The manager, on the other hand, is risk averse and must now bear some project risk. As we saw above, since the risky project's bonus contract implies a loss of expected utility, U_e, to the manager, i.e., a fall of $U_o - U_e$, the project would never be submitted as is—even if the expected return of the risky, effortless project is extremely large. Will the payment of the risk premium, P, overcome this? If $R - P \geq 0$, i.e., if the risk premium is not so large as to make negative the expected profitability of undertaking the project, then the answer is yes; otherwise, there will be no profitable contract the firm could offer that would elicit the risky project from the manager. Clearly, because of the risk premium, there could be borderline, risky projects where the firm could profitably offer an acceptable bonus to the manager when the project entails no risk for the manager, but where there would be no bonus contract attractive enough to be acceptable to the manager and, at the same time, provide positive expected profits to the firm.

To review, we have been discussing an example of an environmentally favorable project with a positive expected profit which, due to manager risk aversion, the manager would not submit to the shareholders for approval because its associated bonus scheme would provide her a lower expected utility than the utility she

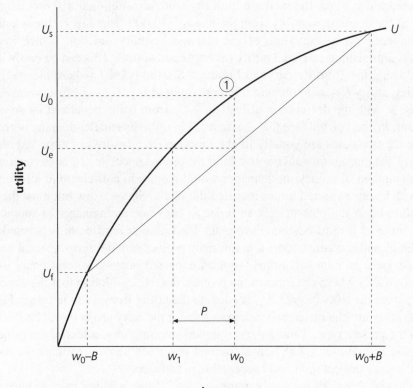

Figure 2. The risk-averse manager has a utility function, U (the same U function as in Figure 1), and a preproject income of w_0. Faced with the opportunity of doing a project where there is a 50:50 chance of getting a bonus, B, or being "fined" by an equal amount, this manager will refuse because her utility would be increased by $U_s - U_0$ if she wins, but be lowered by a *larger* amount, $U_0 - U_f$, if she loses. We call $U_e = (U_s + U_f)/2$ the expected utility of the bonus contract (average of the utilities of the two possible outcomes). Since U_e is lower than U_0, the contract entails a loss of expected utility for the manager. The bet feels to the manager like an income loss, $w_0 - w_1$, since her expected utility, U_e, as a result of the project is equal to that level which she would have by receiving with certainty an income, w_1, as shown in the figure. We call $P = w_0 - w_1$ the "risk premium"; it is the side payment the manager would require to be indifferent between the bet (the risky project) and an income of w_0 with certainty. Thus, for risky projects, the minimum bonus she will accept is increased by such a risk premium relative to the same project with all risk borne by the firm.

would have by ignoring the project opportunity. This is because managers, being averse to risk, seek some insurance in their compensation against bad outcomes that result in spite of having tried hard. Were the same project to have the same expected bonus, but with certainty, it would be submitted by the manager—and approved by the shareholders.

The phenomenon we are describing is, of course, quite independent of our many special assumptions. It is an inherent consequence of the manager's risk aversion, i.e., that a utility function exhibits decreasing marginal utility. The policy implications are clear: the more risk the manager is forced to bear, the fewer profitable projects she will initiate.

Why then doesn't the firm simply shift all of the risk off the manager and onto itself? The reason is that a firm's stockholders can never tell for certain how hard a manager has tried even when they observe the higher profits; the good results may be from her hard efforts or from good luck in spite of managerial shirking (and, similarly, low profits could be due to bad luck with hard efforts by the manager or due to excessive shirking's overcoming good luck). Though we have not proved it here, it may be shown (Tirole, 1988: 35–41) that the firm has an interest in having the manager share risk: the less risk the manager bears, the greater is her incentive to shirk and claim that failure (if it occurs) was due to the uncertainty of the project. The only way open to the firm to ensure that the manager tries as hard as she can is to require her to bear some of the risk. All of these complications in real situations make it unlikely that a manager will share none of the risk of a project with an uncertain outcome.

Most real environmental projects will involve both managerial effort *and* managerial risk. Inducing the manager to submit such projects will require that managers be compensated both for the risk and for their efforts. Computation of the bonus contract in this case is given in Panayotou and Zinnes (1993) and in Tirole (1988).

Other Principal–Agent Problems

Many other issues can act as a wedge between the interests of managers (agents) and shareholders (principals) so that profitable environmental projects are not carried out. In this subsection, we briefly discuss some of these in the framework of principal–agent problems.

- If the manager is faced with a time constraint to evaluate projects, then in her rush she could overlook potentially profitable projects.
- If the manager has a time horizon shorter than the period required to generate the full flow of returns (and, therefore, bonuses) from the project—say, because of the intention to leave for a better job—then she may not undertake the potentially profitable project.
- If there are many managers, each undertaking projects, then it may be hard for the shareholders to identify the contribution to profits of each manager's project. This could lower the bonus offered for successful projects and, thereby, reduce the number of projects carried out.
- A manager could recommend environmental projects that are profitable, but be ignored due to inertia within the firm.
- If the manager or shareholders possess poor information, e.g., undue pessimism about the returns to an environmentally favorable project, then the project may not be undertaken. Moreover, poor information typically increases the riskiness of a project, which, under most incentive schemes, also increases the riskiness of a manager's bonus, thereby lowering the probability that the project would be submitted.

391

- As should be apparent from the foregoing discussion, it is crucial that the shareholders offer the "optimal" incentive structure to the manager. For instance, in our first example (with full information), if the shareholders were to offer the manager a bonus in the form of a share of profits, and that share were less than B_{min}/R, then the manager would calculate her bonus to be less than B_{min} and would refuse to participate; a mutually profitable opportunity would be lost. Bad incentives can leave "free lunches" around to be discovered, but a theory of "free lunches" still requires an explanation of why shareholders would offer bad incentives in the first place. Such an explanation would probably return to some of the other issues mentioned in this article.

- When principal and agent possess different information, or when one of them is lacking critical information about the other, a potentially (mutually) profitable project may not be undertaken. Note that, in the previous full-information case, the shareholders know the manager's utility function and the amount of disutility she attaches to any project, and therefore can calculate her minimum bonus for a given project as accurately as she can. In the absence of such insight into her preferences, how can the shareholders know how much to pay the manager? Asking her may not be adequate, since the manager will have an incentive to misrepresent her true preference—for example, to overstate her minimum acceptable bonus.

- If the firm exists in an oligopolistic market (few firms, noncompetitive prices), it is tempting to argue that some stones may remain unturned in the search for profits. This, however, is not a convincing argument, unless it can also be shown that the shareholders have become saturated with the profits (and the consumption they derive from them) as things stand. Such a state of affairs seemingly violates human nature. Thus, if profitable projects are not undertaken in an oligopolistic industry, it must be due to the presence of at least some other distortion or rigidity, such as those we have just discussed.

Dynamic Considerations

In this section we illustrate how new features of the free-lunch problem emerge when placed in a dynamic context—a context that considers decisions made in some sequence over time. We first present the dynamics of the example developed above. We then consider some pitfalls of regulation which emerge only in a dynamic context. We close by touching upon several additional economic mechanisms best appreciated in a dynamic context.

The Flow of Projects

As described above, incentive schemes to save energy or reduce waste do not simply produce a rash of submitted projects in Year One and none thereafter. Rather, a steady flow of project proposals appears year after year. In this section, we provide an explanation as to why this might occur in the full information case. A formal treatment can be found in Panayotou and Zinnes (1993).

Consider the same framework as in the earlier case, only now assume that several employees of a firm (whom, for consistency, we shall continue to call managers) can submit projects. We assume that every potential project has the same return to the firm, and we further assume that all managers have identical utility functions. This time, however, we assume that what differentiates projects is the amount of effort required of a manager to implement them. We rank the projects in an order that is, equivalently, either from least to most disutility as perceived by the manager, or from smallest to largest minimum bonus as perceived by the manager; or from most to least profitable, $R - B_{min}$, where B_{min} is now project-specific.

As asserted above, regulations can make previously unprofitable projects profitable by lowering the disutility associated with organizational resistance perceived by the manager. We assume that all projects are initially faced with the same organizational resistance and, further, that as this resistance falls over time, the original ranking of projects is preserved. Clearly, then, a regulation that lowers organizational resistance bit by bit over several years will result in a flow of projects following the order just described: beginning with the unprofitable project originally closest to the border of profitability, and moving on through projects originally more and more unprofitable.

This is heuristically illustrated in Figure 3. Recall that since each project was assumed to have the same R, only projects with a B_{min} less than or equal to R would be profitable and, thus, submitted by the manager. Over time, however, the effect of regulation leads L to fall from L' to L'', causing the relevant project utility curve to converge upward from U' to U'' and toward the no-project utility locus, U. In this process, we see how the required B_{min} falls from B' to B'', thereby, creating a flow of projects to be submitted and implemented.

Regulation Dynamics

Industrial ecology carries implications for the dynamics of environmental regulation. Suppose a regulation intended to reduce one form of pollution brings about investments that, directly or indirectly, increase some other form of pollution; suppose, further, that to address the new pollution, a second regulation is promulgated that requires still further investments. In such a situation, the total cost of pollution abatement will depend on the historical sequence of regulation, i.e., the cost of regulation becomes *path dependent*.

To be specific, consider a downstream firm whose production process pollutes. (Here, of course, "upstream" and "downstream" should be taken in the input–output sense, and not in the sense of up or down a river.) New regulations necessitate an additional investment to reduce the level of some pollutant. The firm faces two alternative investments, each involving a capital expenditure and the purchase of an additional input. Investment A requires a fixed cost of $7000 while investment B requires a fixed cost of $10,000. For each unit of the firm's output, investment A requires the purchase of one unit of an additional input X, at a cost of $0.06;

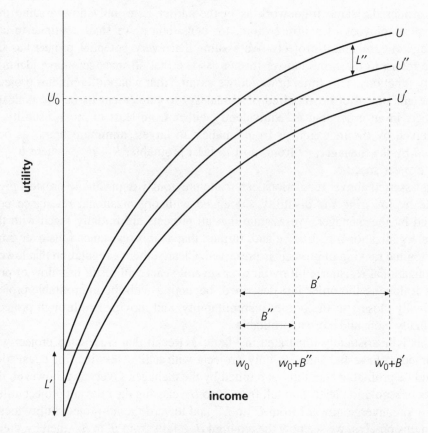

Figure 3. Regulations can lead to a flow of projects over time as, for example, organizational resistance decreases. This is illustrated here by showing how the manager's efforts fall over time from L' to L'', converging toward an L of zero with the fall in this resistance. This corresponds to the project utility curve U' rising to U'' and continuing toward the no-project utility curve, U. As the utility curves rise, the minimum required bonus also falls as shown in the figure by a concomitant fall from B' to B''. Note that U_o corresponds to the manager's utility level prior to considering a project; any project that is submitted would have to achieve at least this level of her utility.

investment B requires the purchase of one unit of a different additional input Y, also at a cost of \$0.06. Investment A will raise the unit price of the firm's output from \$1.00 to \$1.09, while investment B will raise it to \$1.10. The firm knows that if the price of its output reaches \$1.11, it will suddenly lose most of its market. The firm quite naturally decides for investment A.

Unfortunately, a year later, regulators decide to require that the upstream producer of X reduce the level of pollution resulting from its production process. This ultimately causes each firm in the upstream sector to invest \$50,000, which leads the unit price of X to rise to \$0.08. The downstream firm is left with two unpleasant options: it can either raise its output price to \$1.11, with the consequence of a drastic loss of market share, or it can dismantle investment A and invest in alternative B. It does the latter.

In this example, regulating the downstream sector in advance of the upstream sector caused a downstream firm to invest in a technology that, in the long run, was more costly. Rather than the regulations costing $60,000 ($10,000, then 50,000), they cost $67,000 ($7000, then $50,000, then $10,000).

This example makes clear that the regulatory approach may not be cost-effective if applied to sectors of the economy in a random (i.e., noneconomic) order. Some orderings are better than others:

- Best: Multisectoral environmental reforms should be announced (though not necessarily carried out) all at once.
- Second best: Upstream sectors should be regulated before downstream sectors.

It might be asked why the downstream firm did not anticipate the impending regulation on its upstream supplier. One obvious answer is that what is, or will be, going on in the heads of regulators is not always common knowledge. This is especially true when applied to a sector for which a firm is neither a producer nor a long-time consumer, as in the present case. A less obvious answer involves the possibility of collective behavior in sectors containing many firms. Suppose all downstream firms decide to implement investment A at the same time, together substantially increasing the total output of X (and the associated pollution from the production of X) to such an extent as to incur greater regulation of X. An individual downstream firm might not anticipate such collective effects.

Other Dynamic Free Lunches

Other dynamic economic mechanisms may lead to free lunches. We consider two classes here: coordination problems and network externalities.

Coordination Problems

Consider a region (or a country) where there is a preference for a wildlife reserve that does not yet exist. It is likely that a single individual, contemplating taking action on his own to bring the reserve into existence, would discover that his extra disutility for doing the work would exceed his extra utility for enjoying the reserve (and having others enjoy the reserve), once the work were done. Aspects of his disutility include bearing all of the transactions' costs and accepting the uncertainty of success.

Nevertheless, for the group as a whole, the net benefits of establishing the reserve could easily be positive. In such situations, there is a clear role for an agency to coordinate the demands of the group. Typically, this role is filled by a government agency, a nongovernmental organization (NGO), or a trade association. These institutions overcome the coordination problem and provide their members a free lunch, and those outside a free lunch and *free ride*—unless the reserve then charges admission fees.

Coordination issues not only can arise for public goods, as illustrated in the example above, they also are relevant for private goods. Consider an example in the international arena. Suppose all the developed countries want environmentally clean packaging (which for this example we assume is more expensive), but none wants to reduce its (short-run) competitiveness by having its firms be the only ones to sell a more expensive product. International agreements can overcome this coordination problem.

Network Externalities

A network externality occurs when a good generates more utility the more consumers there are that purchase it. A classic example is the telephone or the fax machine, where each unit increases in value as more and more units are deployed. There are costs to being the first on the block to buy some product, or to being the first of a group of mutually competitive firms to make some investment.

Such externalities can affect both the demand for and the supply of new technology. Consider, first, a demand-side network externality. Each of a set of competing firms is considering building, on its own, an auto fleet with very low emissions of traditional air pollutants. Each faces a choice among many alternative technologies. The first firm to act must try to anticipate which technology will be widely used. It could make a costly mistake if it were to choose one technology and, later, the other firms were to choose a different one. The first firm could face higher costs for maintenance and second-generation improvements than its competitors. Though the first firm would like to coordinate its decision with its competitors, the others may have different preferences regarding the best technology, or they may simply prefer to hold back and wait for additional information. (Furthermore, collusion among competitors may be forbidden by antitrust laws.)

One result of this externality is that competing firms often display excess inertia, selecting a technology very late. Alternatively, the externality may lead some firms to display excess momentum, rushing into a potentially inferior technology prematurely. Standard setting either by the market or by regulation may overcome this externality by helping the competitors to make compatible choices.

The story is similar for the effect of network externalities on suppliers. Each of the suppliers of mutually incompatible technologies has an incentive to expand its installed base so as to reduce unit costs. This creates problems for the economy as a whole: in an industry with decreasing costs, incompatibility raises the cost of all products. Standard setting by the government or market to reduce incompatibility would appear to be advisable. However, there is a risk that standardization will repress diversity. Especially when technology is changing rapidly, standardization can result in the adoption of a technology that, with hindsight, is found to be inefficient relative to alternatives.

As an example, consider the case of alternative energy sources for automobiles, such as ethanol, natural gas, diesel fuel, electric batteries, or fuel cells based on hydrogen or methanol. The cost of supply for each alternative falls as the user base expands. On the one hand, standardization on one or just a small number of approaches can reduce the considerable costs sustained by society when it pursues all approaches at once—costs in evidence at a service station offering a multitude of fuels. On the other hand, standardization closes out options. Given that it is still far from clear in this fast-paced sector which approach is best—witness the lack of convergence in the case of battery technology for electric automobiles—one wishes to avoid the problem of "lock-in," where commitment is made prematurely. Interestingly, standardization can actually prevent a more environmentally preferable product from being competitive if it is a late entry into the market and cannot capture the necessary market share to achieve sufficient economies of scale in production to compete with established, more polluting alternatives.

Conclusions

Though industrial ecology concerns more than what goes on inside a firm, its practice ultimately depends on the identification and exploitation of opportunities for minimizing waste and saving energy and materials within the firm. Apparently, there are countless such opportunities, many of them quite profitable, that are not exploited by the managers of profit-maximizing firms for reasons that range from organizational resistance to change, to principal–agent problems, to coordination failure among competitors.

In this article we have provided a few explanations for why *rational* economic agents let such gains go unrealized. We have also described a few ways of inducing actions that achieve these gains. For example, the introduction of environmental regulations, by removing discretion, can reduce resistance to change and catalyze a continuous process of identification and exploitation of hitherto untapped profitable opportunities that also advance industrial ecology. It is the existence of these win–win opportunities which allow industrial firms to do well while they are doing good—free lunches—that offers the best hope that industrial ecology will not remain merely an ideal.

Acknowledgments

Research was supported by a research grant from the Harvard Institute for International Development. We appreciate the helpful discussions with Hafeez Shaikh and Pradeep Srivastava. We also thank Rob Socolow for his extensive and substantive editorial suggestions. All errors and omissions are our own.

References

Panayotou, T., and C. Zinnes. 1993. Incentive structure and regulation dynamics in industrial ecology. Development Discussion Paper No. 454, Harvard Institute for International Development, Cambridge, Massachusetts.

Pearce, D., and E. Turner. 1990. *Economics of Natural Resources and the Environment*. Harvester Wheatsheaf, New York.

Tirole, J. 1988. *The Theory of Industrial Organization*. MIT Press, Cambridge, Massachusetts.

PART 5

INDUSTRIAL ECOLOGY IN POLICY-MAKING

30

Introduction

The Editors

Government interventions in the marketplace are often controversial and problematic, especially for an objective as bold as "eco-restructuring." The five chapters in this part of the book address challenging public policy issues related to industrial ecology. Here we survey the range of policy options and strategies for implementation.

The major instruments by which governments can influence economic activity are tax and regulatory policies: corporate and personal income taxes, excise taxes, subsidies, rate of return regulations, labor laws, and standards for processes, products, and equipment. Governments may also use advertising, education, moral suasion, or signaling. The list of policy options is essentially the same in all countries. But there are differences related to degree and pattern of industrialization that merit some elaboration.

Advanced Industrial Economies

In the advanced industrialized world, the 1970s represented a decade of environmental regulation, spearheaded by passage of the U.S. National Environmental Policy Act in 1970. The 1980s marked a shift in emphasis from command and control regulation toward fiscal incentives and market mechanisms, designed to internalize environmental externalities, as embodied in the sulfur emissions trading provisions of the 1990 Clean Air Act Amendments. The 1990s may turn out to be a decade in which firms and communities become increasingly proactive in seeking environmental improvements, with less micromanagement by government. See the chapter by Andrews, "Policies to Encourage Clean Technologies," for a survey of current policies in key members of the Organization for Economic Cooperation and Development, and that by Griefahn, "Initiatives in Lower Saxony to Link Ecology to Economy," for a closer look at such policies.

Eastern Europe and the Former Soviet Union

Eastern Europe and the former Soviet Union combine the features of both industrialized and less developed countries, while also having their own specific features.

Like the industrialized countries they are characterized by a high level of industri-alization, a high level of education, and significant experience in the development of high-technology industries. Like the less developed countries, these countries are also highly dependent on resource-intensive industries that are often highly polluting.

The historical characteristics of economic and political systems in the former Soviet Union and Eastern Europe have influenced the current economic and environmental situation. The most important characteristics are the strong central-ization of management and limited rights of local authorities; the long-standing priority of producer over consumer; the organization of the economy according to the strategic aims of planned development; the strong monopolization of the econ-omy by the higher-priority ministries; special privileges for the military–industrial complex; emphasis on the production of intermediate products (steel, etc.) without a stimulus for modernization; underdevelopment of downstream industries fol-lowing raw material processing, resulting in structural imbalances; a closed soci-ety resulting in reliance on domestic sources of raw materials, even if of bad qual-ity; an unbalanced system of prices; free use of natural resources and land, and the absence of any stimulus for enterprises to reduce the consumption of energy and resources; a lack of proper environmental information, resulting in a low level of awareness among the population about environmental problems, plus passive social behavior; and little enforcement of environmental laws.

The recent changes in Eastern Europe and the former Soviet Union are creating new conditions for industrial and environmental development. The main strategies for change include increasing the role of market relations, privatizing state prop-erty, integrating these countries into the world economy with a gradual transition to world market prices, wooing foreign investments, and providing a structural and spatial reorganization of the national economy. Industrial ecology can con-tribute to the assessment and analysis of the environmental consequences of such technological restructuring. The chapter by Golitsyn, "Military-to-Civilian Conversion and the Environment in Russia," explores many of these issues.

Developing Economies

There are about 150 developing countries, each differing in its level of industrializa-tion, ecosystem vulnerability, and self-sufficiency. This makes it difficult to general-ize, but several characteristics apply to the policy context in most developing economies.

Developing countries are a fundamental part of a global industrial economy, and they are often the major suppliers of extractive resources to the global economy. The autonomy and sovereignty of the nation-states in the developing countries are often compromised as a result of their weak bargaining position in the international markets for raw materials and industrial goods. The chapter by Bunker, "The Political Economy of Raw Materials Extraction and Trade," gives historical insights into the dynamics of these relationships.

The terms of trade for the developing countries as a whole with respect to the global economy have deteriorated steadily over the past several decades. The large international debts of most developing countries, to industrialized countries directly or through multilateral banks, undermine their capability to deal with environmental issues. Many of the institutions and regulatory mechanisms of the industrial countries are absent or dysfunctional in developing countries. The chapter "Development, Environment, and Energy Efficiency" by Gadgil illustrates the potential for institutional innovations within developing countries to address environmental issues even in the face of such obstacles.

Governmental activities to "internalize the environmental externality" are evolving. The most advanced industrialized countries are already experimenting with incentives, information, and research designed to prevent instead of merely to control pollution. Yet in other parts of the world, the basic foundation of regulatory controls has not been firmly established. This diversity of policy contexts affects our ability to manage global environmental problems, and highlights a need for structured thinking about policies to implement industrial ecology. The five chapters here contribute to the conceptual framework for achieving the goals of industrial ecology from a "top-down" or policy perspective, within specific economic contexts.

31

Policies to Encourage Clean Technology

Clinton Andrews

Abstract

Government policies to promote clean technology in the advanced industrial nations of France, Germany, Japan, the Netherlands, Sweden, the United Kingdom, and the United States are reviewed. Each nation's general environmental policy is characterized, and its use of specific tools for encouraging clean technology is discussed. Acknowledging the complex motivations of targeted firms, some less-used policy options offer promise: financial incentives, training and education programs, consumer information, and especially efforts to reduce firms' risks. Attractive risk-reducing opportunities include activities that develop common visions (plans, goals, and schedules) and increase trust (voluntary agreements, performance monitoring).

Governments are thinking about new environmental policy questions. Following the industrial ecology metaphor, they no longer want only to control pollution, but would prefer to prevent pollution in the first place. They ask: What can governments do to encourage environmentally beneficial innovations in industry? What motivates firms, and how well do policies target firms' motivations?

This chapter explores the questions above by reviewing government policies that encourage clean technology in a set of advanced industrial economies. It first summarizes the general environmental policies of these nations, and identifies tools intended specifically to promote clean technology. It then sketches a conceptual model of what motivates firms, and evaluates how well current policies map onto these factors. The chapter concludes with recommendations for policy options having potential for greater use.

Clean technologies include low-emissions "closed" production systems, internal recycling and reuse of wastes, substitutions away from toxic inputs, improved efficiencies of energy and materials use, and products designed for lower environmental impacts over their life cycle. Clean technologies are important, but still subsidiary, parts of the overall industrial ecology concept. These technologies work to make industrial ecosystems intrinsically more sustainable, and they are being promoted by various government policies.

The basis for this research is a set of structured interviews with government officials in France, the Netherlands, the United Kingdom, and the United States, including a review of government documents (enabling legislation, program plans, and budgets), supplemented with comparative analyses published by others for the remaining countries. These are clearly qualitative comparisons. This chapter includes literature-based discussions of the additional key nations of Germany, Japan, and Sweden. The substantial literature on comparative environmental policy analysis for member nations of the Organization for Economic Cooperation and Development (OECD) is convergent in most respects, and recent major contributions include OTA (1992), OECD (1991a–c, 1990, 1987), IBRD (1991), Baas *et al.* (1991, 1990), Cramer *et al.* (1990), Pearce and Turner (1990), Weidner (1989), and Hohmeyer *et al.* (1988).

Table 1 provides comparative demographic, economic, and environmental statistics for the countries studied. For example, high population density helps explain why Japan and the Netherlands have large relative pollution control expenditures, and why Sweden does not. Countries exhibit great diversity in environmental performance, wherein Japan has low energy intensity and pesticide use relative to the profligate United States, while much of the Japanese population is not yet served by sewage treatment. Economic activity also shows a wide range, with France and the Netherlands appearing stagnant, Sweden and the United Kingdom restructuring, and Japan and the United States innovating.

The underlying conceptual model employed here is that firms have complex profit-maximizing motivations, that government policies seek to influence the behavior of firms, and that the effectiveness of policies depends on their success in targeting those motivations.

General Environmental Policies

Most of today's environmental policies have been put in place during the past 30 years, and have been justified largely by human health and safety concerns. The traditional theory of market failure has suggested several classes of policy options, including regulations, incentives, disincentives, liability, marketable permits, training, and information for firms and consumers (see, e.g., Hartwick and Olewiler, 1986). Surveys such as OECD (1991b) have revealed two additional classes of options: voluntary agreements and plans, or common visions. Table 2 shows these options.

Governments have chosen to emphasize a variety of options in their current policy packages. Table 3 summarizes the policy packages in simple matrix form, but a great deal of additional detail is available from the surveys listed above. Several important patterns are evident in the matrix.

First, every country maintains a floor of command-and-control regulations and of efforts in training and information for firms. Financial incentives are also in widespread use, although the specific mechanisms differ dramatically across countries. Yet each nation then builds a different type of motivational structure upon that base. For example, the United States emphasizes the disincentives of liability and public disclosure, while Germany and France emphasize financial disincentives (taxes).

Table 1: *Selected indicators to facilitate cross-country comparisons*

Indicator	Country							Reference
	France	Germany[1]	Japan	Netherlands	Sweden	U.K.	U.S.	
Demographic								
Population (millions, 1990)	56	61	124	15	8	57	249	WRI, 1992
Population density (per 1000 ha, 1990)	1021	2510	3279	4408	205	2369	272	WRI, 1992
Life expectancy at birth (years, 1990–95)	76.8	76.0	78.8	77.6	77.8	76.1	76.4	WRI, 1992
Economic								
GNP/capita (US$1000, 1990)	17.8	20.8	23.7	16.0	21.7	14.6	21.1	IBRD, 1992
Investment growth (%/yr, 1980–90)[2]	2.6	2.4	5.7	2.3	4.2	6.4	4.4	IBRD, 1992
Labor productivity growth (1980=100, 1987)[3]	109	103	110	107	119	135	125	IBRD, 1992
Patents issued (per 1000 population, 1990)[4]	55	128	168	70	105	53	212	USDOC, 1992
Environmental								
Pollution control expenditures (% of GDP, 1990)	0.89	1.52	>1.25	1.26	0.93	1.25	1.47	OECD, 1991c
Sewage treatment (% of population, 1988±)	52	90	39	89	95	84	80	OECD, 1991c
Municipal waste (kg/capita, 1985–89)	303	318	394	465	317	357	864	OECD, 1991c
CO_2 emissions (kg/$ of GNP, 1989)	0.4	0.5	0.4	0.5	0.3	0.7	0.9	WRI, 1992
Pesticide use (grams/$ of agri GDP, 1983)[5]	5.1	1.1	0.8	1.8	2.1	3.8	5.7	OECD, 1991c
Threatened bird species (% known to exist in region, 1980s)	40	32	8	33	7	15	7	OECD, 1991c

[1] Preunification Federal Republic of Germany.
[2] Average annual percentage growth in gross private investment.
[3] Index of gross output per employee in 1987 standardized on 1980 = 100.
[4] U.S. patents issued in 1990 to residents of each country, per 1000 of that country's population.
[5] Annual grams of pesticide used per dollar of agricultural sector contribution to gross domestic product in 1983.

Table 2: *Classes of policy options*

From the traditional theory of market failure:	
Regulations	ambient standards, emissions standards, technology standards, materials restrictions/prohibitions, employee certifications
Incentives	subsidies for RD&D or implementation, tax relief for investments
Disincentives	output taxes, input taxes, effluent taxes, waste disposal fees
Marketable permits	emissions permits, tradable offsets
Liability	criminal liability for worker and product safety, past environmental damages
Training	employee training, continuing education, worker re-tooling
Information for firms	technical assistance, clearinghouses, publications, seminars
Information for consumers	green labeling, dirty dozen lists, right-to-know legislation
Also observed in practice:	
Voluntary agreements	exhortation, voluntary emissions reductions, negotiated schedules
Plans	national plan and targets, facility plans, participatory planning, schedules

Table 3: *Qualitative comparison of general environmental policies*

Government Policy Actions	Country						
	France	Germany	Japan	Netherlands	Sweden	U.K.	U.S.
Regulations	●	●	●	●	●	●	●
Incentives	●	●	●	●	*	●	*
Disincentives	●	●	○	●	●	●	*
Marketable permits	○	*	○	○	○	○	*
Liability	○	*	○	*	*	*	●
Training	*	*	*	*	*	*	*
Information for firms	●	●	●	●	●	●	●
Information for consumers	*	●	●	*	●	*	●
Voluntary agreements	●	●	○	●	●	○	*
Plans	○	○	*	●	*	*	*

● = much use.
* = some use.
○ = less use.

A second feature is that political culture appears to affect the choice of policies (IBRD, 1991). Figure 1 shows my qualitative mapping of countries' historical behavior along two political dimensions: flexibility of the government's decision-making style, and openness of the decision-making process. Flexibility is defined as an enforcement style allowing negotiation between government and industry on

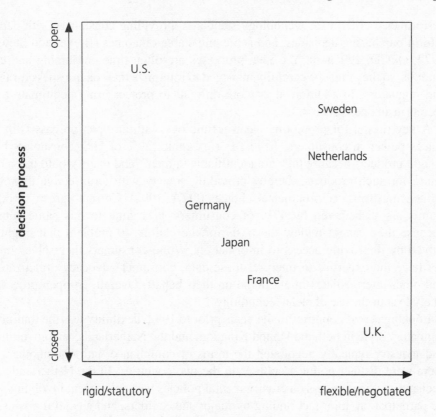

Figure 1. Qualitative mapping along key political variables.

regulatory compliance, rather than rigid interpretation of statutes. Openness is defined as a government decision-making style that provides full and timely public disclosure of data, participants, and methods, and provides opportunities for public input into the decision.

Hawkins and Thomas (1989) argue that flexibility encourages the use of clean technology because it (in contrast to rigid enforcement) allows innovation and permits solutions to be tailored to the individual circumstances of each firm; in short, it promotes economic efficiency. For example, a flexible regulator would allow a firm to meet environmental standards when replacing obsolete capital rather than forcing immediate (and therefore more costly) compliance.

Yet effective regulation also depends upon maintaining the political legitimacy of the decision-making process. The requirements for legitimation differ across countries and change through time, with some emphasizing well-credentialed decision makers and others credible decision processes (IBRD, 1991). Where the process is the source of legitimation, it is often open or reviewable. Regarding clean technology, an open decision process provides opportunities for testing the credibility of regulators, monitoring the level of compliance by industry, substituting performance criteria for technology standards, providing consumers with data to inform purchasing decisions, and achieving stable outcomes (Breyer and Stewart, 1979; OECD, 1991a–c). Yet open processes are often time consuming and contentious, so they must be carefully managed to minimize their onerousness for firms and regulators. In addition, it is more difficult to protect firms' legitimate trade secrets in an open process.

A key reason for promoting clean technology with an open process is that it places power in consumers' hands. For example, 29% of U.S. consumers have bought products because they are identifiably "green," and many would pay a premium for such products, thereby providing society with an effective means of influencing firms' environmental choices (OTA, 1992). Current "green" product claims are disbelieved by 47% of consumers, according to this same source, because they cannot review the basis for the claims—a problem that might be solved by improving access to information. While consumers themselves might not have the expertise to interpret these data, consumer advocacy organizations and other intermediates might do so on their behalf. Overall, an open process is likely to spur the use of clean technology.

Looking across countries in the years prior to 1992, flexibility was the traditional regulatory style in both the United Kingdom and the Netherlands, and government and industry typically negotiated the terms of compliance with one another. Yet there were distinct political cultures in the two countries. In the Netherlands, the decision-making process on environmental policies was quite open, involving public participation, joint fact finding by major stakeholders, and a careful process of consensus building on major issues. The United Kingdom, on the other hand, placed a great deal of decision-making power in the hands of government inspectors, who were then expected to strike a fair bargain with industry (i.e., best available technology not entailing excessive costs) during closed negotiating sessions that involved only modest public consultation and parliamentary oversight.

The United States offered a third combination: rigidity and openness. Like the Netherlands, it had a very open decision process. Both legislation and rule making enjoyed significant public participation, and performance data such as the Toxic Release Inventory allowed independent monitoring of progress. However, the relationship between government and industry was traditionally adversarial instead of cooperative in the United States. Once given statutory authority, government typically specified nonnegotiable compliance requirements for industry, and resolved any disputes in court.

Most nations shown in Figure 1 historically fell short in terms of either flexible compliance or open process. Sweden and the Netherlands, whose environmental policies were widely admired, performed relatively better along these dimensions. Innovative policy-makers recognized these differences and recently have been testing ways to change the relationships among government, industry, and the public, as discussed in the next section.

Policy choices are clearly evolving as government officials, industry, and advocates share ideas and experiences across international borders. In recent years, for example, there has been an increasing interest in financial incentives. The United States has been increasing its flexibility of enforcement by relying on tradable emissions permits for sulfur dioxide emissions. The United Kingdom is proposing to increase the openness of its decision-making process by requiring firms to publicly report toxic releases. With international trade representing an ever larger share of the world's economic activity, we can expect some convergence on policy mechanisms among different countries, as both nations and firms seek to level the competitive playing field. The harmonization efforts of the European Community provide an illustrative example, in which all member nations are moving towards similar environmental standards (IBRD, 1991).

Policies to Encourage Clean Technology

All of the advanced industrial nations have supplemented their environmental protection policy packages with special efforts to encourage clean technology, typically promoting specific types of innovation rather than generic investment. In the United Kingdom and the Netherlands the clean technology promotion effort has been led by economic development agencies, while elsewhere it has most often been led by an office within the environmental protection bureaucracy. Table 4 summarizes the current use of policies to encourage clean technology. Technical assistance is the most widely used policy, and mandatory recycling/takeback and goal setting the least used policies. Usage is measured using the policy's sum of scores across countries (see appendix for details).

While each country is unique, the most broadly used policies have addressed problems of imperfect information for firms (technical assistance, information for firms) plus their tendency to underinvest in research (research support, financial incentives). Both of these targets flow from the classic theory of market failure (Harmon, 1980). To think prescriptively about what policies should be employed, however, we need to develop a logic for evaluating the effectiveness of policies. Ultimately, effectiveness depends upon the extent to which firms' motivations are addressed, as discussed below.

What Motivates Industry?

Before evaluating programs to encourage clean technology, it helps to explore a prior question: What makes business tick? A brief look at the variety of motivations may suggest an expanded range of policy options. Profit maximization is the obvious driver, but we need to dig more deeply if we seek policy-relevant insights.

Many useful ideas about the factors affecting innovation appear in the evolving literature on organizational behavior. Classical theories, such as Taylorism, have emphasized engineering concepts of scientific management, unambiguous goals, and rationalization activities (Taylor, 1947; Koontz and O'Donnell, 1976). Revisionists have emphasized sociological factors such as friendship networks and other informal group processes (Locke and Schweiger, 1979). Neoclassical

Table 4: *Policies currently used to encourage clean technology*

Policy	Examples	Score
Regulations		
Regulatory adjustments	Flexible enforcement, interagency coordination, multimedia permitting	16
Mandatory recycling	Product take-back	11
Incentives		
Research support	Grants, loans, government labs	20
Financial incentives	Planning grants, investment tax credits, subsidies, loans	19
Disincentives		
Financial disincentives	Fees, fines	14
Marketable Permits		7
Liability		7
Training		
Employee training	Employee seminars, university curricula, training conferences	14
Information for government	Inspectors' manuals, training	14
Information for Firms		
Information	Telephone hot lines, fact sheets, manuals and guides, newsletters, clearinghouses, waste exchanges	17
Technical assistance	On-site consultations, basic information on options, generators' planning manual, referral services	21
Information for Consumers		
Information	Awards, eco-labels, education and outreach	13
Reporting requirements	Waste audits, toxic release inventories, trend analysis	14
Voluntary Agreements		7
Plans		
Facility plans	Waste reduction steps and schedules, reduction targets, capital replacement cycle, citizen advisory boards	13
Set national (or state) goals		11

Policies are scored by their usage, measured as the sum of country scores (3 = much use, 2 = some use, 1 = little use) across seven countries for each policy. Maximum possible score is 21. Note that the clean technology policies were scored separately from the general environmental policies in Table 3.

economists have contributed a contingency approach, suggesting that different organizational structures are a rational response to the nature of the external environment and the types of tasks being performed (Lawrence and Lorsch, 1967). Economists have also focused upon "principal–agent" issues that cause individual managers to act against an organization's best interest due to their personal incentive structures (see Panayotou and Zinnes, this volume). The systems approach has attempted to pull many of these strands together by considering firms to be work-performing, problem-solving, learning, goal-directed systems with both individuals and objects as elements, all of which are subject to exogenous forces that have strong or weak interactions with one another (Churchman, 1979). See Tornatzky *et al.* (1983) for a pertinent review of this literature.

A relatively simple systems approach will suffice for an illustrative discussion that emphasizes the components of the firm's profit-maximizing motive. In the standard theory of the firm, an enterprise usually achieves profitability by targeting two subobjectives: lower costs and higher revenues. These in turn have multiple components and distinct time frames, as shown in Figure 2. The specific business targets shown cover most key motivations but are not meant to be exhaustive. Typical company actions listed next to each target are likewise illustrative rather than comprehensive.

Cost minimization depends in large part upon efficiency improvements. Business decision-makers widely recognize innovation as an effective way to reduce the costs of regulatory compliance. Public policy that decreases the burden and increases the flexibility of regulations is also a popular recommendation for improving industrial efficiency. Pollution taxes can induce firms to reoptimize their production processes to reflect society's valuations of inputs and outputs.

Risk reduction is an important, if less familiar, cost reduction tool, one that reduces the cost of capital and allows management to focus on constructive activities. Government can play a role in helping firms manage marketplace risks by legislating liability caps for socially desirable but risky activities, or by allowing precompetitive research collaborations. Regulatory uncertainty imposes important risks that can be reduced by government, through developing clearer goals and timetables, and by improving the stability of decisions. Firms can also reduce regulatory risks by communicating their intentions to government, such as with pollution prevention planning documents.

Another key to cost management is employee involvement. Both the gospel of total quality management and old-fashioned common sense suggest that direct employee participation can improve worker satisfaction, safety, and productivity, while also uncovering cost reduction and pollution prevention opportunities invisible to process design consultants (see Nelson, this volume). Government can support training programs for employees and improve the educational level of the general population, making employee involvement more likely to be productive.

Governments may reduce firms' costs of production by reducing their costs of regulatory compliance, or may affect their relative costs by improving the uniformity of regulations across national and global markets. The search for a level

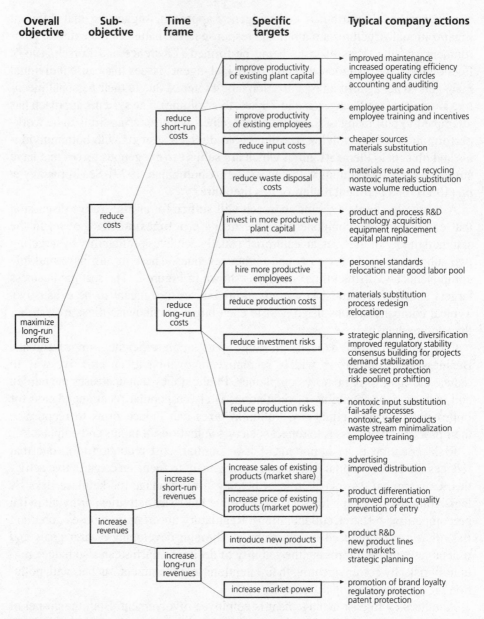

Overall objective	Sub-objective	Time frame	Specific targets	Typical company actions
		reduce short-run costs	improve productivity of existing plant capital	improved maintenance increased operating efficiency employee quality circles accounting and auditing
			improve productivity of existing employees	employee participation employee training and incentives
			reduce input costs	cheaper sources materials substitution
	reduce costs		reduce waste disposal costs	materials reuse and recycling nontoxic materials substitution waste volume reduction
maximize long-run profits		reduce long-run costs	invest in more productive plant capital	product and process R&D technology acquisition equipment replacement capital planning
			hire more productive employees	personnel standards relocation near good labor pool
			reduce production costs	materials substitution process redesign relocation
			reduce investment risks	strategic planning, diversification improved regulatory stability consensus building for projects demand stabilization trade secret protection risk pooling or shifting
			reduce production risks	nontoxic input substitution fail-safe processes nontoxic, safer products waste stream minimalization employee training
	increase revenues	increase short-run revenues	increase sales of existing products (market share)	advertising improved distribution
			increase price of existing products (market power)	product differentiation improved product quality prevention of entry
		increase long-run revenues	introduce new products	product R&D new product lines new markets strategic planning
			increase market power	promotion of brand loyalty regulatory protection patent protection

Figure 2. Decomposing firms' motivations.

playing field (or a favorably tilted one) is a central feature of industry–government interactions. It represents a key theme in the negotiations over the North American Free Trade Agreement, the General Agreement on Trade and Tariffs, and similar arrangements (Rees, 1993).

Revenue increase is a path to profitability that complements cost reduction. Products that better satisfy consumer preferences are likely to result in increased sales or higher prices; hence the popularity of "green" product strategies. Of course, environmental compatibility cannot substitute for other product attributes; products must remain functional, cost-competitive, attractive, and of high quality. Demonstrated corporate citizenship may also result in increased sales. Government may recognize firms for contributing to societal and ecological sustainability in both product design and facility operations. Conversely, government may publicly place negligent firms on a "dirty dozen" list.

Entry into new markets is another way to increase revenues. Thus, any activity that increases a firm's options, opportunities, and adaptability is likely to be attractive. Government support of research and development is one classic way to bring new products to the marketplace. Support of a more effective system of process improvement and early adoption of innovations may also achieve this goal.

Mapping Policies onto Firms' Motivations

Table 5 shows government policy actions that are relevant to specific business targets. By comparing the policy emphases of OECD countries shown in Table 4 with business motivations from Figure 2 one can see in Table 5 what pressure points have potential for greater use. The appendix describes a systematic approach used for the comparison, but the discussion here only highlights key findings.

Potential exists in most countries for additional policy efforts to increase employee productivity, enhance revenues, and reduce risks. This does not mean that greater government efforts in other areas would have no value; rather, it suggests that there is particular potential in these three areas.

Government may affect employee productivity in the short term by encouraging training/retooling programs, and in the longer term by ensuring an educated, adaptive work force. An emphasis on training and education is not unique to clean technology policy, but appears in most broad technology policy recommendations (e.g., Rivlin, 1992; Markusen and Yudkin, 1992). Special to clean technology would be a focus on environmentally sensitive design skills (for examples, see Graedel, this volume; Paton, this volume). Firms are probably better positioned to provide such training than is government, although government could play a role in supporting this activity.

Governments may enhance firms' revenues by providing consumers with information to help them differentiate "green" products from others. For example, the U.S. Environmental Protection Agency's (EPA) Toxic Release Inventory provides a basis for interfirm environmental comparisons. Government eco-labels such as the White Swan (Scandinavia), Blue Angel (Germany), and Eco-Mark (Japan) help consumers identify "greener" brands. In the United States, this task is left to private sector organizations, one of which is moving beyond the simple "green seal" concept to produce environmental report cards for products (OTA, 1992). Yet credible execution of this task is not easy (see Graedel *et al.*, this volume). The labeling must be done in a credible way that avoids oversimplification of tradeoffs

Table 5: *Mapping policies onto firms' motivations*

Specific Business Target from Figure 2 *Relevant Government Policy Actions[1]*	Coverage of Business Target by Government Policies	
	Average of Policy Scores[2]	Comparison to Median Coverage
Reduce short-run costs		
Improve productivity of existing plant capital *Technical assistance, information for firms, regulatory adjustments, reporting requirements, information for government, facility plans*	16	Median
Improve productivity of existing employees *Training*	14	Worse
Reduce input costs *Technical assistance, research support, information for firms*	19	Better
Reduce waste disposal costs *Technical assistance, research support, information for firms, financial disincentives, mandatory recycling*	17	Better
Reduce long-run costs		
Invest in more productive plant capital *Technical assistance, research support, financial incentives, information for firms, regulatory adjustments, financial disincentives*	18	Better
Hire more productive employees *Training*	14	Worse
Reduce production costs *Technical assistance, research support, information for firms*	19	Better
Reduce investment risks *Financial incentives, information for firms, regulatory adjustments, information for government, information for consumers, facility plans, set national goals*	15	Worse
Reduce production risks *Technical assistance, research support, financial incentives, information for firms, training, information for government, information for consumers*	17	Better
Increase short-term revenues		
Increase sales of existing products *Information for firms, reporting requirements, information for consumers*	15	Worse
Increase price of existing products *Reporting requirements, information for consumers, mandatory recycling*	13	Worse

Table 5: (*contd.*)

Specific Business Target from Figure 2 *Relevant Government Policy Actions*[1]	Coverage of Business Target by Government Policies	
	Average of Policy Scores[2]	Comparison to Median Coverage
Increase long-run revenues		
Introduce new products	17	Better
Research support, financial incentives, information for firms, regulatory adjustments, information for consumers		
Increase market power	13	Worse
Reporting requirements, information for consumers, mandatory recycling		

[1] Note that these assignments are subjective and interested readers may want to revise them to understand how they affect the conclusions.
[2] From Table 4.

among multidimensional environmental impacts. Further, it is only likely to be effective for products that compete on quality and other nonprice characteristics.

Potential also appears to exist in the area of risk reduction. Risk allocation has not been given the attention it deserves from regulatory economists, and is often overlooked for its role in market failure. Risks related to investment and production are significant factors in business decision-making, and students of finance know that firms are willing to trade off higher expected returns on investment for lower risk (e.g., Levy and Sarnat, 1984). Mitigating those risks could arguably unleash increased capital investment that could be channeled towards clean technologies. Policy options that address risk, such as voluntary agreements and planning exercises (see Table 2), are not widely used at present.

While many nations use the risk of fines or prosecution as a negative "stick" to encourage cleaner production, few employ reinforcing "carrots," or risk reduction incentives. In part, this is because perverse outcomes can occur, as was the case in the United States with the Price Anderson Act of 1957, which legislated a cap on accident liability exposure for nuclear power plants, freeing utilities to invest aggressively in this relatively unproven technology (Lee *et al.*, 1990, p. 58).

Yet there are alternative government actions that could help reduce firms' investment risks without increasing societal exposure. First are government incentives for firms to exceed minimum regulatory standards, in order to garner good will or avoid emissions taxes. Second are concerted efforts to build trust among firms, consumers, and governments. For example, voluntary agreements allow different parties to build context-specific credibility (a "track record") with one another, and to avoid surprises. Opportunities to demonstrate trustworthiness using reporting systems and performance monitoring could become a larger part of the regulatory tool kit (Barnett, 1986).

417

Another risk-reducing government action is to seek regulatory stability. The technique of developing a common vision, or plan, for each jurisdiction helps to do this. Long-term planning provides a mechanism for defining targets and scheduling milestones. This technique applies to both national environmental goal-setting and firm-level planning. However, stable results may depend upon inviting broad participation in the planning process and seeking to build a consensus in support of each policy decision (Kelman, 1992).

Regulatory flexibility can also reduce investment risks. Firms seek flexibility that allows coordination of environmental efforts with capital replacement cycles, so that undepreciated equipment is not constantly at risk (OECD, 1987). Market-based regulatory mechanisms such as tradable emissions permits also offer such flexibility.

Risk management strategies such as pooling of precompetitive research investments are gaining popularity. For example, France (this volume) discusses a U.S. partnership on recyclable automobiles. Yet without government encouragement, many such arrangements would be precluded by antitrust laws.

Conclusions

This quick tour of clean technology policies in advanced industrial countries shows that a significant amount has been done, but much more could be done. All countries studied have a strong foundation of regulations to improve environmental protection. They all likewise make some efforts in training and providing information for firms, although greater efforts may be warranted. Most countries employ financial incentives, but they choose very different specific mechanisms.

Countries have historically differed in other aspects of political culture: the United States has emphasized an open decision-making process and rigid enforcement of regulations; the United Kingdom has had a relatively closed process and flexible enforcement of regulations; and other nations have covered a spectrum in between. The flexibility allowed by market mechanisms such as tradable emissions permits is just beginning to enter use, and may serve as a substitute for flexible enforcement.

A crucial goal of evolutionary (as opposed to revolutionary) industrial ecology is to develop a sustained dynamic of technological improvement. This research suggests that, in pursuit of that goal, OECD governments have opportunities to give more emphasis to education and training, information for consumers about the environmental attributes of firms and products, risk management for firms, and the use of flexible enforcement mechanisms.

References

Baas, L., M. van der Belt, D. Huisingh, and F. Neumann. 1991. Cleaner production: What some governments are doing and what all governments can do to promote sustainability. *European Water Pollution Control Journal 1(3/4)*.

Baas, L., H. Hofman, D. Huisingh, J. Huisingh, P. Koppert, and F. Neumann. 1990. *Protection of the North Sea: Time for Clean Production.* Erasmus Center for Environmental Studies, Erasmus University Rotterdam, Rotterdam, Netherlands.

Barnett, S. 1986. A New York state consumer energy mindset. In *The Future of Electrical Energy* (S. Saltzman and R. Schuler, eds.), Praeger, New York, 260–266.

Breyer, S. G., and R.B. Stewart. 1979. *Administrative Law and Regulatory Policy.* Little, Brown and Co., Boston, Massachusetts, p. 1059.

Churchman, C. 1979. *The Systems Approach and Its Enemies.* Basic Books, New York.

Cramer, J., J. Schot, F. van den Akker, and G. Maas Geesteranus. 1990. Stimulating cleaner technologies through economic instruments: possibilities and constraints. *Industry and Environment 13(2),* 46–53.

Davis, B., and B. Greer (eds.) 1992. *State Leadership in Pollution Prevention.* Proceedings of symposium at the Woodrow Wilson School of Public and International Affairs, Princeton University, Princeton, New Jersey.

Foecke, T. 1991. State pollution prevention programs: Technical assistance and promotion. *Pollution Prevention Review,* 313–321.

Foecke, T. 1992. State pollution prevention programs: Regulatory integration. *Pollution Prevention Review,* 429–441.

France, Ministère de l'Environnement. 1989. Service des Technologies Propres et des Déchets. *Clean Technologies?,* Paris.

France, Ministère de l'Environnement. 1990a. Les Technologies Propres 10 Ans apres. *Etat de l'Environnement,* Paris.

France, Ministère de l'Environnement. 1990b. Service des Technologies Propres et des Déchets. *Rapport d'Activité,* Paris.

France, Ministère de l'Environnement. 1990c. Agence Nationale pour la Récupération et l'Elimination des Déchets. *The Disposal of Industrial Waste,* Paris.

Harmon, A. J. 1980. Industrial innovation and governmental policy: A review and proposal based on observation of the U.S. electronics sector. *Technological Forecasting and Social Change 18,* 15–37.

Hartwick, J. M., and N. D. Olewiler. 1986. *The Economics of Natural Resource Use.* Harper and Row, New York, p. 473.

Hawkins, K., and J. M. Thomas (eds.). 1989. *Making Regulatory Policy.* University of Pittsburgh Press, Pittsburgh, Pennsylvania, 263–278.

Hohmeyer, O., W. Schäfer, and A. Sieber. 1988. Systematic Presentation of the Clean Air Policy of Individual OECD Member States: Country Reports for France, Germany, United Kingdom, and United States. Prepared on behalf of the (German) Federal Environmental Service, Fraunhofer-Institut für Systemtechnik und Innovationsforschung, Karlsruhe, Germany.

IBRD (International Bank for Reconstruction and Development) (World Bank), Environment Division, Europe, Middle East and North Africa. 1991. *Patterns of Environmental Management.* World Bank, Washington, D.C.

IBRD. 1992. *World Development Report 1992.* Oxford University Press, New York.

Innes, A. 1991. Scope and approach of state pollution prevention legislation: An overview. *Pollution Prevention Review,* 369–387.

Kelman, S. 1992. Adversary and cooperationist institutions for conflict resolution in public policy-making. *Journal of Policy Analysis and Management 11(2),* 178–206.

Koontz, H., and C. O'Donnell. 1976. *Management: A Systems and Contingency Analysis of Managerial Functions*. McGraw-Hill, New York.

Lawrence, P., and J. Lorsch. 1967. *Organization and Environment*. Harvard University Press, Cambridge, Massachusetts.

Lee, T. H., B. C. Ball, and R. D. Tabors. 1990. *Energy Aftermath*. Harvard Business School Press, Boston, Massachusetts.

Levy, H., and R. Sarnat. 1984. *Portfolio and Investment Selection: Theory and Practice*. Prentice-Hall International, Englewood Cliffs, New Jersey.

Locke, E., and D. Schweiger. 1979. Participation in decision making: One more look. *Research in Organizational Behavior 1*, 265–339.

Markusen, A., and J. Yudkin. 1992. *Dismantling the Cold War Economy*. Basic Books, New York.

Netherlands, Ministry of Economic Affairs and VROM (Ministry of Housing, Physical Planning, and the Environment). 1991. *Technology and Environment*, The Hague, Netherlands.

Netherlands VROM. 1989. *National Environmental Policy Plan*. The Hague, Netherlands.

Netherlands VROM. 1990a. *National Environmental Policy Plan Plus*. The Hague, Netherlands.

Netherlands VROM. 1990b. *National Environmental Policy Plan Plus, Annex 1*. The Hague, Netherlands.

Netherlands VROM. 1990c. *Report Recommendations on and Responses to the NEPP*. The Hague, Netherlands.

OECD (Organization for Economic Cooperation and Development). 1991a. Policy Options to Encourage Environmentally Friendly Technologies in the 1990s. Note by the Secretariat on Technology and Environment, Paris, September 27, 1991.

OECD. 1991b. Policy Tools and their Applications in Member Countries. Note by the Secretariat, prepared for a workshop on Technology and Environment, Paris, June 17, 1991.

OECD. 1991c. *The State of the Environment*. Third edition, OECD Publications, Paris.

OECD, Environment Directorate. 1987. The Promotion and Diffusion of Clean Technologies in Industry. Environment monograph, OECD, Paris.

OECD, Environment Directorate. 1990. Procedure for the Notification of Financial Assistance Systems for Pollution Prevention and Control: Results of the 1978–1979, 1981–1982, 1987–1988 Notifications. Environment monograph, OECD, Paris.

OTA (U.S. Congress Office of Technology Assessment). 1992. *Green Products by Design*. U.S. Government Printing Office, Washington, D.C.

Pearce, D., and R. Turner. 1990. *Economics of Natural Resources and the Environment*. Harvester Wheatsheaf, New York, 23–25.

Rees, P. 1993. NAFTA brings changes to Mexico and Central America. *Geography: A Newsletter for Educators 3(1)*, 10.

Rivlin, A.M. 1992. *Reviving the American Dream: The Economy, the States and the Federal Government*. The Brookings Institution, Washington D.C.

Taylor, F. 1947. *Scientific Management*. Harper Brothers, New York.

Tornatzky, L., J. Eveland, M. Boylan, W. Hetzner, E. Johnson, D. Roitman, and J. Schneider. 1983. *The Process of Technological Innovation: Reviewing the Literature*. Report prepared for the Productivity Improvement Research Section, Division of Industrial Science and Technological Innovation, National Science Foundation, Washington D.C.

UKDOE (United Kingdom Department of Environment). 1989. *Clean Technology*. London.

UKDOE. 1990a. The Environmental Protection Act of 1990. Mimeo, London.

UKDOE. 1990b. Her Majesty's inspectorate of pollution. In *Integrated Pollution Control: A Practical Guide*. London.

UKDOE. 1992. Summary of U.K. Environmental Protection Situation. Mimeo, London.

UKDTI (U.K. Department of Trade and Industry). 1991a. *Cleaner Technology in the UK*. Prepared by the PA Consulting Group, London.

UKDTI. 1991b. Innovation: Competition and Culture. Transcript of speech by Rt Hon Peter Lilley, MP, London.

UKDTI. 1991c. *Environmental Contacts: A Guide for Business*. London.

UKDTI. 1991d. *Cutting Your Losses: A Business Guide to Waste Minimisation*. London.

USDOC (U.S. Department of Commerce). 1992. *Statistical Abstract of the United States 1992*. Bureau of the Census, Washington, D.C., Tables 856 and 1390.

Weidner, H. 1989. A Survey of Clean Air Policy in Europe. Report No. FS II 89-301 of the Veröffentlichungsreihe der Abteilung Normbildung und Umvelt des Forschungsschwerpunkts Technik-Arbeit-Umvelt, Wissenschaftszentrum Berlin für Sozialforschung, Berlin, Germany.

WRI (World Resources Institute). 1992. *World Resources 1992–93*. Oxford University Press, New York.

WRITAR (Waste Reduction Institute for Training and Applications Research, Inc.). 1992. *Survey and Summaries: State Legislation Relating to Pollution Prevention*. April 1992, Minneapolis, Minnesota, Survey Section updated September 1992.

Appendix

Seven structured interviews involving 25 government officials and others provided much of the primary data used for this analysis. Each interview lasted about two hours and covered 23 questions on the origins, emphasis, methods, strengths, weaknesses, and degree of integration of governmental clean technology programs.

In France, the key contact was the engineer in charge of implementing the Clean Technology program within the Department of the Environment. Major documents reviewed included France (1990a–c, 1989). In the Netherlands, there was a governmental session including representatives of the Ministries of Economic Affairs and of Housing, Physical Planning and the Environment. There was also a session with members of the faculty in Environmental Studies at Erasmus University, Rotterdam. Major documents reviewed included Netherlands (1991, 1990a–c, 1989). In the United Kingdom there were two sessions, one with representatives of the Department of Trade and Industry, and another with those from the Department of the Environment. Major documents reviewed included UKDOE (1992, 1990a–b, 1989) and UKDTI (1991a–d). In the United States, there were discussions with representatives of the EPA Office of Pollution Prevention and the New Jersey Department of Environmental Protection and Energy Office of Pollution Prevention, plus participation in a government/industry/academia symposium on pollution prevention (Davis and Greer, 1992). Major documents reviewed included WRITAR (1992), Foecke (1992, 1991), and Innes (1991). For Germany, Japan, and Sweden I relied on secondary sources, including

OTA (1992), OECD (1991a–c, 1990, 1987), IBRD (1991), Baas *et al.* (1991, 1990), Cramer *et al.* (1990), Pearce and Turner (1990), Weidner (1989), and Hohmeyer *et al.* (1988).

Respondents were asked which environmental protection activities were favored by their government, as measured by percentage of budget and staff, and about the specific emphasis of their clean technology program. Based on my understanding of their (sometimes oblique) responses, and a review of documents, I assigned a country score of 3 = much use, 2 = some use, or 1 = little use to each policy in each country. The overall usage of policies to encourage clean technology (shown in Table 4) was based on the sum of the seven country scores for each policy. The maximum possible overall score was 21.

This research developed a subjective but systematic method for mapping policies onto firms' motivations. Taking the usage measure described above for each policy (sum of scores across countries), it estimated their current usage for specific target motivations of firms. Business targets having below-median policy usage scores were designated as having worse coverage. For each target motivation it calculated two alternative measures of policy emphasis: average and "best" scores. The average score, shown in Table 5, was a measure of the average usage of the policies relevant to that particular business motivation. The "best" score, not shown, measured the usage of the most widely used policy relevant to the particular business motivation. One was not conceptually preferable to the other. These measures tracked one another quite closely, improving the robustness of the analysis.

These quantitative measures should not mask the subjective basis of the comparisons: the original assignment of policy usage within each country into categories of "much," "some," or "little use" was a judgment call. Also, the scoring method involved strong additivity/transitivity assumptions, such as giving equal weight to each country. Given these caveats, the method provided some significant, if subjective, insights, discussed above.

32

Initiatives in Lower Saxony to Link Ecology to Economy

Monika Griefahn

Abstract

A key theme of industrial ecology is that environmental and economic policy linkages must be established for real progress to be made on both fronts. The government of Lower Saxony is integrating economic and environmental policy using targeted economic development funds, environmental levies, public sector procurement, corporate environmental accounting requirements, and trade fairs.

One of the primary aims of the government of Lower Saxony is the ecological reorganization of industry. We are abandoning the principle of remedial environmental protection, which has proved to be more and more expensive and inefficient. Instead of "end-of-the-pipe" technologies, which only cure the symptoms, we require integrated technologies and processes which prevent environmental damage from occurring in the first place.

This means, however, that economy and ecology should no longer be viewed as conflicting issues. Only an economy which switches over to environmentally friendly products and processes secures the basis for its own existence in the long term. The current economic principle of production, consumption, disposal, and rehabilitation is a vicious circle that has to be broken. A simple "Carry on as usual!" is irresponsible and finally results in ecological disaster.

Thus the responsibility for the protection of the environment must be transferred more than at present to companies and should not—as is still often taken for granted—be borne by the public.

Ecology Funds

The government of Lower Saxony has unanimously decided to provide direct financial support for worthwhile approaches to an ecological restructuring of industry. Thus, associated with a fund for promoting economic development, we have set up a special ecology fund, jointly managed by the Ministry of Economic Affairs and the Ministry of the Environment. A sum of 280 million Deutsch marks (about $165 million U.S.) has been made available for the ecological restructuring of the economy in 1992 and 1993. Although the program has been especially designed for small and medium-sized companies, the money is also available to communities and larger corporations. The ecology fund supports projects that:

- Apply and use new and renewable forms of energy and energy-saving ideas
- Avoid, reduce, and recycle waste and residual materials
- Develop and test innovative and environmentally friendly products and production processes
- Provide specific environmental counseling and environmental education
- Stimulate investments in environmentally and socially compatible tourism.

The response to our offer has been extremely encouraging. By mid-1992, 5500 applications were filed. From a total of approximately 3000 which have been dealt with, 2500 have been accepted. Most of the applications that have been filed relate to the energy sector, especially the promotion of energy efficient boiler technology.

Environmental Levies

By increasing the number of environmental protection laws, directives, and decrees, we shall never arrive at the heart of ecological restructuring. Few companies and consumers are ready to do more than is required by law. If they did, they would be sneered at in our society as being mad, or "martyrs to ecology." Most of us are realists, and realists think only in the short term and do only what the law requires.

The aims of a profound ecological restructuring can only be achieved through the concept of an ecological market economy. This concept is based on a very simple and plausible idea: the regulation of structural changes via price—scarce goods are expensive, plentiful goods are cheap.

The motivating force for technological development in the past was the fact that the labor force was becoming increasingly expensive; technologies succeeded when they increased labor productivity. However, the key force in the future will be energy and raw material productivity. An intact environment is a rare commodity; nobody can seriously doubt this. But the environment is considered to be a public commodity and therefore not tradable. This means that the real costs for water consumption, air pollution, or the use of the soil as a rubbish dump are hardly noticeable in industrial cost accounting. The costs are "externalized," which means that the public is expected to bear them as a matter of course.

The economy can only be ecologically restructured when real environmental costs have been included in industrial cost accounting—in other words, by internalizing them. In this context, environmental policies must define framework conditions. The market alone is not capable of making an accurate assessment of the public commodity, "environment." The price of using natural resources will have to increase noticeably.

Consider two illustrative environmental policies of Lower Saxony:

- By increasing water extraction prices, we shall regulate the use of water, our rare natural resource. The private and industrial use of water will become more expensive, and more visible in industrial and private cost accounting. Income from water charges will be purposely used to protect natural water resources.
- By imposing a special waste tax (in January 1992) we shall reduce waste generation. Past experience has clearly proved that low waste disposal prices lead to excessive waste production, with consequent long-term problems for our environment. Our tax will draw the attention of companies to waste-avoiding technologies. The income from this tax will be used, especially, for the promotion of commercial recycling projects and methods of avoiding waste, as well as for the establishment of a waste disposal consulting agency.

Public Sector Procurement

In addition to subsidies and taxes, the awarding of public contracts by government can significantly influence the ecological restructuring of the economy.

With annual expenditure by governments at the federal, state, and regional levels of about 80 billion Deutsch marks ($50 billion) in the Federal Republic of Germany, considerable pressure can be exerted to promote environmentally friendly products, building materials, and services. By using contract award procedures that give importance to environmental protection, the contract-awarding authority can initiate and support the development, manufacture, and marketing of energy-saving and environmentally friendly products and processes.

To be a model for public authorities, the local government of Lower Saxony has a new environmental directive, intended to encourage the use of environmentally friendly products and to promote ecologically advantageous proposals from those seeking building contracts and contracts for other supplies and services. Where formerly economic criteria were applied exclusively in choosing among proposals, we have now added an ecological evaluation. It often turns out that when all social costs are taken into consideration, the most ecological solution is also the cheapest.

These are some of our new decision-making criteria:

- Reduction of water consumption by using water-saving devices and sanitary equipment
- Reduction of emissions, for example, by using low-emission, energy-saving gas boilers or short- and long-distance district heating
- Reduction of noise, for instance, by equipping private vehicles with long-life silencing systems
- Avoiding waste by using returnable bottles, containers, food crates, trays, and similar industrial packaging
- Salvaging rubbish by using recycled products (made from synthetic or plastic materials, cardboard and paper, used tires, rubber, glass, and building site rubble) in preference to natural raw materials, in order to conserve natural resources

- Using solar-powered appliances
- Using zinc–air batteries rather than batteries using cadmium or mercury
- Avoiding, whenever possible, polyvinylchloride (PVC) and products containing its copolymeric substances
- Using rare types of wood from tropical rain forests only if it can be proven that they come from long-term protected forests (secondary forests).

Before a construction permit for a new building is issued, we require an examination of the design to see whether any of the planned construction measures will have a harmful effect on the environment. Depending on the type of construction and the local circumstances, the following criteria may be applied:

- Avoidance of excessive sealing of buildings and their infrastructure
- Optimum heat insulation and the use of efficient methods of heating
- Use of passive solar energy
- Use of renewable energy sources
- Collection and use of rain water.

The implementation of these new assessment criteria is placing great demands on all the authorities we can influence, both the federal government above us and the local governments below us. Specialized and advanced training of employees is required. We are well aware that proper assessments can only be made when sufficient information is available concerning the environmental compatibility of products. This will necessitate the setting up of a product data base that provides information clearly and accurately.

Corporate Ecological Control

An ecological restructuring of the economy will also be accelerated by the adoption of "ecological control" at the level of the corporation. In Lower Saxony a pilot experiment is being carried out in conjunction with the Institute for Ecological Economic Research in Berlin, entitled Environmental Control: The Active Use of Environmental Balances by Companies within the Framework of Preventative Environmental Policies. The project is motivated by our awareness that although many companies in Germany are assigning ever greater importance to active and preventative environmental protection, many of these companies are hampered by a severe lack of basic concepts. A shortage of experience and tools constrains the ecological assessment of industrial practices and the inclusion of ecological considerations in company decision-making. This applies to research and product development as well as to manufacture and sales.

One way to plan, regulate, and control all corporate activities relevant to the environment is to apply the concept of ecological control. This concept, analogous to traditional financial control, is being worked out in detail in the course of a project we are running, involving three medium-sized companies. The companies chosen to participate in this pilot experiment have both a management open-minded about ecological matters and a set of products with special environmental problems. Examples include manufacturers of foil packaging and wallpaper, and suppliers for the paint and varnish industry.

The first step in corporate ecological control is to analyze all current activities of a company relevant to the environment. All material and energy flows involved in production must be identified, and an ecological balance must be developed. In an ecological balance, the inputs of raw and processed materials are compared quantitatively with the resulting products and emissions, including waste materials, waste water, fumes, waste heat, and noise. To make the data useful for company decision-making, the data must be systematically processed and assessed for relevance to environmental issues. The result is a "weakness analysis," which forms the basis for formulating ecologically oriented company objectives as well as for establishing concrete remedial measures. Weakness analysis is designed to help a company identify and assess alternative products, technologies, processing materials, and production methods—including the zero option, in which a product line or a particular product is abandoned altogether.

An essential part of corporate ecological control, which is an ongoing process, is the constant scrutiny of measures adopted earlier. Advances in ecological research may reveal that well-intended modifications or substitutions have larger environmental effects than were earlier understood.

A proper ecological balance will identify and assess as many environmental effects as possible relating to industrial activities: effects occurring not only in the factory but throughout the entire product life cycle, including extraction of raw materials, transport, usage, and disposal. Assessment criteria will include legal, social, and scientific considerations.

Our pilot experiment has shown that corporate ecological control is able to improve significantly the quality of information about ecological conditions available to a company. With improved recognition of a company's impact on the environment, a company finds many new ways to convert its production to more environmentally friendly processes and to develop measures to conserve raw materials, energy, and water. For example, one company participating in the experiment has succeeded in saving substantial quantities of energy by constructing a cogeneration plant with coupled waste-heat utilization.

A formal integration of ecological control into corporate procedures is necessary to ensure that the concept of protecting the environment is permanently embedded in a company's daily decision-making process. It is beneficial for a company to give a single corporate officer responsibility for coordination of all environmental matters. Ensuring the participation of the entire staff is equally important because it is the only way to ensure that a plan for corporate ecological control will succeed in overcoming internal resistance and that the abilities of all persons involved can be put to good use.

We are preparing commercial business handbooks for classes of businesses, to empower companies to apply and implement the concept of ecological control on their own. We have begun with the furniture industry, and we will move next to the carpet industry and the electroplating industry.

World Exposition, Year 2000, Hanover

Whether the establishment of ecological principles for industrial production can be achieved in a reasonably short time depends on the level of worldwide effort in this direction. A successful worldwide effort requires common aims, implemented regionally. Ecologically important technical approaches will be developed in one location that have broad applicability. The global communication of information has to be improved to disseminate such ideas.

An important event, designed to assist in global information flow and to give the world's public answers to the questions of economy and ecology in the next millennium, will be the World Exposition in Hanover in the year 2000 (Expo 2000). Its theme is "Mankind—Nature—Technology."

Expo 2000 should create an awareness of the fact that coexistence between mankind and nature can only be sustained if there is a change in the use of technology. It is true to say that many global environmental problems are a direct result of technology. Yet we cannot abandon technology. Instead, we must redirect technical progress according to ecological criteria. We are in need of a new set of technologies that are superior from an ecological point of view.

We are inviting the industrial nations of the Northern Hemisphere to outline perspectives on reducing currently excessive levels of consumption of resources to sustainable levels and on terminating their exploitation of Third World countries. In addition to the question of introducing suitable technologies, there is above all the question of changing attitudes.

We are inviting Third World countries to propose ways to achieve a progressive improvement in living conditions with new and adapted technologies. This includes, of course, strategies for controlling the population explosion.

The history of previous world expositions illustrates that in most cases these events convey a superficial euphoria regarding technology. In contrast, Expo 2000 should serve as a confirmatory event, to ascertain whether the objectives of the U.N. Conference on Environment and Development—embodied in Agenda 21 and the Earth Charter—have been implemented and to initiate exemplary projects worldwide. As minister for the Environment of Lower Saxony, one of the organizing states, I am seeking international contacts to realize these objectives for Expo 2000.

33

Military-to-Civilian Conversion and the Environment in Russia

George Golitsyn

Abstract

In the former Soviet Union, defense-oriented industry and its research and development organizations could play a large role in addressing environmental problems. Examples include monitoring the environment from ground-based and space satellite systems and mitigating the impact of supersonic transport on stratospheric ozone. The prospects of cooperation with foreign national and multinational corporations are also considered.

Introduction

To write an essay about linkage of military conversion and environmental quality would seem bizarre to almost anyone in Russia, or elsewhere in the former Soviet Union (FSU), or anywhere. In principle, the essay can be produced—by digging through diverse materials and talking to knowledgeable people. But both the economic and the political scene change so rapidly now in the FSU that by the time one understands some feature, it has taken a new shape. To make the preparation of such an essay even more difficult, for many decades the true financial and material state of the defense-related industries was classified. There are no open detailed official statistics even now, in mid-1992. What is worse, the Russian Committee on Statistics has ceased to publish detailed quarterly reviews of the state of the economy and society. Its predecessor in the FSU, the State Committee on Statistics, regularly published such reviews in all major national newspapers. Although data in the reviews were often distorted, at least there was something official available. In 1992 only one newspaper, the weekly *Economy and Life*[1] (*E&L* hereafter), has published a review of the performance of the economy in Russia in 1991, and this yearly (not quarterly) review considered many fewer items than earlier reviews.

---

[1] This author must confess that he had not heard of this newspaper before starting to gather the material for this paper.

One can only guess whether these shortcomings are due to lack of data in deteriorating conditions, a lack of workers, or a decision not to upset people by publishing large amounts of oppressing detail about our impoverishing society.

This essay draws on documents and materials published in *E&L*; some documents from a forum, The World Experience and Economics of Russia, held in Moscow in May 1992; talks with people working in military-related industries and research and development (R&D) institutions; and talks with people working on environmental problems. There is only a small hope that in the present fluid situation, with everything changing and so much changing for the worse, some of the observations here may preserve their significance for a while.

The Decline of the Defense Industries

The area of the Russian Federation (RF), 17.1 million km^2 (11.8% of the world land, while the FSU occupied 15.5%), is still almost twice that of the United States. The RF population was 148.8 million people on January 1, 1992 (*E&L*, 4^2), only about 52% of the population of the FSU and about-two thirds of the population of the United States.

Very recently an active Soviet foreign policy has brought about a substantial reduction of the tension between the two superpowers and between the North Atlantic Treaty Organization (NATO) and the Warsaw Pact. Agreements to reduce rockets, nuclear arms, and conventional arms; the general improvement of the political climate; and the whole process of *perestroika* posed the question: Does the country need armed forces of about 4 million people and a huge military industry? Between 15 and 30% of the gross national product (GNP) of the FSU was devoted to military purposes. In the United States the comparable number is about 6% of GNP. These numbers explain why the USSR, with only about half the production of the United States, could successfully match the U.S. military capability; in items like land-based rockets, armor, artillery, and chemical weapons could attain numerical superiority; and in numbers of nuclear weapons could attain parity. But to spend such a large fraction of GNP on military production was too heavy a burden on the Soviet economy and was one of the main reasons for its slow deterioration, already distinctly observable in the first half of the 1980s and even earlier.

The Soviet military industry was always in a better position than the other sectors of Soviet industry. It has had preferential funding, better materials, better-trained workers, and strong and plentiful R&D. Moreover, competition nourished by the Soviet government (there were several firms for aircraft, rockets, and satellites) allowed the USSR to be on a par with the West, or sometimes ahead of it—especially in areas dominated by mechanical engineering rather than electronics. In communications, control systems, and computers the USSR was considerably

---

[2] The number after the comma (here and hereafter) is the issue number of *E&L*.

behind, mostly due to a less developed general culture of industrial production and some wrong strategic decisions back in the 1960s.

In about 1990, with the first signs of economic freedom and of markets, traditional economic linkages began to break. The process was especially painful and destructive, because in the centrally planned economies thousands of factories had unique production capabilities. For example, four factories in Latvia and Lithuania produced the parts—the absence of which can stop all of the FSU's agricultural machinery production and operation.

The result has been a steady contraction in the economy. Total industrial output in the RF in January 1992 was 15% below that of January 1991. For the whole year of 1992, compared to 1991, the production of consumer goods is expected to drop by 13%, food products by 18%, cereals by 16%, and sugar by 20%. Even more disturbing, capital investments in physical terms are expected to decrease by 30–40%, postponing any hopes for light at the end of a tunnel of steadily increasing length.

What is the situation with military-oriented industry? There are no precise figures for military production alone, because most military plants have also been producing civilian consumer goods. In 1991 military plants produced 100% of all TV sets, photocameras, video systems, and sewing machines; 98% of refrigerators; 95% of computers; 83% of equipment for producing medicine; 76% of equipment for food processing; 72% of vacuum cleaners; and 66% of washing machines (Golovachev, 1992).

On March 20, 1992, President Yeltsin signed a law on the conversion of the defense industry in the Russian Federation (Yeltsin, 1992). This law establishes the basic principles governing the operation of defense plants. It proclaims that the high-technology capabilities of these should be directed toward the production of goods that are competitive in international markets. And each converted plant is required to preserve its ability to return to military production under mobilization. But since there is no agreed-on way to assess the mobilizing capability of a converted industry, it is hard to see how such a requirement can be implemented. Much needed for a well-reasoned course of conversion would be a doctrine of "sufficient defense," still virtually absent in Russia.

On June 26, 1992, the Supreme Soviet adopted the law on the Defense of the Russian Federation. An important constraint in the law is an upper limit on the size of the RF Army: in times of peace it should not exceed 1% of the population of the federation. Inasmuch as the RF military personnel is currently about 2.5 million, at least a 40% reduction is expected in the near future, along with accompanying reductions in arms and munitions. Such a large reduction in personnel poses serious problems, including how to house officers and their families, and how to retrain at a time of growing unemployment.

The Sorry State of the Environment

The current environmental problems in Russia and the rest of the FSU are those experienced in the West 20 or more years ago (see Golitsyn, 1992; National

Report, 1991). They include urban air pollution, acid deposition, and pollution of water and soils by municipal sewage, industrial waste, agriculture, and animal husbandry. Especially heavy damage is to be found around metallurgical complexes (Norilsk, Bratsk, etc.), where thousands of square kilometers of forests have died out completely. An increase of about 17% in cancers from 1980 to 1990 in the RF, and of 12% in the whole FSU, has been observed, as well as a 6.5% increase in deaths in the RF from heart and blood diseases (see National Report, 1991; *E&L*, 12[3]). Though the life expectancy did not change appreciably during the same period, these numbers suggest an increased deterioration of the environment, i.e., the quality of the air we breathe, water we drink, and food we eat. There has also been a large drop in birth rates: in 1991 the natural increase of the population in RF was only 128,000 compared to a total population of 149 million (*E&L*, 13), and in 1992 the population fell for the first time since the end of World War II.

These environmental problems of Russia and the other newly independent states of the FSU originate in an inefficient economy with a structure disproportionately skewed toward heavy industry (largely because of the defense sector) and the manufacture of products of poor quality and low reliability. As a result the repair industry is a far larger fraction of the FSU economy than of the economy of any other developed country, and spare parts are a chronic deficit and a constant headache for managers (see Golitsyn, 1992). Production that is very energy, material, and resource intensive generates large stresses on the environment.

In the 1970s and early 1980s some real efforts were undertaken in the FSU to purify water wastes at certain especially dangerous industrial facilities and to treat the municipal sewage of some large cities. Replacing coal with natural gas at many power plants allowed the FSU to fulfill the goal of a 30% reduction in sulfur dioxide emissions: in 1990 these emissions were 32% lower than in 1980 (see National Report, 1991; Golitsyn, 1992). The amount of untreated water waste was declining, and an increasing fraction of industrial water has been recirculated—in 1985, about three-fourths of total industrial water.

However, with the advent of economic liberties and a market approach, environmental control was weakened. The links between enterprises started to loosen, production slowed, shortages of everything increased sharply, and prices went up. With a decrease of central funding, the first to suffer were cleaning facilities. The emerging new private enterprises, most of which are still small, pay little attention to the environment. With environmental laws in Russia undeveloped, unfinished, and unsuited to market conditions, these private enterprises are theoretically exempt from any interference from the state in their production processes (except taxes). No environmental officer is allowed to inspect their air, water, or solid waste discharges.

---

[3] Unfortunately, the USSR National Report on the State of Environment was not distributed at the U.N. Conference on Environment and Development in June 1992. Completed in August 1991, it contained a wealth of material, most of which had never been widely published or made easily available. Instead, much shorter reports hastily compiled by not all 15 states were presented.

Another important polluter is agriculture. Outdated machinery, poor discipline in the workplace, and, above all, giant losses of agricultural products are to blame. Losses occur at every stage: at harvest, transport from the fields, storage, and distribution. Losses range from 30% to more than 70% for various crops and products. Especially large losses occur during storage, due to poor facilities, lack of refrigeration, and a weak industry of agricultural product treatment. There are now some incentives from the state to raise crop yields, but no matching incentives to treat and store food. The domestic food needs of the country could be met by a considerably smaller quantity of agricultural production if products were better treated and conserved—substantially easing the stress on the environment from agriculture and removing the need of the RF to spend billions of dollars to buy food from abroad.

Military-to-Civilian Conversion in the Service of the Environment

Inasmuch as military activities themselves produce considerable stress on the environment, simply decreasing the size of the armed forces will be beneficial for the environment. The reduced Russian military forces will need considerably fewer tanks, aircraft, ships, and other materials-intensive products. Furthermore, if conversion proceeds as its proponents in the RF dream and desire, then it may lead to improved quality and reliability throughout the civilian economy, including consumer goods. Product durability will increase, reducing demand and, consequently, production—again decreasing the stress on the environment.

What can conversion do directly to solve, or at least mitigate, environmental problems? Directly, retrained scientists and engineers can measure the amount and composition of waste, monitor the environment, and invent, develop, and commercialize new waste treatment technologies. Military R&D in Russia has developed many promising new materials and technologies. These could revolutionize manufacturing, including the construction of buildings, machinery, and ships, and the improvement of safety. Energy, material, and capital intensity per unit of product could substantially decrease. Most of these materials and technologies were classified, and many still are. Now, however, faced with great reductions in funding, there is a frantic effort to declassify, to obtain patents, and to seek commercial sponsors, including abroad. Multibillion-dollar industries based on new products can be envisioned, with very diverse applications.

Soviet remote-sensing techniques are not bad at all and should be helpful. What have always been bad in the FSU are computing facilities, information and control systems, and electronics. These weaknesses have prevented operations with large data flows.

In 1992 the Ministry of Environment and Protection of Natural Resources of the Russian Federation initiated a new program: Conversion for Ecology. The idea is to use, adapt, and improve the potential of the defense industry and its R&D to make sensors, measuring devices, communication and control systems, and information facilities for monitoring the state of the environment. The main contractor for each project in the program is supposed to be an organization from the defense-related part of the economy, and one of the subcontractors must be from civilian science, e.g., from the Russian Academy of Sciences or from a university. As one example, in 1992 our Institute of Atmospheric Physics—well known in Russia and abroad for its studies of solar and thermal radiation transfer in the atmosphere—was asked to work formally under an unfamiliar military R&D organization, to rebuild its first-class photopolarimeter as a meteorological instrument to measure aerosols.

During the last three years, three major Soviet (even in 1991 it was still Soviet!) firms developing and manufacturing satellites asked our institute to help them develop the concept of a space system for environmental monitoring. Unfortunately, all three initiatives came from below; as yet the Russian Ministry of the Environment has adopted neither a concept nor a specific plan for a monitoring system. It is time for the recently formed Russian Space Agency, together with the Ministry of the Environment, to develop such a concept and program. To be sure, this effort must be closely coordinated with the efforts of the National Aeronautics and Space Administration (NASA), the European Space Agency, and other international and national organizations.

Space is a field with very large conversion potential. Due to less reliable electronics (e.g., solar batteries of shorter life and smaller communication capacity), the Soviet Union has had to launch many more satellites than the United States—especially military reconnaissance satellites. Thus the Soviet Union developed an experienced and capable rocket-launching infrastructure, now available for the launching of satellites such as those needed for an ecology-oriented global observing system.

In August 1991 the Soviet Union launched a U.S.-made instrument, the total ozone monitoring spectrometer (TOMS), that measures the total ozone content in an atmospheric column. It works successfully, on the spacecraft *Meteor-3*. Much red tape from the U.S. side had to be overcome, because the United States has a law against transferring high technology to potential enemies. This launch, evidently for an environmental purpose, was done for free. The Russian Space Agency cannot allow itself such a luxury on a regular basis, but it is ready to do launches considerably more cheaply than NASA.

In 1989–90 hundreds of Soviet medium-range rockets were destroyed in accordance with a U.S.–Soviet treaty. Our military chose a spectacular method of destruction: each rocket was shattered with large amounts of explosives. Our institute took the opportunity to record atmospheric pressure signals from the explosions up to 1000 km away. An analysis of the signals at various distances has given us insights into the fine structure of the stratosphere and lower thermosphere (Kulichkov, 1992). One high official in our space administration said to this author in 1991, that these findings were a really good example of conversion! On the contrary, it was clear then and is especially clear now that the mode of destruction was not optimal: the rockets could have been deprived of their military capabilities while still remaining useful for launchings of civilian satellites up to about 500 kg in weight. Such satellites can be launched from almost any place, not only from a special launching site.

As for its special launching sites, Russia may be in trouble. For many years the main site was Baikonur, in the desert in Kazakhstan. Now an independent state, Kazakhstan has already claimed its rights to the site but has not yet confiscated it. Russia's other launching site, Plesetsk, a few hundred kilometers south of Archangelsk, can be used for launching satellites of considerable weight, but not for manned missions. When President Yeltsin visited Plesetsk in April 1992, he announced plans to expand its launch and control capabilities to permit manned missions, but no one knows where the funds can be found.

Another major dual-purpose large institution is the Central Aerohydro-dynamical Institute near Moscow, the oldest of its kind in Russia. It has one of the world's largest wind tunnels, well equipped and with experienced personnel in both science and technology. Because all kinds of aircraft and rockets can be tested there, it is now in substantial demand from U.S., French, and German aircraft companies. This allows the institute to stay afloat in these difficult times.

The same institute will play a role in the future of supersonic civilian transportation, which is economically sound only if planes fly in the stratosphere. It is now quite clear that the Anglo–French Concorde was far from an optimum aircraft, both from the environmental and the economic point of view. The main constraint on next-generation design is that the aircraft cause minimal destruction of the stratospheric ozone layer; this constraint will determine the altitude of aircraft operation, which in turn will dictate what kind of aircraft it should be. Such work, about to start with French funding, is another example of making conversion work for the environment.

A large sector of the military industry deals with nuclear weapons. Russia has several large laboratories similar to the U.S. national laboratories in Los Alamos, New Mexico, and Livermore, California, with experienced, highly qualified scientific and technical personnel. These laboratories also seek their ecological niche under the changing conditions. Some of their scientists have started working on global problems, like energy and climate change, safe nuclear reactors, and transforming bomb plutonium into a dilute nuclear fuel. These laboratories already have good computer models of pollution dispersion, and they are in a position to take the lead in environmental modeling. Especially after the nuclear accident at Chernobyl, these laboratories seek international cooperation. The potential benefit is large.

Many major multinational corporations have already understood the potential of the defense industry of Russia and its significance for the international arena. A few real collaborations have started, but in many cases the multinational corporations simply pay Russian organizations for information about actual or potential markets, or ask their own governments to exert political pressure on the RF to prevent commercial contracts with other countries involving high technology (as was the case with nitrogen-cooled rocket engines for India in the spring of 1992). Thus, multinational corporations can help transform the Russian defense industry into something useful, but they can also be unhelpful, as when they try to extinguish potential competitors.

Conclusion

I have tried to present a quick glimpse of the state of the economy in Russia as of 1992. In the process of converting a major part of its defense industry to civilian purposes, the Russian Federation may be able to improve the environment. The potential effect is large, but much larger are the problems in its realization: economical, financial, political, organizational, geographical, and personal. Particularly critical is the availability of capital, because conversion needs large investments.

With a deteriorating economy and worsening life conditions in the RF, and throughout the FSU, social apathy is widespread. Most of the population does not understand the economic reforms. The degradation of discipline in the workplace, the increase in crime and corruption, and many other negative phenomena are all consequences of today's economic conditions. How and when we will be able to scramble out of this, only God knows.

Acknowledgments

I want to thank Robert Socolow, who proposed that I write on this subject. At first his request astonished me, but then my fragmentary thoughts on the subject materialized, aided by several talks with the people mentioned below. V.L. Makarov, director of the Central Economics and Mathematics Institute of the Russian Academy of Sciences (RAS), directed me to official sources: Yu.V. Yaremenko, director of the Institute of Economics and Prognosis of Science and Technology Progress, RAS, and especially K.V. Frolov, vice-president, RAS, and director of the Institute of Machinery, RAS, helped me to clarify my ideas.

References

E&L. 1992. Social-economic state of the Russian Federation in 1991. Prepared by the Ministry of Economics of RF. *Economy and Life (4),* January.

E&L. 1992. Causes of death in the Russian Federation. Editorial. *Economy and Life (12)*, March.

E&L. 1992. Population of the Russian Federation. Editorial; and Prognosis of the social-economic development of the Russian Federation for the second quarter and up to the end of 1992. Prepared by the Ministry of Economics of RF. *Economy and Life (13)*, March.

Golitsyn, G. S. 1992. Environmental aspects of transformation of Eastern Europe and the former Soviet Union. In *Science and Sustainability: Selected Papers from IIASA's 20th Anniversary Conference*, International Institute for Applied Systems Analysis, Laxenburg, Austria, 123–156.

Golovachev, F. 1992. Collapse of the defense production complex can cause the break-up of the country's economy. *Economy and Life (18)*, May 1992.

Kulichkov, S. N. 1992. Long-range sound propagation in the atmosphere (a review). *Izvestia–Atmospheric and Oceanic Physics 28 (4)*, 339–360.

National Report of the USSR to UNCED, Moscow, August 1991 (unpublished draft report).

Yeltsin, B. N. 1992. The Law: On conversion of the defense industry in the Russian Federation. *Economy and Life (18)*, May 1992.

34

The Political Economy of Raw Materials Extraction and Trade

Stephen Bunker

Abstract

A central feature of the industrial era has been the development of global markets for raw materials. Industrialized countries have employed a variety of strategies to ensure their supplies of key mineral and plant resources, resulting in significant economic and environmental effects in host countries, which are typically less developed in economic terms. The industrial ecology and strategic significance of policy-driven changes in global mineral markets are examined, using the starkly-contrasting examples of copper and aluminum.

Secure access to an expanding, cheap, and secure supply of raw materials is critical to economic growth and stability under industrial capitalism. Industrial firms and states of industrial societies therefore act strategically as well as economically to assure access. Their strategies may contravene the sovereignty of nations and the environmental and social well-being of communities in areas rich in natural resources. At the same time, their strategies are unlikely to be effective without the acquiescence and cooperation of powerful economic and political agents—including states—in the resources-rich areas.

Access strategies have changed over time and space because of changes in technology, in markets, in transport capacity, and in world political and economic organization. A particularly important feature has been the role of the rising economy in a competitive world system of production and trade. In seeking to challenge the dominance of already well-established national industrial economies, rising economies often devise particularly aggressive access strategies. The consequences for global flows of natural resources are highlighted in this chapter.

The concept of industrial metabolism developed by Ayres (1989) provides a useful frame for analyzing the complex interactions between social, economic, political, geological, chemical, and physical processes that drive raw materials extraction, processing, trade, transformation, consumption, and disposal. Ayres pays close attention to natural transformation cycles of minerals, as well as to the interactions of minerals with soils, water, and biota. His tools of analysis are helpful in assessing both the environmental constraints on and the environmental effects of social uses of minerals. However, his methods are incomplete. After all, the extraction of raw materials directly changes not only the physical environment, but also the demography and the social organization of the regions where the deposits are located and of the regions where ancillary transport and energy infrastructures are constructed (Bunker, 1985, 1989, 1992, in press). Much is to be gained by expanding our attention so as to include the ways that social organization and process interact with physical characteristics and technologies. A physical ecology of material flows in industry and in nature is well complemented by a political economy that takes into account the close interactions between the material requirements of industry and the social organization that drives industry.

The Dynamics of Raw Materials Flows

Historically, increased diversity of materials consumed, economies of scale in extraction and processing, and the progressive depletion of the resources most accessible to industrial centers have combined with the absolute spatial fixity of most mineral resources to increase the mean distances between natural resource extraction and industrial production. Longer mean distances have increased the significance of scale economies in transport. These scale economies, in turn, have reinforced the technologically driven increases in the scale of extraction. Investments sunk into larger ships, larger ports, and longer rail lines—frequently dedicated to a single extractive enterprise—can only be recouped with larger shipments of minerals sustained over longer periods of time.

These dynamics restrict greenfield mining projects to large deposits, of which there are relatively few. Such selectivity further reinforces the tendency toward longer distances between extraction and industrial production. With fewer, larger sites of extraction also comes a higher proportion of raw materials transported across national boundaries prior to transformation (see Table 1), and a higher likelihood that extractive enterprises will dominate the economy and politics of the region, and sometimes of the nation, in which the resource is located.

As distance and scale increase, mines tend to locate in areas with sparse populations and little effective integration into political, economic, and legal systems. Communities in these areas have only limited access to the economic and technical information required for effective rent bargains or for environmental or social regulation. Isolated exporting nations compete against each other in contracts with informed importing firms and states, so raw materials rents and prices remain low and damages to environments are omitted from contract costs. Profits move further from the mine, exacerbating the tendency of raw materials prices to decline (Drucker, 1986) and reducing the bargaining power and revenues which exporting states need to offset the adverse consequences of extraction and processing. At the same time, increased scale and distance also raise the strategic stakes for the firms and the states in the industrialized countries.

Table 1: *Changes in seaborne materials trade*

	Petroleum	Coal	Iron Ore	Bauxite and Alumina
Seaborne raw materials trade volume (million metric tons per year)				
1960	366	46	101	17
1990	1190	342	347	52
Seaborne materials trade activity (billion ton-miles per year)				
1960	1650[1]	264	34	99
1990	6261	1849	1978	205

[1] This refers to the trade activity in 1962 rather than 1960.

Source: Calculations by Stephen Bunker and Paul Ciccantell.

Resource Access Strategies of Rising Economies

A rising economy elaborates especially aggressive strategies to secure access to raw materials. Its challenge to dominant economies usually begins during periods of global economic expansion, when demand approaches or surpasses capacity and prices rise. The raw materials industries reflect not only expanded commodity production, but also intense building of plants, roads, ports, warehouses, and energy networks. The established economies have usually captured most of the sources and may still dominate extractive and transport technologies. Sustained ascent therefore requires that rising economies exploit opportunities and weaknesses in the established political and economic relations that govern raw materials extraction and export.

The strategies of rising economies have a long history. Britain's early colonial expansion was driven in part by the need for timber to build ships. British strategies of expansion were conditioned by the demographic and political condition of the Americas, as well as by competition from France and Spain. Two centuries later, Britain used very different strategies to secure access to the minerals that were rapidly replacing fibers as the most critical raw materials. Britain exploited rifts in the Spanish and Portuguese colonial system and among the newly independent Latin American dominant classes to break down trade barriers and to secure concessions for resource extraction. Britain used its power to induce other states to assume part or all of the costs of building the transport systems required for raw materials exports.

In the late 19th century, Britain used newly developed botanical technologies to transplant the cultigens of rubber, cinchona, and sisal from Latin America, where it controlled neither land nor labor, to Asia, where it controlled both and could move these raw materials from extractive to plantation regimes (Brockway, 1979). To assure a steady flow of raw materials from around the world for its burgeoning urban industrial economy, British capital backed by the British state financed and built railroads and canals in Canada, Latin America, Africa, and Asia, many of them under agreements with local governments that guaranteed a certain rate of profit. British and U.S. firms and their states competed and collaborated in establishing the forms of government and the physical infrastructure that effectively secured access to Middle East oil. In each case, changing technologies, new demands and opportunities, and evolving geopolitical conditions drove the choice of strategy, but every strategic aim was the same: to secure access to raw materials considered crucial for industrial growth and military security.

U.S. strategies to secure access to metals during its rise to industrial and military preeminence responded to the very different political organization of the world in the early and middle 20th century. In the 1920s and 1930s, the Council on Foreign Relations (CFR) formulated a geopolitical program for securing access to critical and strategic materials. The CFR was particularly concerned with manganese, critical for hardening steel and available in relatively few and distant locations. Together with the U.S. Department of State and U.S. Steel, it devised an elaborate scheme to induce the Brazilian government to build a railroad from its iron deposits to a mill designed by U.S. Steel; the trajectory of the railroad would facilitate the export of manganese as well. Then, U.S. Steel withdrew its participation in the steel mill, once the railroad was secured and manganese exports were assured. Both the State Department and CFR were concerned about the risk that individual Brazilian states would restrict exports; a centralizing coup during this period conveniently rearranged Brazilian federal authority and removed that risk. Under this regime, the Department of State was able to secure a manganese mine for U.S. Steel by providing arms to the military (Priest, 1993).

The CFR also proposed the creation of new international financial institutions to stabilize investment flows and costs. World War II and the Bretton Woods Conference created the conditions for the concrete realization of these proposals. As Britain had in the 1820s, the United States in the 1940s and 50s manipulated nationalist aspirations to secure access to natural resources as these resources emerged from European colonial control. Through the threat of withholding Marshall Aid, the United States also used Europe's desperate need for capital to coerce access to minerals still under European colonial control (Bunker and O'Hearn, 1992). The actions of the World Bank and the International Monetary Fund have served both to palliate the economic distortions of U.S. political and economic intervention in raw materials exporting countries, and to ensure that exports continue despite economic crises.

The Japanese confronted still another geopolitical structure as they began their industrial ascent in the 1960s. The United States had faced a weakened imperial system run by decapitalized industrial powers already in relative decline. The Japanese faced a politically and economically vibrant and still expansive U.S. economy that directly controlled the majority of natural resources through direct investment and ownership. The Japanese had far less capital at their disposal than the United States had after World War II. In the intervening years the scale of extractive, processing, and transport technology in most minerals had increased, and many underdeveloped but resource-rich nations had become more exigent in negotiating their resource contracts. The Japanese also had to confront the most dramatic manifestation of these exigencies in the middle phase of their ascent when the Organization of Petroleum Exporting Countries (OPEC) pushed up oil prices in 1973 and 1979, greatly raising energy costs for industrial processes and the transport of raw materials.

The Japanese turned some of these obstacles to their own advantage. They solved the twin problems of finding capital in amounts matched to the increased scale of minerals facilities and placating the demands for host-country control by channeling a portion of the large global surpluses of finance capital that were created by the oil shocks of the 1970s into new forms of investment—long-term contracts, loans, and joint ventures with poor and indebted states willing to assume even greater levels of debt.[1]

A comparison of Japanese strategies to secure access to two very different, but partially substitutable, metals—copper and aluminum—illuminates some of the interactions between physical and social processes. Copper and aluminum are both high-volume metals in world production and trade.

A Comparison of Copper and Aluminum

History, Technology, and Industrial Organization

The copper and aluminum industries differ along multiple important dimensions. Copper has been a major raw material for thousands of years. Due to its relative scarcity and to an early spurt in its rate of consumption, it was traded internationally in high volume before many other major minerals. The international copper trade grew most rapidly at the beginning of this century, and, by the 1920s, there were significant investments of U.S. and European capital in African and Latin American mines.

The three principal stages of copper production are mining, milling, and smelting.[2] The product of milling, *concentrate*, is still not chemically reduced to metal, but, on the one hand, it is greatly reduced in bulk relative to ore, while, on the other hand, it is still only about 30% copper by weight. The reduction of copper ore (generally, copper sulfide or copper oxide) to copper metal takes place in smelters.

---

[1] Many resource economists have explained Japanese actions in specific metals in terms of normal market operations in a recessive world economy. The comparison of Japanese actions across two metals in this paper suggests that there were strategic Japanese actions as well.

[2] There is a fourth stage, refining, where copper blister (99% copper) becomes pure copper, in one form known as anode copper. We will omit reference to this step, and will use "smelted copper" or copper metal where the professional literature refers to "refined copper." We do so in order to minimize the confusion with the "refining" of aluminum, which occurs at an earlier stage in processing. It is unfortunate that "refining" has such different meanings for aluminum and copper.

Depletion has steadily reduced the grade of commercial ore (the percent of copper by weight in the ore). A typical commercial ore grade today is less than 1%. A wide variety of technologies have been developed to deal with leaner grades and with the great diversity of ores of copper. The diversity of ores means that older, more polluting mines and smelters generally return to production whenever copper prices are high.

The grade of commercial copper ore is so low that the location of milling is always as near as possible to the mine itself. Between them, mining and milling generally account for over half of the total capital and operating costs of copper production. This means that the major capital barrier to entry has been at the mine itself.

For much of its history the copper industry was tightly integrated within national territories. Copper smelters were almost always located near mines, in part because smelting reduces so much bulk and because most mines are remote from ocean transport. Until recently, it was presumed that concentrate would not be traded internationally; rather, concentrate would first be smelted to metal, and only the metal would move long distances. As we will see, this presumption was successfully challenged after 1970 by the Japanese, and to a lesser extent by the Germans.

The aluminum industry is much newer. The metal was hardly used commercially until a hundred years ago, and major growth occurred only after World War II. There are still vast reserves of the types of bauxite for which the original technologies were developed. This has meant that technologies have remained more uniform, and that innovations for efficiency and pollution control can be adopted far more generally than in the case of copper. In contrast to copper, currently used bauxite deposits are all close to 45% ore.

The three principal stages of aluminum production are mining, refining, and smelting. Mining of bauxite constitutes less than 2% of total production costs, and refining (the conversion of bauxite to aluminum oxide, or alumina) constitutes only about 16% of total production costs. By far the most capital intensive step is the smelting, where alumina is reduced to aluminum metal. The amount of electricity required to smelt aluminum is large because oxygen bonds particularly tightly to this metal.

As a consequence, for aluminum in contrast to copper, the greatest barrier to entry is at the smelter rather than at the mine. The ease of transporting bauxite and alumina and the large electricity requirements of smelting have tended to separate extraction, refining, and smelting in space but to integrate them in terms of corporate control (Stuckey, 1983). The costs and conditions of electricity supply are a major factor in location decisions and plant operations at the smelting stage.

Technologies for each stage of aluminum production are relatively homogeneous, so patent ownership and control over energy sources are an important aspect of corporate control. The relative youth of the aluminum industry has enhanced corporate control over information; it was still possible in the 1940s and 1950s for aluminum companies to buy ore deposits (O'Hearn, in press) and to secure riparian rights for hydroelectric dams at prices far below their potential value (Barham, in press).

Not surprising in view of the geological realities, the aluminum industry has long been organized internationally and has been highly concentrated. Four firms—Alcoa, Alcan, Pechiney, and Alusuisse—were able to control most of the world's trade in aluminum until World War II; with the addition of Kaiser and Reynolds after World War II, the six major firms controlled 83% of the market in 1955, 72% in 1965, and 64% in 1972. The copper industry, though still highly centralized in comparison to most non-mineral industries, is far less concentrated than the aluminum industry.

Environmental Impacts

The association of copper ores with numerous other metals, including zinc, lead, cadmium, and arsenic, assures that copper mining and milling will involve chemical processes with great potential for pollution. Leaching from slag at mines and accidental spills of the cyanide used for separation of metals during milling have severely polluted river systems, estuaries, and bays from Chile and Peru to the Philippines and Papua New Guinea (OTA, 1988; Moody, 1992). Copper smelting releases large amounts of sulfur dioxide, which is recaptured only where there is a proximate market for sulfuric acid. The particulate matter released from copper smelting contains lead, cadmium, arsenic, zinc, and nickel.

In contrast, the principal chemical step at the bauxite mine is the removal of chemically bound water: the bauxite deposits exploited commercially contain 40–60% water. Recent technical advances have further reduced pollution at the pit (the site of extraction) by facilitating the reconstitution of liquid wastes for return to the pit, and subsequent revegetation.

The key step in the refining of bauxite into alumina involves the removal of reactive silica in a bath of caustic soda (sodium hydroxide) under heat and pressure. The total production cost for aluminum metal is sensitive to the cost of caustic soda. Recently, when the cost of caustic soda rose, a prior cost advantage of the higher grade Brazilian ore relative to lower grade Australian ore disappeared: the Brazilian ore had greater amounts of reactive silica. The mixture of silica and caustic soda that today is a waste product of refining is a red mud that drastically reduces soil fertility when it enters ground water.

During smelting the alumina is reduced to aluminum in an electrolytic reduction process, where electric current is passed through a fluoride bath in a carbon-lined steel pot. While fluoride emissions used to be the most significant direct environmental problem related to aluminum smelting, some newer plants capture and recycle most of the fluoride. Other direct environmental problems of smelting include managing the carbon solid wastes associated with the carbon anode and potlinings and controlling the emissions of sulfur dioxide.

The indirect environmental costs of aluminum smelting, stemming from its intensive use of electrical energy, often dominate the direct costs. In Australia and

Spain the electricity for smelting comes from power plants that burn low grade, high sulfur coals. In Venezuela, Canada, Surinam, and Brazil, the sources of electricity, instead, are large hydroelectric dams constructed primarily for these smelters; their negative environmental effects include releasing mercury from soils and producing methane as the flooded forests decompose.

For geological reasons, deposits of commercial bauxite and sites suitable for low-cost hydropower production tend not to occur at the same locations. Trihydrate bauxite, the lowest cost aluminum ore, was formed when the silica bonding aluminum to surrounding rock was dissolved and the aluminum percolated down to form a horizontal stratum just above some ancient water table. This geological process requires high levels of rainfall on older soils on relatively flat ground, conditions most conspicuously present in the humid tropics. The location of the best bauxite deposits (i.e., lowest cost extraction with present technology) on relatively flat land means, in turn, that only low-head dams can be built nearby. As a result, the generation of large amounts of power requires either reservoirs nearby that flood large areas, or reservoirs at a considerable distance where the terrain is mountainous and the hydropower is cheaper.

Recent Trends in Markets

Between 1950 and 1988 copper production expanded 3-fold, but aluminum production 12-fold. In tons of annual production, aluminum surpassed copper in the 1960s and nearly doubled copper by 1988.

Not surprisingly, given these different rates of expansion, the rank order of producers has been far more stable for copper than for aluminum. For copper, all six of the countries with the highest output in 1988 had been among the top ten producers in 1950. For aluminum, by contrast, the four countries with the highest bauxite production in 1988 (accounting for 68% of world production) were producing little or no bauxite in 1950.

The international aspects of the structure of both of these industries has changed dramatically since 1970. Where copper smelting had largely been restricted to locations near the mine, unsmelted copper concentrate is now transported around the world. Where bauxite used to be exported from Guyana, Jamaica, and a few African countries for refining and smelting almost entirely in the United States, Canada, and Europe, aluminum smelting is now located near the mine. I argue below that these changes are in large part the results of Japanese strategies to secure access to materials they considered critical to their industrial development, just as earlier changes in the flows of rubber, sisal, iron, manganese, and other then-critical industrial materials resulted from British and North American strategies during their earlier rise to economic dominance.

Japanese Access Strategies in Copper and Aluminum

The Japanese started to invest in copper in the 1960s, earlier than in aluminum.

They focused at first on long-term purchase contracts and loans for mines in poor nations with little experience of industrial development. In most cases, Japanese support was for the mine and the mill alone. Generally, the host countries had little access to capital, little managerial expertise, and little experience with copper. Indeed, the only early Japanese investment in an established copper-mining nation, Zaire, was abandoned as a failure. In the Philippines, Malaysia, Indonesia, Australia, and Papua New Guinea, however, the Japanese stimulated either the initiation of mining or its expansion from a very small base. Since most of the existing copper smelters in the rest of the world were already tied to other mines, new mines were left dependent on Japanese and West German smelters. The Japanese set very low tariffs on the import of concentrate and very high tariffs on the import of copper metal. As the Japanese developed their domestic capacity for copper smelting, they expanded their reach to established overseas mines and induced them to produce more copper concentrate than could be processed in host-country smelters.

Japan, almost single-handedly, established a world trade in copper concentrate: in the 12-year period from 1970 to 1982, the share of the world copper trade associated with ores and concentrates doubled, from 15% to 30% (Wagenhals, 1984: 36–37), and Japan became the importer of 60% of the total concentrate traded. Consequently, the concentrate-exporting countries became dependent on the Japanese market. In that same 12-year period, U.S. exports of concentrate increased by a factor of four (while exports of refined copper decreased by a factor of six); Chile increased its concentrate exports by a factor of five; Zaire initiated concentrate exports; and Papua New Guinea, the Philippines, and Indonesia, all of whom had not participated in the copper trade before 1970, entered the export market, but only in concentrate.

Since the 1970s, the Japanese have formed joint ventures in copper mines in Mexico, Chile, Canada, and the United States. Particularly in the United States, they have bought into mines tied to smelters that had been closed due to the cost of complying with U.S. Environmental Protection Agency (EPA) restrictions on sulfur dioxide emissions. The concentrate now goes to Japanese smelters located close to ports and industrial plants, and 95% of the sulfur dioxide is captured at low cost. The clustering of smelters near each other and other industries also facilitates the capture of metal byproducts (such as cadmium and zinc) on an economically viable scale. This not only reduces the effective cost of the copper ore, it also reduces pollution. The low price of sulfuric acid and the large distance between the tied smelters and industrial markets had made the capture of sulfur dioxide prohibitively expensive in the United States.

Turning to aluminum, we find that the Japanese campaign to secure access to aluminum started in earnest only in the early 1970s. The Japanese sought joint ventures where their equity was as small as possible. The joint venture has proven to be a particularly effective way of surmounting the barriers to entry that the aluminum oligopoly had constructed. Alcan and Alcoa had tied up some of the best bauxite reserves and hydroelectric sites, leaving potential competitors dependent on higher-cost ores and electricity. As demand promised to exceed reserves through the 1960s and 1970s, exploration was stimulated, leading to the discovery of major new bauxite reserves in Guinea, Australia, and Brazil. The growth of demand also assured that host countries would be willing to develop new hydroelectric sites for smelting.

The aluminum oligopoly had kept prices high by restricting capacity. The Japanese countered by diversifying their sources, and by building a large number of hydroelectric dams, transmission systems, and ports. This would have been a costly and risky undertaking, especially since a successful diversification of supply was likely to bring down prices, and therefore profits. The Japanese reduced their risks by encouraging selected countries to make their own investments in infrastructure and share equity in smelters. Unlike their copper strategy, where poor exporting countries were made dependent on Japanese smelters, their aluminum strategy required partners who were at least partially industrialized. Japan courted the relatively well capitalized developing nations, but ignored Guinea, the poorest and least industrialized of the major bauxite exporters, altogether.

Convincing a less developed nation to assume the costs of a hydroelectric dam and transmission lines is not easy. In Brazil and Indonesia, the initial Japanese proposals for bauxite production were sweetened with promises of investments in dams, infrastructure, and refineries. After the respective governments were committed to the project, however, the Japanese withdrew their participation in the dams and refineries. In the Brazilian case, they also significantly reduced the size of the smelter, so the Brazilians were left with a huge, largely idle dam. The unreturned costs of this dam have been a major factor in the financial and administrative difficulties of the host electric company. The Japanese aimed at realizing profits further down in the fabrication stream, rather than in the mining and processing, as the large oligopolists had done. The bauxite-exporting countries, which had invested in large dam projects and infrastructure in order to support domestic smelters, faced disastrous reductions in aluminum prices.

Between 1970 and 1980, the Japanese initiated joint ventures in aluminum smelting in New Zealand, Australia, Indonesia, Venezuela, Brazil, and Canada. These six countries have played a major role in the creation of excess aluminum capacity in the world market, and thus have contributed to the loss of market control by the majors. Significantly, rather than reducing aluminum capacity during the downturn in aluminum price in the early 1980s, all six of these countries expanded capacity, presumably because of the debt pressure from its investments in infrastructure. This further depressed prices. An expanding spot market had reduced the majors' market share in the 1960s, but they were still dominant. The excess supply on the world market not only broke the majors' control of international trade, but also broadened the development of a spot market. The loss of market control was formalized in 1977, when aluminum was quoted on the London Metals Exchange for the first time.

Since the 1960s, as seen in Table 2, the trend in the aluminum industry has been toward less trade in ore and more trade in refined products. (Recall that for copper

Table 2: *The changing world aluminum trade (millions of tons, at three levels of processing)*

	1962	1975	1988
Bauxite	17.0	28.0	27.8
Alumina	1.1	9.5	12.2
Aluminum	1.0	2.9	8.8

Source: United Nations Trade Tables.

the trend has been the opposite: toward more trade in concentrate, less trade in metal.) In terms of contained aluminum, bauxite constituted over 70% of total trade in 1962, less than 45% by 1975, and less than 30% in 1988; aluminum metal constituted about 20% in 1962, just over 20% in 1975, and over 40% in 1988. Since, per ton, alumina prices are about five times higher than bauxite prices, and aluminum metal prices are more than ten times higher than alumina prices, the changes in trade in favor of more highly processed aluminum are even more dramatic in terms of dollar value than in terms of tonnage, as seen in Table 3.

The aluminum industry, with the more transportable ore, is trading it less (in percent terms, relative to metal), while the copper industry, with the less transportable ore, is trading it more. Japanese lending, buying, and investments have moved aluminum toward territorial integration and copper toward territorial de-integration.

To review, Japanese firms first sought copper mines in poor countries, with no previous ties to smelters and refiners, while developing their own smelting and refining capacity. Once they had achieved scale economies of transport, smelting, and refining, they followed this same strategy in more developed countries. In so doing, Japan has created a world market in copper concentrate, dependent on sales to Japan. On the other hand, Japan has fostered joint ventures in aluminum in countries anxious to integrate forward from bauxite mines and willing to provide cheap hydroelectricity for smelting aluminum. As venture partners it has chosen countries with sufficient capital and management skills to assume majority equity and national management responsibility. In both instances, the effect of reversing earlier patterns has been to break down earlier U.S.-dominated trading regimes. In both cases, Japanese strategies turned established physico-spatial logic upside down, but even so the success of their actions depended on their manipulation of physical properties of the different minerals and of the socio-economic characteristics of the host countries.

In other minerals, Japanese access strategies have also manipulated physical properties. In iron and coal, both high-bulk minerals, these strategies have focused on stimulating massive rail–port–shipping complexes rather than on primary processing. Scale economies in transport are critical to Japan because of its distance from raw material suppliers. Japanese financial, technical, and infrastructural initiatives in shipping have led to the massive increases in raw material transport

Table 3: *The changing world aluminum trade (millions of current U.S. dollars at three levels of processing)*

	1962	1975	1988
Bauxite	220	670	1700
Alumina	80	1200	4100
Aluminum	760	4000	40000

Source: United Nations Trade Tables.

shown earlier in Table 1, and have consequently restructured both global markets and the global environment.

Japanese firms and the state collaborated in these strategies. Together, they responded to unanticipated changes in global demand, as well as to changes (some of which they engendered) in the economic and political conditions of the countries with which they were dealing. They were able, in most cases, to maintain strategic advantage by understanding these changes more clearly than either the host countries where they were investing or the hegemonic firms, whose control they were undermining. For both copper and aluminum, they succeeded in substantially restructuring the world markets and material flows at limited cost and risk, despite importing less than 20% of total world imports.

Conclusion

Global mineral markets are driven by the needs of industrialized countries to obtain secure access to raw materials. Successful strategies for securing access respond to physical and topographical properties of specific raw materials and significantly affect environmental and economic conditions in host countries, while substantially altering world patterns of material flows. An understanding of these market forces provides an important complement to the existing literature on industrial ecology.

The physical interdependence of extraction and industrial transformation, and the increasing spatial scale of markets, can best be understood as the result of both economic and ecological dynamics, each molded by the collusive and competitive behaviors of individual states and firms. Industrial ecology is enriched when it incorporates human agency and intention into models that are firmly grounded in material processes. One may anticipate that similar powerful constraints will operate through global markets to affect other materials policies, such as those bearing on reduction, substitution, and high-proportion recycling of major industrial materials.

Earlier forms of subordination of less developed countries to industrialized countries were based on foreign direct investment in mineral extraction and oligopolistic control of world minerals markets. Today, new forms of subordination are appearing, based on competition between indebted suppliers in a market destabilized by uncoordinated excess capacity. Japan has greatly weakened the established oligopolists, whose capital was largely tied to fully-owned, self-financed primary extraction, by moving the locus of surplus profit forward in the commodity chain. The result has been upward pressures on the costs of new mining projects and downward pressure on the prices of basic metals (Oman, 1989: 43, 45, 46, 67, 68).

The costs that states and firms are willing to assume, and their capacity to impose some of those costs on the economies and environments of the resource-holding nations, vary with the organization of state–firm relations in dominant industrial countries, with the political organization of specific resource-holding nations, and with the political and economic organization and condition of the world system of nations. Free trade, imperialism or colonialism, foreign direct investment, and joint ventures or shared responsibility are modes of resource access that have characterized different periods of world resource trade. Each mode has called for quite different strategies, and has had quite different effects on the resource-exporting societies.

Historical data show that no matter what the particular commodity or particular market structure, actual strategic behavior is driven by a blend of physical principles and political economy. Among the issues that are rooted primarily in physical principles one might list a commodity's absolute physical scarcity, those physical characteristics that drive its use in particular commercial and military technologies and products, and the requirements for infrastructure that result from responses to physical location (both physical and national-territorial), ore grade and type, energy requirements, and scale. Among the issues that are rooted primarily in political economy are a commodity's perceived scarcity (requiring a judgment of acceptable political and economic cost for each possible source of supply), its perceived importance (even, criticality) as a raw material in industry, and the costs of substitution. The problem of analysis is to integrate systematically both physical and political economic considerations.

References

Ayres, R. U. 1989. Industrial metabolism. In *Technology and Environment* (J. Ausubel and H. Sladovich, eds.), National Academy Press, Washington, D.C., 23–49.

Barham, B. Strategic capacity investments and the Alcoa-Alcan monopoly, 1888 to 1945. In *States, Firms, and Raw Materials: The Political Economy and Ecology of the World Aluminum Industry* (B. Barham, S. Bunker, and D. O'Hearn, eds.), University of Wisconsin Press, Madison, Wisconsin, in press.

Brockway, L. 1979. *Science and Colonial Expansion: The British Royal Botanical Gardens*. Academic Press, New York.

Bunker, S. 1985. *Underdeveloping the Amazon: Extraction, Unequal Exchange, and the Failure of the Modern State*. University of Illinois Press, Champaign, Illinois.

Bunker, S. 1989. Staples, links, and poles in the construction of regional development theories. *Sociological Forum IV(4)*, 589–609.

Bunker, S. 1992. Natural resource extraction and power differentials in a global economy. In

Understanding Economic Progress (S. Ortiz and S. Lees, eds.), University Press of America, 61–84.

Bunker, S. Flimsy joint ventures in fragile environments. In *States, Firms, and Raw Materials: The Political Economy and Ecology of the World Aluminum Industry* (B. Barham, S. Bunker, and D. O'Hearn, eds.), University of Wisconsin Press, Madison, Wisconsin, in press.

Bunker, S., and D. O'Hearn. 1992. Strategies of economic ascendants for access to raw materials: A comparison of the U.S. and Japan. In *Pacific Asia and the Future of the World-System* (R. Arvind Palat, ed.), Greenwood Press, Westport, Connecticut, 83–102.

Drucker, P. 1986. The changed world economy. *Foreign Affairs 64*, 768–791.

Moody, R. 1992. *The Gulliver File: Mines, People, and Land: A Global Battlefield*. Minewatch, London.

O'Hearn, D. Producing imperialism anew. In *States, Firms, and Raw Materials: The Political Economy and Ecology of the World Aluminum Industry* (B. Barham, S. Bunker, and D. O'Hearn, eds.), University of Wisconsin Press, Madison, Wisconsin, in press.

Oman, C. 1989. *New Forms of International Investment in Developing Country Industries*. OECD Development Center, Paris, France.

OTA (U.S. Congress Office of Technology Assessment). 1988. *Copper: Technology and Competitiveness*. U.S. Congress Office of Technology Assessment, Washington, D.C.

Priest, R. T. 1993. The most essential projects: Manganese and U.S. development assistance to Brazil, 1948 to 1953. Paper presented to the Society for Historians of American Foreign Relations Annual Conference, Charlottesville, Virginia, June 19, 1993.

Stuckey, J. 1983. *Vertical Integration and Joint Ventures in the Aluminum Industry*. Harvard University Press, Cambridge, Massachusetts.

United Nations. Various years. *World Trade Tables*. United Nations Organization, New York.

Wagenhals, G. 1984. *The World Copper Market: Structure and Econometric Model*. Springer-Verlag, Berlin.

35

Development, Environment, and Energy Efficiency

Ashok Gadgil

Abstract

Technology choice is a central feature of economic policy in many developing countries, where progress in building an energy infrastructure, for example, is often constrained by lack of financing. Investments in energy efficiency may be more economical than a traditional supply-side-only energy strategy. An efficiency-based strategy provides both economic and environmental benefits, illustrated here by a potential project to build a compact fluorescent lamp manufacturing plant in India. However, significant market and institutional barriers hinder the implementation of such as strategy. Policy innovations such as a Global Collaborative on Energy Efficiency may be needed to overcome these barriers.

Introduction

In most developing countries, national priorities are focused on development. Development, as commonly used, does not mean merely growth in per capita income. Rather, inherent in this term is also a vision of structural change of the economic base, from one centered primarily on agriculture and natural resources to one emphasizing industrial production. A population shift from rural to urban is also envisioned. While this sounds unattractive to those who have an idyllic vision of rural life, we know of no other pathway to satisfy the needs and priorities of developing countries. Illustrative priorities are: reduced unemployment, lower infant mortality, improvements in industrial and agricultural productivity, accessible modern health care, improved nutrition, increased literacy, housing availability, access to good communication, development of a transport infrastructure, food security, industrialization, and good public hygiene. The ranking of these priorities changes in the course of development from the satisfaction of basic needs to higher-order needs (e.g., from food and shelter to vehicles, television, refrigerators, etc.). In any case, the problems of the environment, particularly global ones, rank relatively low on this list.

This chapter proposes, however, that the large-scale use of energy-efficient technologies can simultaneously support the efforts of developing countries to meet their priorities and the global efforts to reduce human impact on the environment. The likelihood of large-scale acceptance of energy-efficient technologies by the developing countries is based on a variety of conditions internal to the developing countries, including (1) the fact that demand for electricity continues to outstrip supply, despite large expansions in power systems that have supported 8% annual growth during the past two decades (Levine *et al.*, 1991; Meyers *et al.*, 1989); (2) the impossibility of financing the expected demand for capital of more than $100 billion (throughout the paper all dollars are U.S. dollars) annually by 2010 for the planned power sector expansion in the developing world; (3) the rapidly increasing share of developing country budgets absorbed by the energy sector; (4) the much cheaper availability of increased energy services through efficiency improvements, compared to supply side options; and (5) the rising local concerns for the protection of the environment.

In discussing energy systems, a distinction is made between supply-side technologies (e.g., refineries, gas turbines, power plants) and end-use technologies (e.g., motors, compressors, lights). A corresponding distinction must be made between raw energy (kWh) and energy services (which may have units appropriate to the service desired). Raw energy is what is supplied to end-use technologies, which then deliver specific energy services, such as lighting, heating or refrigeration, transport, or motive power. The energy appliances consume useful energy and provide an energy service to the user. The energy efficiency perspective focuses on technologies that provide a comparable energy service while consuming less useful energy.

Energy efficiency is often incorrectly perceived by developing country policymakers as a luxury affordable and essential only to the rich industrialized countries, because of both their wealth and high energy use. In reality, energy efficiency is a vital constituent of a successful development strategy that developing countries can ill afford to ignore. The promotion of high-efficiency appliances and equipment in developing countries is attractive to the industrialized countries because of environmental concerns regarding carbon emissions. For developing countries, such improved energy end-use efficiency is attractive because it provides a rare example of a cost-effective, no-regrets policy to speed the development process, especially for the world's poorest citizens, while simultaneously reducing local and regional environmental problems and the risk of potential global climate change.

In the next section, we illustrate how new energy-efficient technologies can provide energy services to the developing economies much more cheaply than conventional technologies powered by electricity from new power plants. We show that investments in factories for producing the new technologies are far more favorable than investments in new power plants. Barriers to the adoption of such technologies are then discussed, and examples of experience deploying these technologies in the United States and in developing countries are provided. Finally, I outline a possible action plan for accelerating the implementation of energy efficiency in the developing countries.

Economics of Energy-Efficient Technologies in Developing Countries

Building plants to manufacture energy-efficient technologies such as compact fluorescent lamps (CFLs) is more than a hundred times cheaper than building new power plants per unit of electricity produced or conserved. The analytical framework allowing such comparison is summarized in the following illustrative example using the Indian energy system as a specific context for a CFL and a CFL factory (for a detailed analysis, see Gadgil and Jannuzzi, 1991; Gadgil *et al.*, 1991a).

Compact Fluorescent Lamps

Today, India's utilities (almost all of which have evening peak demand) subsidize residential electricity for lighting and other services while many industries are unable to obtain enough power to meet their needs. Electricity for lighting now represents approximately 34% of Indian peak power and roughly 17% of the electrical energy consumed in the country. Incandescent lighting is estimated to constitute at least 17% of the peak demand and roughly 10% of the national electricity consumption (210 TWh[1] in 1990–91). Incandescent lighting consumption and its contribution to peak electric demand could grow rapidly because only about 30% of India's 155 million households are currently electrified.

CFLs last 10 to 20 times longer than incandescent light bulbs and provide the same high-quality light using less than one-quarter the electricity. In the United States, one 16-watt CFL replaces a 60-watt incandescent bulb and conserves 44 watts at the meter. When transmission and distribution (T&D) losses (8% in the United States) are factored in, the replacement achieves a savings of 47.8 watts at the bus bar of a U.S. power plant. Incandescents designed for the vagaries of developing countries' power systems are more robust and have lower luminous efficacy. Thus, in India a 16-watt CFL replaces a 60- or 75-watt incandescent (weighted average wattage 65.5 watts), and the consumers obtain a little more light with the CFLs than they did with the incandescents. Ignoring the value of the additional lighting, the savings at the meter from this replacement are 49.5 watts. When India's high T&D losses of 20% are taken into consideration, bus bar savings increase to 62 watts. Since CFLs replace only the most heavily used incandescent bulbs, they have a peak period coincidence rate that is significantly higher than the average incandescent. Consequently, replacing a heavily used incandescent in India with a CFL conserves 42 peak-coincident watts at the power station.

Assume that a 16-watt CFL always has a lifetime of 10,000 burning hours (the life can vary somewhat depending on the on–off cycle). Over its lifetime it will save 478 kWh and 440 kWh respectively at a U.S. bus bar and meter, and 619 kWh and 495 kWh at an Indian bus bar and meter.

---

[1] 1 TWh = 1×10^9 kWh

Energy Economics of a Single CFL: Cost of Conserved Electricity and Cost of Avoided Peak Installed Capacity

The cost of conserving a kWh by replacing incandescents with long-lasting CFLs can be compared to the cost of producing a kWh of energy. Calculating an energy-efficient technology's cost of conserved energy (CCE) is straightforward. It is given by the annualized net cost of the energy-saving technology divided by the annual energy savings. This can be written as:

CCE = [(investment) × (capital recovery rate) +
    (net annual increase in operation and maintenance costs)] / (annual energy
    saved)

A CCE is a useful measure of an energy-efficient technology because it is independent of any supply-side variables (e.g., energy prices), and depends only on the incremental cost, incremental efficiency, and use-intensity of the new technology. However, a technology's net benefits can be different from different perspectives. From a societal perspective, the net benefit is the difference between the cost of new generation and the CCE. For a residential application in a lamp socket used about 1000 hours annually, calculations show that the societal CCE of a 16-watt CFL ($0.02/kWh) is one-sixth the long-run marginal cost (LRMC) of electricity (approximately $0.12/kWh). From the consumer's perspective, the annual benefit derived from a CFL is the difference between the annual savings from avoided energy consumption and avoided purchases of incandescent light bulbs, and the annualized cost of the CFL. From the utility's perspective, the net benefits of energy-efficient technologies are calculated by adding avoided generation expenditures to avoided electricity subsidies and subtracting the lost revenues from reduced electricity sales and subsidies (if any) of the energy-efficient technologies. In most developing countries the benefits to utilities of installing CFLs are large enough that subsidizing the technology heavily is a remunerative proposition in almost every case. For most Indian utilities a 50% subsidy of CFLs (to ensure their market success) would yield returns of approximately 250% (on an annualized basis). A scenario for the introduction of CFLs in India by a small transfer of subsidy from residential electricity to residential CFLs has been illustrated (Gadgil and Jannuzzi, 1991). The authors show that in ten years, at 20% saturation, CFLs would save the Indian utilities the equivalent of $1 million per day.

Major policy decisions related to power system expansion in the developing countries are often based on the availability of capital resources for initial investments. The cost of avoided peak installed capacity (CAPIC) can be used to inform such decisions. CAPIC refers to the net present value of an energy-efficient technology (to be operated for the duration of the life of a power plant) that renders unnecessary the installation of a kW of peak generating capacity. In India, the avoided peak demand at the bus bar (42.38 watts) divided by power plant availability at peak hours (0.573) equals 74 watts, the installed capacity avoided with use of one residential CFL. The investment cost of new installed capacity (which lasts 30 years for a typical Indian utility) is about \$867/kW. In comparison, from a societal perspective, one CFL operated for that period in India costs (in net present value) \$10.13. Hence the CAPIC is \$137/kW, six times less than the cost of new installed capacity.

Energy Economics of a CFL Manufacturing Plant

In this section, we compare a CFL production plant with an electricity generation plant. We do not ignore the "free" additional light that the consumers may obtain with the CFLs; instead, we assume that the 16-watt CFL will replace a 75-watt incandescent lamp in India and will provide the same amount of illumination (900 lumens).

A modern automated CFL plant costs about U.S \$7.5 million: \$5 million for the machinery and \$2.5 million for the buildings and land.[2] It produces 3 to 7 million CFLs annually, depending on the number of shifts it operates. For this calculation we assume that the CFL plant operates four shifts like a power plant, producing 6 million lamps annually. We credit each production year with all the energy saved by the 6 million CFLs over their lifetime. Socket life (as opposed to burning life) of the CFL may vary from 1.1 years to 10 years or more, depending on the duty cycle of use.

A 16-watt CFL producing 900 lumens replaces a 75-watt incandescent, saving 59 watts at the meter, or 74 watts at the bus bar. Assuming lamp use of 4 hours/day (1460 hours annually) in a residential setting and a burning life of 10,000 hours, the socket life of the CFL is 6.85 years. If a factory produces CFLs only for residences, production for the first 6.85 years goes to increase the number of lamps in the sockets. Afterwards, each year's production just replaces the lamps that have burned out after 10,000 hours of use. So, in the steady state there are:

(6 million CFLs/year) (6.85 years) = 41.1 million lamps in use.

Assuming only a 70% peak coincidence, these will save a peak demand of:

(41.1 million CFLs) (74 W/CFL) (0.7) = 2129 MW at the bus bar.

To produce this amount of peak power, the installed capacity needed in India is 2129/0.573 = 3715 MW.

If the capacity were to consist of coal-fired thermal power stations (common in India), an investment of \$5.6 billion would be required (\$1500/kW) of which 40% (or \$2.2 billion) would be in foreign exchange. If the peaking capacity is obtained from cheaper (but more expensive to operate) gas turbines, the total investment

---

[2] To be conservative, we use typical land costs in industrialized countries; costs for land near urban areas in developing countries (e.g., Mexico City) would probably be less.

would still be $2.8 billion, with at least $1.1 billion in foreign exchange. In comparison, investment in a CFL plant requires $5 million in foreign exchange and a total investment of $7.5 million, more than 350 times cheaper than the power plant investment.

Barriers

For a variety of reasons, households, businesses, manufacturers, and government agencies in both industrialized and developing countries fail to fully exploit cost-effective energy-conserving opportunities. The result is a significant gap between current and optimum levels of energy efficiency. The reason this occurs is that there are a number of barriers in the research and development, production, commercialization, acquisition, and use of energy-efficient systems. These barriers include a lack of international and national funding for efficiency investments, a lack of effective collaboration between industrialized and developing countries, weak national commitments to efficiency, weak national institutional capability, supply-oriented and centralized utilities, energy subsidies, the payback gap, a lack of energy-efficient products, a lack of information on energy-efficient products and practices, and capital constraints.[3] In addition to these market barriers, other reasons for the failure to use energy economically and as efficiently as technically possible include inadequate monitoring and program evaluation, a shortage of trained building designers and tools, a shortage of trained auditors and retrofitters, and inadequate research and development funding to keep the pipeline full of more efficient products.

Barriers that are unique to developing countries are discussed in a later section.

Experience in the United States (Where Markets and Institutions Work Reasonably Well)

Technologies and policies for improving energy efficiency have primarily originated in the United States and other industrialized countries, especially in response to high energy prices following the 1973 Organization of Petroleum Exporting Countries (OPEC) oil embargo.

Figure 1 shows United States energy and electricity consumption (E) and gross national product (GNP) trends since 1960 (Gadgil *et al.*, 1991b). For the United States, we use the conventional definition of energy intensity as energy (primary, excluding biofuels, since they are renewable) consumed per dollar of gross national product (E/GNP).

As a nation industrializes, E/GNP first rises as energy-intensive technology is used to build an infrastructure, then falls as more of a service economy prevails

---

[3] For a review of literature on barriers to increased energy efficiency, see Hirst and Brown (1989), Blumstein *et al.* (1980), Stern and Aronson (1984), and Fisher and Rothkopf (1989). Reddy (1991) provides an excellent discussion of these barriers with particular reference to developing countries and proposes ways to overcome them.

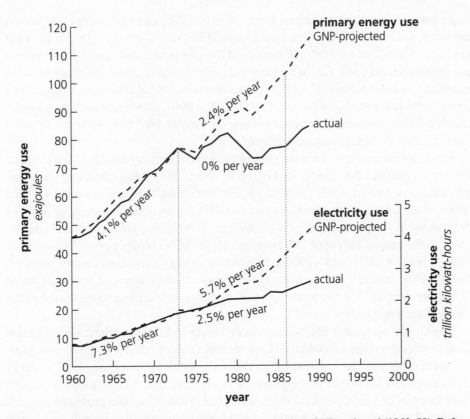

Figure 1. Total U.S. primary energy and electricity use: actual vs. GNP projected (1960–89). Before the 1973 oil embargo, total primary use was growing at about the same rate as GNP, and electricity use was growing about 3% faster. By 1986, projected primary energy use was 36% higher than actual use, indicating a savings of 28 EJ. Increased energy efficiency is credited with two-thirds of these savings, or about 19 EJ, with the remainder resulting from structural change. Electricity use followed a similar pattern, growing at a much smaller rate than projected until 1986. GNP projected values are based on 1973 efficiency and GNP; electricity projections include an additional 3%/yr to account for increasing electrification. (Energy data from Energy Information Administration, U.S. Department of Energy; structural change data from Schipper *et al.*, 1990, and OTA, 1990).

and as energy intensities decline with time (Reddy and Goldemberg, 1990). This pattern was especially prevalent for the United States and other Organization for Economic Cooperation and Development (OECD) member countries; countries that are now industrializing can use more efficient products and methods for both infrastructure building and energy services. Between 1960 and 1973, energy was inexpensive and no distinction was made between providing raw energy and providing the energy services of space heating, lighting, motor shaft power, etc. Accordingly, little explicit attention was paid to improving energy efficiency, which remained "frozen." Primary energy use and U.S. GNP were thus linked, and each climbed about 4% per year. In 1973, the OPEC oil embargo introduced a powerful incentive to conserve energy. During the 13 years of high oil prices and more progressive energy policies from 1973 to 1986, national energy use stayed constant, while U.S. GNP grew by a total of 35%, i.e., 2.4% per year. Of the total savings of 28 EJ (the difference between GNP-projected and actual consumption during this period), one-fourth to one-third is attributed to structural changes in the economy and the remaining two-thirds to three-fourths is attributed to improved energy efficiency (Schipper *et al.*, 1990; OTA, 1990). Efficiency measures implemented during this period avoided a sharp increase in coal use and hence avoided a rise of 50% in U.S. greenhouse gas emissions.

Even more impressive than the past reductions in primary energy is the electricity conservation, also shown in Figure 1. Until 1973, total electricity use was growing at a rate of 7.3% per year (3.2% faster than GNP). Between 1973 and 1986, electricity use grew only as fast as GNP, for an annual savings of 3.2% or 50% in the 13-year period. This 50% savings, 1160 billion kWh per year, is equivalent to the annual output of 230 baseload (1000 MW) power plants. In late 1985, when OPEC's oil prices collapsed, gains in energy efficiency nearly stopped. Since 1986, primary energy consumption has climbed again at a rate of about 2.5% per year vs. 3% per year for GNP, directly contributing to increased emissions of carbon dioxide.

The United States has enjoyed success with several policies about which a great deal has been written elsewhere. These include: electric utility demand-side management programs (Krause and Eto, 1988; EEI, 1990; Levine *et al.*, 1992); energy use labels and standards for appliances, equipment, and buildings (OTA, 1991; McMahon *et al.*, 1990; CEC, 1988); and "golden carrot" rebate programs for new product development, and revenue-neutral fee/rebate programs for efficient buildings, appliances, and automobiles (Rosenfeld and Price, 1992).

Experience in Developing Countries (Where Markets and Institutions Do Not Work Terribly Well)

Several major characteristics are common to utilities in developing countries. First, because availability of electricity has been seen by developing country governments as a key aspect of the development process, the governments have historically taken a lead (and often legislated an exclusive) role in investments and operation of the electricity sector. In most developing countries, the government (or a government-owned and -financed corporation) is the sole owner and operator of electricity services. The electricity sector owns and controls a very large amount of capital stock; sometimes it is the financially most powerful sector of the government. Large and powerful bureaucracies, whose thinking and operations are supply dominated (often exclusively so), operate this sector. Like all large and mature bureaucracies, they are resistant to challenges to their traditional ways. In practical terms, this means that most do not have a strong commitment to identify and implement the least costly solutions to supplying energy services. Often there is only an understaffed and powerless office that is responsible for conservation.

Another major common characteristic of utilities in developing countries is the subsidized electricity tariffs, particularly for domestic sectors (and sometimes also for small commercial and small-scale industrial sectors, because these are operated in conjunction with the residence of the sole proprietor). The justification for subsidy is societal; access to affordable electricity is considered essential to an improved standard of living. If the electricity tariff equaled the true marginal cost it would be unaffordable to a large majority of the population in the poorer developing countries. This logic is also extended to subsidize electricity supply to agricultural pumping, for example, in India. The agricultural sector is seen as essential for national self-sufficiency in food supply. As a result, the tariffs for the agricultural sector in India can be as low as $0.01/kWh; in some parts of the country there is a small flat monthly charge, no meter, and no tariff.

A third common characteristic of utilities in developing countries is that most are losing money as a result of low tariffs and rising costs of electricity expansion. The shortfall is made up by "loans" from the government to the (government-owned) utilities. It is not clear how and when these can be repaid. There is substantial political pressure on the utilities to keep the tariffs low. As a result, the utilities are fed up with subsidies and are struggling to keep themselves from going bankrupt. They are thus deaf to any suggestion that a subsidy from the utility to an energy-efficient appliance may be economically viable and a better alternative to system expansion.

Oil and gas are treated much like electricity. In oil or gas exporting countries, the problems are commonly further complicated by assigning low energy prices to the fossil fuels in the domestic markets, which undercuts any efficiency improvements.

Despite the difficulties described above, several developing countries have made significant and substantial improvements in the efficiency of their energy use. Several members of the Association of South East Asian Nations (ASEAN) countries (in particular Thailand, Singapore, and Malaysia) have experienced very rapid growth in office buildings, and in the associated electricity consumption for cooling. These countries have adopted building standards aimed at limiting the rapid growth of electricity demand for cooling from new construction. Commercial buildings in the ASEAN countries already consume more than 30% of the total electricity generated in the region. The demand from this sector may account for 40% of new electricity demand in the near future (Levine and Deringer, 1987). With scientific support from researchers at Lawrence Berkeley Laboratory, teams from these countries developed their own standards for new buildings based on the computer program DOE-2. Some of the standards are based on simplified methods for estimating energy consumption that were calibrated with the use of DOE-2.

A national electricity conservation program named PROCEL was initiated in Brazil in 1985. PROCEL primarily engaged in technology R&D, demonstration projects, educational and promotional campaigns, and direct installation of conservation measures. As of 1990, PROCEL had undertaken projects worth about $20 million, with matching contributions from utilities and research institutions.

One of PROCEL's large projects has been outright replacement of incandescent street lights with mercury vapor or high-pressure sodium lamps. About 300,000 lamps were replaced by 1989, with most of the lamp cost paid directly by PRO-CEL. This activity has continued, with utilities bearing the cost of replacement. This is necessary because the municipalities that own and operate the lamps are financially too strapped to make the investments in efficient lamps themselves. The utilities have an incentive to replace the lamps because they sell electricity for those lamps to the municipal councils at a lower tariff than the utilities' cost of generation and distribution. Although it is difficult to estimate the magnitude of total energy conservation that can be attributed to PROCEL's efforts, the officials operating the program attribute electricity savings of at least 1.1 TWh annually as of 1989 to their programs. Some estimates of savings are as high as 2.5 TWh annually (Geller, 1990).

Beyond Technology, Economics, and Policy: The Need to Address Institutional and Organizational Behavior

Improved end-use efficiency holds the promise of increasing the level of electrical services provided to consumers while reducing the requirements for new generation plants. Recent increases in cost-effectiveness and availability of efficient technologies are argued as evidence that improving end-use efficiency is an effective development strategy, even (or especially) for the poorest nations. Based on the current low levels of electrification and small penetration of appliances, developing countries are said to possess the opportunity to "leap-frog" over the development stage experienced in the United States and other industrialized countries in which low-efficiency appliances and equipment achieved wide-scale penetration. With such leap-frogging, developing countries could proceed directly to the increased use of higher-efficiency appliances.[4]

The cost–benefit framework used in the CFL example above represents one approach for designing implementation programs. In such an approach, the goal is to quantify the economic dimensions of options that have been shown to be technically feasible. Policy analysts, utility planners, and other decision-makers frequently use such technical-economic evaluations as the basis for recommending implementation of a technology or program. This traditional methodology—which has been developed and used in the United States and other industrialized countries in the two decades since the oil crisis—is not always sufficient for achieving successful implementation in a developing country. In this section, we draw upon our experience in India to explain why, and point to the need for a wider approach. For a detailed discussion see Gadgil and Sastry (1994).

There is a growing recognition that technical and economic analyses alone do not account for all the factors likely to be important in planning and assessing the

---

[4] The idea of leap-frogging appears to have gained currency in international development circles (see, for example, Goldemberg *et al.*, 1988).

prospects for energy efficiency (Cebon, 1992). Most of the research in this area has arisen from the perspectives of researchers from the industrialized countries, and argues that behavioral dimensions should be added to the traditional approach (see, for example, Robinson, 1991; Katzev and Johnson, 1987; Stern and Aronson, 1984; and Kempton and Nieman, 1987). While accounting for the behavioral aspects of consumer response to policy instruments is certainly important, and energy-efficiency programs that are designed to be economically attractive indeed may fail if they neglect the social, psychological, and behavioral aspects of consumers' decision-making, a recognition of institutional and organizational behavior must be added to the analysis. However, there are few examples both in the literature and in practice of a given energy-efficiency proposal being tackled simultaneously at the technical, economic, behavioral, and institutional levels. It is exactly this lack of an integrated approach that has resulted in the poor record of energy-efficiency implementation in developing nations. Thus, if we are to understand which policy options work and which do not, the "top-down" approach (such as the World Bank's pricing reform strategy) must be combined with a program-specific, "hands-on" approach.

The Indian situation provides a useful example. Governmental organizations, of one form or another, have been central to decision-making in many walks of public life in India for decades. Like governments throughout the developing world, the Indian government perceives itself as the active promoter of change in the country, and exerts unparalleled influence in the economy both directly and through government-owned corporations and enterprises. The elites that run the institutions of the national and state governments can influence decisions in both public and private sectors, since enterprises in the latter are governed by a large number of government regulations (such as requirements for licenses and permits, quotas for scarce raw materials, connections for electricity and water, environmental restrictions, and access to loans from the large banks). While economic activity is overregulated on paper, the actual enforcement of the regulations is erratic (Crook, 1991; Bardhan, in press) and can depend on the disposition of any of a number of influential officials. This fact of life in India explains why it is difficult to start up a new enterprise without having access to a member of this power elite to enable one to negotiate the myriad rules and regulations. Despite recently initiated efforts to cut the red tape (Gargan, 1992), reducing the traditional influence of the bureaucracy is a slow and difficult process (Bardhan, in press).

Three points about the role of the government officials are important to note. The first is that connections to these decision-makers can be important to anyone attempting to implement change, within the government as well as outside of it. The second is that the influence of government administrators is not necessarily malignant: bureaucrats can (and do) play a positive role both in promoting change and in reconciling perspectives of different interest groups. Lastly, because any individual or organization that needs to get things done must draw on a limited stock of personal goodwill and informal connections to get around the various

461

bureaucratic obstacles, there is a reluctance to risk this limited capital on ventures not perceived to be major.

Another feature of the institutional environment in developing countries is also important. Relative to the advanced industrial countries, developing countries have a low level of institutional diversity in terms of the number of organizations which may be authorized, capable and willing to carry out a given project. For example, in the United States, support for a cost-effective energy-efficiency demonstration project may be obtained from one of the numerous offices of the Department of Energy, or any of its large institutions, or the appropriate state energy commission, or even the U.S. Environmental Protection Agency, within which there is a multiplicity of channels by which the project may be funded. This is rarely the case in developing countries, where there are only a few opportunities for obtaining support for a given project. As a result of the relative paucity of "niches" in a given area, the institutions that occupy a niche guard it jealously and, often, by all means, fair or foul. If an organization regards projects in a certain area as within its purview, it can be very reluctant to allow others to undertake similar projects and may call on the connections it has already established as niche occupier to block potential entrants. Consequently, there is little competition within each niche and low mobility of organizations between niches. A further characteristic of the Indian environment is that the pyramid of hierarchical authority narrows extremely fast as one goes up the institutional ladder. As a result, the personal opinions of the people at or near the top have a disproportionately large influence on organizational outcomes. Thus, three factors—steeper hierarchical structure of institutions, fewer opportunities for project support, and more intense infighting for niches—make it possible for top managers to block implementation of a new idea more effectively in India than in industrial countries.

In summary, large-scale implementation programs for energy efficiency in developing countries must pay attention not only to technology, economics, and policy, but also to a complex set of institutional and organizational problems. A possible structure for addressing such problems is proposed in the next section.

Possible Action Plan: Global Collaborative on Energy Efficiency

The action plan outlined here is intended to facilitate an alternative growth path for the inevitable future expansion of developing country economies that is more capital intensive on a first cost basis, but is ultimately less costly and less energy intensive. This action plan addresses the two related hurdles that developing countries face in implementing energy efficiency aggressively. First they are short of capital, and thus commonly opt for the option with the lower first-cost, and secondly, they do not have the institutional capacity for analyzing the options in the first place, and then do not have the institutions for implementing them once the better ones are known. The action plan proposes the establishment of a project tentatively called the Global Collaborative on Energy Efficiency (GCEE).[5]

The key elements of this plan call for the establishment of international-level central office and individual national-level offices in developing countries. These offices will implement two distinct but related functions: financing and information. The national-level offices will strongly emphasize collaboration among the numerous players and stake-holders involved in energy efficiency: international, regional, national, and local funding sources; institutions; agencies; businesses; universities; technical schools; advocacy groups; etc.

For the financial functions, the central office will include an energy-efficiency bank that collects funds from multilateral and bilateral agencies, passes them on to national offices, and receives returns on energy-efficiency investments from national offices for passing them on to the primary funding agencies. The national offices receive investment funds from the central office, and have a projects division that disaggregates these funds into several smaller projects (less than $100 million each) that the World Bank-sized organizations find so difficult to deal with. The projects division also provides grants for project preparation, conducts national-level analyses, and supports demonstration programs and policy experiments.

The second function, that of information, is similarly vertically integrated between the central office and the national offices in individual countries. The national-level activities will include access to technical and programmatic international experience, support for conduct of national surveys and analyses, information support to government and business decision-makers and opinion-makers in the country, assistance for curriculum development in universities and technical institutions of learning, and conduct of multi-level training programs. It will also include a fellowship program that supports exchange of activists between various non-governmental organizations (NGOs) in industrial and developing countries to improve their effectiveness.

To be successful, it is essential that the GCEE will simultaneously address the constraints faced in the areas of information and awareness, technical issues, policy, and institutions. An approach addressing only some of these elements while ignoring others is not likely to succeed in the long run. The last constraint, of adequate local institutional support, is often the most difficult and is the barrier on which many well-intentioned assistance programs flounder.

One could avoid the creation of a large bureaucracy for GCEE by housing it within an international organization such as the World Bank or the United Nations Development Program (UNDP). The success of the recently founded Global Environmental Facility (GEF) in the World Bank is an encouraging move in this direction.

---

[5] The term "collaborative" has been recently used in the United States to describe a process in which a diverse group of institutions and individuals (including representatives from utilities, independent power producers, industry, local and state government agencies, consumer advocacy groups, and environmental advocacy groups) work together to devise regulatory reforms to encourage utilities to promote energy efficiency. A similar global concept is suggested here.

Concluding Remarks

The long-term prospects for industrialization and development in the developing countries to the current levels of affluence in the industrial world are clouded by the untenability of the resource and pollution burden it will impose on the planet's ecosystem. In the short term (on the order of a couple of decades), however, there appears to be considerable scope for incorporating an accelerated introduction of energy efficiency as an integral part of the development process. There are various barriers to such an introduction, some arising from markets, others from institutions. My proposal for a Global Collaborative on Energy Efficiency is aimed at addressing the institutional barriers.

Acknowledgments

I would like to thank Lynn Price for her thoughtful review and editorial assistance. I would also like to thank Vicki Norberg-Bohm and George Golitsyn for their helpful comments.

References

Bardhan, P. A political-economy perspective on development. In *The Indian Economy: Problems and Prospects* (B. Jalan, ed.), Penguin, New Delhi, India, in press.

Blumstein, C., B. Krieg, L. Schipper, and C. York. 1980. *Energy 5(4)*, 355.

Cebon, P. B. 1992. Organizational behavior, technical prediction and conservation practice. *Energy Policy 20(9)*, 802–814.

CEC (California Energy Commission). 1988. *Conservation Report.* CEC, Sacramento, California.

Crook, C. 1991. India: Caged. *The Economist.*

EEI (Edison Electric Institute). 1990. *State Regulatory Developments in Integrated Resource Planning.* EEI, Washington, D.C.

Fisher, A. C., and M. H. Rothkopf. 1989. Market failure and energy policy. *Energy Policy 17*, 397.

Gadgil, A. J., and G. DeM. Jannuzzi. 1991. Conservation potential of compact fluorescent lamps in India and Brazil. *Energy Policy 19(6)*, 449–463. Also Lawrence Berkeley Laboratory Report LBL-27210.

Gadgil, A. J., and M. A. Sastry. 1994. Stalled on the road to the market: Lessons from a project promoting lighting energy efficiency in India. *Energy Policy 22(2)*, 151–162.

Gadgil, A. J., A. H. Rosenfeld, D. Arasteh, and E. Ward. 1991a. Advanced lighting and window technologies for reducing electricity consumption and peak demand: Overseas manufacturing and marketing opportunities. *Proceedings,* IEA/ENEL Conference on Advance Technologies for Electric Demand Side Management, April 4–5, 1991, Sorrento, Italy. Also Lawrence Berkeley Laboratory Report LBL-30389.

Gadgil, A. J., A. H. Rosenfeld, and L. Price. 1991b. Making the market right for environmentally sound energy-efficient technologies: U.S. building sector successes that might work in developing countries and eastern Europe. *Proceedings*, ESSETT'91, Environmentally Sound Energy Technologies and Their Transfer to Developing Countries and European Economies in

Transition, Milan, Italy, October 21–25, 1991. *Proceedings* to be published by ENEL, Italy. Also, Lawrence Berkeley Laboratory Report LBL-31701.

Gargan, E. 1992. A revolution transforms India: Socialism's out, free market in. *New York Times*, Sunday 29 March, p. 1.

Geller, H. S. 1990. *Electricity Conservation in Brazil: Status Report and Analysis*. American Council for an Energy-Efficient Economy.

Goldemberg, J., T. B. Johanssen, A. K. N. Reddy, and R. H. Williams. 1988. *Energy for a Sustainable World*. Wiley Eastern Limited, New Delhi, India.

Hirst, E., and M. Brown. 1989. Closing the Efficiency Gap: Barriers to the Efficient Use of Energy. Oak Ridge National Laboratory, Oak Ridge, Tennessee.

Katzev, R. D., and T. R. Johnson. 1987. *Promoting Energy Conservation: An Analysis of Behavioral Research*. Westview Press, Boulder, Colorado.

Kempton, W., and M. Nieman. 1987. *Energy Efficiency: Perspectives on Individual Behavior*. American Council for an Energy-Efficient Economy, Washington, D.C.

Krause, F., and J. Eto (eds.). 1988. *Least-Cost Planning Handbook for Public Utility Commissioners, Volume 2: The Demand Side: Conceptual and Methodological Issues*. NARUC.

Levine M. D., and J. Deringer. 1987. Implementation strategies for achieving energy-efficient buildings in ASEAN. *Proceedings* Workshop on Energy Conservation Policy and Measures for Energy Demand Management, October 12–16, 1987, Bangkok, Thailand. Also Lawrence Berkeley Laboratory Report LBL-24134.

Levine, M. D., A. Gadgil, S. Meyers, J. Sathaye, J. Stafurik, and T. Wilbanks. 1991. *Energy Efficiency, Developing Nations, and Eastern Europe: A Report to the U.S. Working Group on Global Energy Efficiency*. Secretariat, Global Energy Efficiency Initiative/U.S. Working Group on Global Energy Efficiency, IIEC, Washington, D.C.

Levine, M. D., H. Geller, J. Koomey, S. Nadel, and L. Price. 1992. *Electricity End-Use Efficiency: Experience with Technologies, Markets, and Policies Throughout the World*. American Council for an Energy-Efficient Economy, Washington, D.C. Also Lawrence Berkeley Laboratory Report LBL-31885.

McMahon, J. E., D. Berman, P. Chan, T. Chan, J. Koomey, M. D. Levine, and S. Stoft. 1990. Impact of U.S. appliance energy performance standards on consumers, manufacturers, electric utilities, and the environment. *Proceedings of the 1990 ACEEE Summer Study on Energy Efficiency in Buildings*. American Council for an Energy-Efficient Economy, Washington, D.C.

Meyers, S., J. Sathaye, and A. Ketoff. 1989. *Plans for the Power Sector in Thirteen Major Developing Countries*. Report No. LBL-27764, Lawrence Berkeley Laboratory, Berkeley, California.

OTA (Office of Technology Assessment, U.S. Congress). 1990. *Energy Use and the U.S. Economy*. OTA, Washington, D.C.

OTA. 1991. *Changing By Degrees: Steps to Reduce Greenhouse Gases*. OTA, Washington, D.C.

Reddy, A. K. N. 1991. Barriers to improvements in energy efficiency. *Energy Policy 9(12)*, 953–961. Also see the longer version available with the same title, as Report No. LBL-31439, Lawrence Berkeley Laboratory, Berkeley, California.

Reddy, A. K. N., and J. Goldemberg. 1990. Energy for the developing world. *Scientific American 263 (3)*, September.

Robinson, J. B. 1991. The proof of the pudding: Making energy efficiency work. *Energy Policy 19(9)*, 631–645.

465

Rosenfeld, A. H., and L. Price. 1992. Incentives for efficient use of energy: High prices worked wonders from 1973 through 1985, what are today's alternatives to high prices? Presented at *POWER (Program on Workable Energy Regulation) Conference: The Economics of Energy Conservation*, University of California, Berkeley.

Schipper, L., R. B. Howarth, and H. Geller. 1990. United States energy use from 1973 to 1987: The impacts of improved efficiency. *Annual Review of Energy 15*, 455–504.

Stern, P. C., and E. Aronson (eds.). 1984. *Energy Use: The Human Dimension*. National Research Council, Washington, D.C.

END PIECE

END PIECE

36

The Industrial Ecology Agenda

Clinton Andrews, Frans Berkhout, and Valerie Thomas

All human activity, from the most basic (our individual metabolism), to the most industrialized (energy infrastructures), is embedded in the earth's environment and leads to some transaction with it. This relationship works in both directions. Human activity is bounded by environmental conditions while also influencing the environment, by preempting a part of it and by emitting waste residuals into it. Indeed, the history of human activity can be seen as a history of overcoming environmental limitations in the pursuit of personal, political, or social goals. Throughout this history one finds expressions of a fear that certain absolute limits would be reached in the ability of the earth's environment to absorb human influences, whether material or spiritual. These fears have to a large extent conditioned the environmental values of human cultures. Today we are better able than in any previous time in human history to investigate the validity of such fears.

It is commonly believed that mass consumption society and the industrial structure on which it rests are depleting and overwhelming the earth's resources. Current levels of resource use and intervention in natural systems are already producing serious environmental impacts, and with growing economies and populations these problems are unlikely to diminish. Although industrial and regulatory systems have already demonstrated an enormous capacity to learn and adapt to perceived environmental constraints, these adaptations have not radically altered the basic environmental problems faced by industrial societies. For instance, although a trend toward "dematerialization"—cars are smaller and lighter, computers are smaller, and the service sector of the economy is growing—has been noted by some authors (Larson et al., 1986), the amount of matter and energy processed by global industrial systems continues to increase. Moreover, increases in atmospheric carbon demonstrate that fundamental global cycles have been significantly altered by human activities.

Human beings are recreating the biosphere in their own image. Agriculture long ago deforested Asia, Europe, and much of North America. The regeneration of these immense lost wildernesses is not even a point of reference. Likewise, we are now changing the atmosphere in ways that will change the global climate. Returning to preindustrial levels is considered to be practically and politically impossible; the debate is bracketed between stabilization and no action at all.

469

Industrial ecology is not just about managing industrial change; it is also about managing change of the earth. We are faced not with the task of reforming our activities to spare nature, but with the task of managing the transformation of nature into an extension of ourselves. In this process, human society itself will be fundamentally transformed.

Even with the knowledge that human activity is producing serious and irreversible damage to the environment, uncertainty and lack of focus inhibit our understanding of this damage and our attempts to reduce it. The principal claim of industrial ecology is that vigor and focus can be provided by viewing human activity and the environment as embedded in one another.

The Analytical, Critical, and Prescriptive Aspects of Industrial Ecology

Industrial ecology is an ambitious attempt to recast the discussion about the long-term future of human industrial activity. As represented in this book, it is analytical, critical, and prescriptive. In these various guises it can be an approach to a systematic analysis of nutrient flows in the environment, but also a vision of a sustainable industrial economy.

The "analytical" aspect of industrial ecology is to investigate the flow of materials and energy through industrial systems, and to catalogue how the discards and emissions of these flows affect natural fluxes of materials and energy. In other words, how do industrial economies sit within the biosphere?

Human activity has produced significant changes in the flows and reservoirs of some materials (for reviews see: Ayres, 1989; Turner *et al.*, 1990; and Houghton *et al.*, 1990). Raw materials are extracted or cultivated, processed and often transformed, used, and then emplaced or dissipated as waste residuals back into the biosphere. Study has shown that in many industrialized countries, point source pollution from industrial plants has been falling steadily under the impact of environmental regulation and structural change. On the other hand, dissipative sources of pollution associated with the consumption and use of products have continued to increase. In this book much space has been devoted to human disruption of the grand nutrient cycles and to the emission and impacts of toxics in the environment. To stabilize these impacts will require adaptations of production and consumption activities, but it may extend further into active management of the biosphere as well.

The "critical" aspect of industrial ecology seeks to understand which industrial fluxes of material and energy are causing serious disruptions to natural ecosystems, or generating human health risks. Such assessment is already common for some types of material: radioactivity, chlorofluorocarbons, and the greenhouse gases, to name a few. But there remain many substances used in industry for which systematic mass balances and environmental impact assessments have not yet been done.

Direct impacts on human health have driven the great majority of efforts to reduce pollution and prevent environmental damage. A shift is now occurring, which industrial ecology would encourage, toward a more inclusive treatment of impacts to include a range from specific human impacts to global environmental impacts. Today the significance of environmental impacts is judged primarily in terms of final human health impact. We can at least imagine a "Copernican" revolution in environmental impact assessment which moves away from fixing the human subject at the center of its universe. For one thing, the timescale of a human life is not a good metric for thinking about sustainable development.

The third, "prescriptive" aspect of industrial ecology is to implement change. Here we are immediately faced with the problem of agency. Where will the principal impulse for change come from—the regulator or the regulated? Industrial ecology builds on the observation that much regulation has failed to deliver qualitative improvements in the environment, and argues that regulation of emissions can no longer be seen as the primary agent for the achievement of sustainable development (Commoner, 1990; Stavins and Whitehead, 1992).

In response, industrial ecology places on industry a primary responsibility for reducing and eliminating environmental impacts, with governments and consumers supporting the process. Industrial firms alone have the technological and organizational capabilities to accomplish the broad changes envisaged by industrial ecology. But virtue will not descend on firms unprompted, although there is strong evidence that an environmental consciousness already exists in some board rooms—witness the industrial interest in industrial ecology. A powerful set of economic, social, technological, and legal forces will push them in this direction. Once this stimulus has been given, then perfectly ordinary profit-seeking companies may find it in their own enlightened interests to pursue environmental improvements in their products and activities.

The Ecological Metaphor

Natural ecosystems provide the central metaphors of industrial ecology: nutrients and wastes become raw materials for other processes, and the system runs almost entirely on solar energy. The analogy suggests that a sustainable industrial system would be one in which nearly complete recycling of materials is achieved. The agenda of industrial ecology, simply put, is that all wastes will find an economic use, or be placed into the environment without disruption. It further suggests that current industrial systems, equipped with rather primitive metabolisms, will be forced by environmental and social constraints to evolve more sophisticated metabolisms which exploit the nutrient or energy value of matter more completely.

Like most analogies, this one should not be taken too literally. It is clear that human societies are governed by quite different rules and conditions than other natural systems. Indeed, the unbridled use of natural analogues in defining social or economic policy can produce dangerous results, like Social Darwinism. Moreover, industrial systems apply quite different principles to the processing of materials and energy, depending in general on fast, high-temperature, high-pressure processes. Nevertheless, it is instructive to explore in some detail what an industrial ecosystem could involve.

Energy and Materials in an Industrial Ecology

Three aspects of the simple industrial ecology metaphor will be highlighted: the use of renewable energy, the recycling of limited material resources, and the robustness of systems. Natural ecosystems rely chiefly on inexhaustible energy inputs, specifically solar energy, and this ensures their long-term sustainability. Today's industrial systems rely instead mostly on energy stored in fossil fuels. Likewise, on a system-wide basis, natural systems are highly efficient users of the materials consumed, unlike today's "linear" industrial systems (see Graedel, this volume). This is primarily achieved through recycling of nutrients. Finally, healthy natural ecosystems exhibit a species diversity that make them robust to exogenous shocks. Inexhaustibility, recycling, and robustness are central themes in the industrial ecology agenda.

Renewable Energy

Internalizing, physically and economically, the environmental impacts of energy resource use will require significant, and as yet imperfectly understood, changes in the global energy infrastructure. Several paths are open. Fossil fuel combustion products could be captured and sequestered, rather than being allowed to enter the atmospheric sink. While commonplace for some constituents (e.g., particulates, sulfur, nitrogen oxides), the large quantity of carbon dioxide emitted would require unfamiliar underground or deep ocean storage. Researchers have yet to demonstrate the cost-effectiveness and stability of such CO_2 sequestration. Nuclear power is attractive for its lack of emissions, and is in some respects very close to the industrial ecology ideal in that it was conceived of as an integrated recycling system. But its wastes are toxic and not readily available for re-use, and the nuclear fuel cycle produces nuclear weapons-usable materials (Berkhout, this volume). Alternatively, one can imagine a renewable energy future that relies on solar, biomass, wind, hydropower, and geothermal energy. Its problems are also well known: low power density, large scope of energy generating and conversion facilities, environmental externalities, the need for new transmission infrastructures, and so on.

Williams (this volume) takes on the challenges of renewables and spells out a detailed global scenario for their development over the next 60 years. This scenario projects that nearly one-half of the world's energy could be supplied from renewable sources by the year 2025, with biomass providing the largest share. While fossil energy use siphons carbon from the long (geological) cycle to the short (biochemical) cycle, biomass energy manages carbon stocks entirely within the short cycle. Although this scenario requires radical change in energy sources and energy use, and is optimistic in its projection of the development and adoption of renewable technologies, the result is a world that still depends on fossil fuels for one-half of its energy. Even to achieve this goal will require important changes in public policy; significant institutional and economic barriers would need to be crossed.

Turner, B. L., W. C. Clark, R. W. Kates, J. F. Richards, J. T. Mathews, and W. B. Meyer (eds.). 1990. *The Earth as Transformed by Human Action*. Cambridge University Press, Cambridge, U.K.

Uutela, E., and N. P. Black. 1990. Recycled fiber use expected to grow by 41% and reach 130 million tons yearly by 2001. *Tappi Journal*, 50–52.

Organizing Committee Members

Robert Socolow, Director
William Clark
Robert Frosch
James Galloway
Thomas Graedel
Franklin Harris
John Harte
Robert Kates
Henry Kelly
Jessica Mathews
William Moomaw
Michael Oppenheimer
Kumar Patel
Stephen Peck

Working Groups

<table>
<tr><td>

Grand Nutrient Cycles

Moderators:

William Moomaw
Bruce Paton

Rapporteurs:

David Angel
Daniel Lashof

Members:

Robert Ayres
Robert Chen
Daniel Deudney
Elizabeth Economy
George Golitsyn
Arnulf Grübler
Jiang Zhenping
James McNeal
Kenneth Nelson
Theodore Panayotou
V. Ramanathan
William Schlesinger
Valerie Thomas
Robert Williams

</td><td>

Group II

Exotic Intrusions

Moderators:

Wayne France
Thomas Spiro

Rapporteurs:

Frans Berkhout
Steven Fetter

Members:

Stefan Anderberg
Susan Anderson
Stephen Bunker
Robin Cantor
Monika Griefahn
Bette Hileman
Saleemul Huq
Peter Jaffé
Vladimir Kotlyakov
Jerome Nriagu
Jerald Schnoor
Richard Sonnenblick

</td><td>

Group III

**Implementation of
    Industrial Ecology**

Moderators:

James Galloway
Thomas Graedel

Rapporteurs:

Clinton Andrews
Robert Harriss

Members:

Michael Braungart
Ashok Gadgil
Olga Gritsai
Inge Horkeby
Yuri Kononov
Vicki Norberg-Bohm
Jackton Boma Ojwang
Kumar Patel
Steve Rayner
Lowell Smith
Robert Socolow
Richard Somerville
William Stigliani

</td></tr>
</table>

Index